Concrete Durability

Concrete Durability

Thomas Dyer

CRC Press
Taylor & Francis Group
Boca Raton London New York

CRC Press is an imprint of the
Taylor & Francis Group, an **informa** business

A SPON PRESS BOOK

Cover Artwork: Courtesy of Bill Revie

CRC Press
Taylor & Francis Group
6000 Broken Sound Parkway NW, Suite 300
Boca Raton, FL 33487-2742

First issued in paperback 2019

© 2014 by Taylor & Francis Group, LLC
CRC Press is an imprint of Taylor & Francis Group, an Informa business

No claim to original U.S. Government works

ISBN-13: 978-0-415-56474-5 (hbk)
ISBN-13: 978-0-367-86585-6 (pbk)

Library of Congress Cataloging-in-Publication Data

Dyer, Thomas (Civil engineer)
    Concrete durability / Thomas Dyer.
        pages cm
    Includes bibliographical references and index.
    ISBN 978-0-415-56475-5 (hardback)
    1. Concrete--Deterioration. 2. Concrete--Service life. I. Title.

TA440.D894 2014
620.1'3692--dc23

2013046812

Visit the Taylor & Francis Web site at
http://www.taylorandfrancis.com

and the CRC Press Web site at
http://www.crcpress.com

Contents

Preface

The increasing importance placed on the whole-life performance of structures means that there is a growing demand for long service lives with minimal maintenance requirements. Furthermore, the operation of infrastructure beyond the originally intended service life is becoming an increasingly common scenario. Thus, the durability of construction materials is of more concern to civil engineers than ever before.

Concrete is a highly durable material that is also capable of imparting protection to steel embedded within it. However, concrete structures are frequently required to function in a wide range of aggressive environments for long periods of service. Moreover, measures to optimize the durability performance of concrete structures often find themselves in conflict with structural and aesthetic design requirements.

Over the past decade, the introduction of new U.K. and European standards has sought to readdress the issue of the durability of concrete structures in a comprehensive manner. However, negotiating the resulting body of standards and guidance can be a daunting prospect for anyone unfamiliar with their content.

This book individually examines all of the major physical and chemical mechanisms that threaten the durability of concrete and addresses the options available for achieving appropriate durability, with emphasis on the approaches addressed by standards. It also provides a coverage of procedures for durability assessment, testing of structures, and repair and rehabilitation methods.

This book has been written with an audience of graduate students and young professionals in mind.

Author

Dr. Thomas Dyer is a materials scientist working in the field of civil engineering. He is a lecturer within the Division of Civil Engineering at the University of Dundee in Scotland and a member of the university's Concrete Technology Unit.

His research interests center around the chemical interactions of cement with other constituents of concrete and substances from the external environment and the use of recycled materials as concrete constituents. He has published more than 30 articles in academic journals and contributed to a number of chapters in books in the fields of concrete construction and sustainability.

Chapter 1

Introduction

Among the many things manufactured by humans, the built environment is notable in the magnitude of longevity required. Although the reader can probably think of many functional articles that have survived for hundreds or even thousands of years, this was, in most cases, not the intention. Such articles are now usually no longer used but are rather kept for display purposes, possibly in museums under conditions intended to preserve them.

In contrast, many structures that have survived such periods of time and, in many cases, are still in use today. Admittedly, most of these structures will have undergone repair and renovation and may be subject to measures intended to preserve them. Nonetheless, the human race invests unusual efforts in making sure that the structures we build last for periods often exceeding the lifetimes of those designing and building them.

Part of this is driven by practical considerations – the expense, effort and inconvenience of building a structure (and the same factors associated with demolition and replacement) make a short service life undesirable. From a sustainability perspective, a long service life will maximise the benefit obtained from the materials used in a structure with positive benefits with regards to resource use. Additionally, a structure may establish itself in the collective conscience of a population as a fundamental component of that region, city, or settlement, making people reticent to lose it. The argument that such sentimentality represents a barrier to progress and, possibly, to a better quality of life is, in some cases, not entirely without base. On the other hand, it should surely be the objective of all architects, designers and engineers to create structures that instil these feelings, because they are evidence of success.

The end of the life of a structure is reached when it no longer satisfies its function – it ceases to be serviceable. Although this may be rectifiable by repair, replacement, or retrofitting of components, eventually a point will be reached where this becomes uneconomic or impractical, or the desire to prolong the life of the structure is lost. The design service life – the period that a structure is intended to remain serviceable – is very much dependent on its nature and function. *EN 1990: Eurocode – Basis of Structural*

Table 1.1 Indicative design service lives in *EN 1990* [1], with modifications made in line with the *U.K. annex* (in Brackets) [2]

Category	Structure type	Design service life (Years)
1	Temporary structures, not including structures or parts of structures that can be dismantled with a view to being reused	10
2	Replaceable structural parts, for example, gantry girders and bearings (10–25 years)	10–25 (30)
3	Agricultural and similar buildings	15–30 (25)
4	Building structures and other common structures	50
5	Monumental building structures, bridges, and other civil engineering structures	100 (120)

Design provides suggested service lives for a range of different structures, as shown in Table 1.1.

Countless introductions to textbooks on concrete have argued the case for the material's importance in the modern built environment. It is not the author's intention to bore the reader by restating these points, but it is sufficient to say that concrete is ubiquitous in modern construction and has earned this position through characteristics that no other construction material possesses.

Concrete and the materials we used in combination with it are no exception. One of the main factors in the evolution of construction technologies has been driven by the need to use materials that are capable of lasting for long periods of time – that are durable. This is one of the reasons for concrete's success – it is a strong and largely chemically inert material that can potentially last for centuries. However, the relative immaturity of concrete construction as a technology has meant that much of the concrete building stock has experienced unexpected problems with inadequate durability performance within the design service life. Moreover, there has been a growing trend among governments and the operators of structures to extend the service life of structures for economic and practical reasons: The emergence of unexpected durability problems has meant that the last few decades have been a learning process for engineers involved in concrete construction: it has been estimated that the annual cost of repair of concrete structures in Europe is in excess of $20 billion [3].

This book intends to provide an understanding of how concrete elements in structures deteriorate, what must be done to protect structures from unacceptably rapid deterioration, and how existing durability problems can be identified and rectified.

In the United Kingdom, *BS 7543* [4] provides guidance on the durability of buildings and their components. It points out that, when stating the design life, it is not enough to simply define how long a structure is intended to function – it is essential to recognise that different components will remain serviceable for different periods of time. Thus, a definition of service must also make it clear what schedule of maintenance would be required to sustain serviceability and what criteria will be used to decide when a component has ceased to function correctly and will require repair or replacement. Although it is reasonable to expect, for instance, that a flooring material may become unserviceable or obsolete within the design service life of a structure, it is the norm to expect the components that make up the fundamental substructure and superstructure (including concrete structural elements) to remain serviceable throughout the design service life.

BS 7543 outlines the main agents that can cause deterioration:

- Temperature
- Radiation: infrared, visible, ultraviolet and thermal radiation
- Water: in both solid and fluid forms
- Normal air constituents: oxygen, carbon dioxide and sea spray
- Air contaminants
- Freezing/thawing
- Wind
- Biological factors: bacteria, insects, fungal attack, rodents and birds, surface growths and plants and trees
- Stress factors: sustained and intermittent stresses
- Chemical incompatibility: leaching, solvents, contaminated land and expansive materials
- Use factors: normal wear and tear or abuse by the user

Concrete is affected by many of these agents, and so, it is best to first eliminate the ones that are less relevant or whose influence is outside the scope of this book.

Concrete is not vulnerable to the types of radiation listed in the standard. Other forms of radiation can potentially cause more significant damage, such as that induced by exposure to high neutron fluxes in nuclear reactors [5]. However, this aspect lies outside the remit of this book. Temperature change will be induced by exposure to infrared radiation, and this can have implications for concrete with regards to the volume changes it will induce. Aspects of this are discussed in Chapter 2, which looks at physical deterioration mechanisms, including cracking from volume change.

The design of structures to withstand the differential stresses that can be induced by wind action is very much a structural issue, which cannot be dealt with here. However, wind can act as a means of bringing rainwater in contact with concrete surfaces and will also act to propel solid particles against a structure, leading to abrasion. Abrasion is another physical process dealt with in Chapter 2.

Most of the biological factors listed in the standard are of limited concern to a concrete structure, although it is conceded that surface growth of moss and lichen can present problems relating to appearance in some instances and that the growth of tree roots can potentially cause significant damage. Bacterial activity, however, can damage concrete through the production of by-products that can cause sulphate attack and other types of chemical attack, which are discussed in Chapter 3.

Stresses on a structure can take many forms: sustained or intermittent, cyclic or random in nature. Stresses are significant from a durability perspective because they lead to cracking, whether caused by overload of a structural element, creep under sustained loading (which leads to the formation of microcracks) or fatigue under cyclic loading. These cracks offer a route for harmful substances to make their way into the interior of concrete with relative ease. This aspect of concrete deterioration is not addressed in detail here (with the exception of cracking resulting from volume change). However, the influence of cracking on the ingress of water and gases, and the impact on durability, is addressed.

Water plays probably the most important role in the deterioration of concrete structures. Contact with moving water can lead to erosion through interaction with sediment particles or through a process known as 'cavitation'. An unusual characteristic of water – its expansion as it freezes – is the reason why exposure to cycles of freezing and thawing can have serious implications for concrete durability, unless appropriate measures are taken, a topic covered in Chapter 2. Water will permeate the pores of concrete, carrying with it substances that may produce deterioration of concrete or, in the case of chloride ions, corrosion of the steel reinforcement. The presence of water may be necessary to permit damaging processes, such as alkali–silica reaction, to progress. Moreover, moisture plays important roles in other processes that can lead to deterioration, such as plastic and drying shrinkage, and carbonation. Thus, aspects of water's influence on concrete durability appear throughout this book. Indeed, although the author has not carried out such an analysis, it is likely that the word 'water' features almost as frequently as the word 'concrete'.

Concrete does not react with the two main gases in our atmosphere – nitrogen and oxygen. However, the presence of oxygen plays a fundamental role in the corrosion of steel reinforcement. Furthermore, the reaction between constituents of the cement matrix of concrete and carbon dioxide in the atmosphere – carbonation – can lead to a change in the chemical environment around steel reinforcement, which renders steel considerably more prone to corrosion. Sea spray carried by the air in marine environments will also threaten steel through the introduction of chloride ions. All of these processes are examined in Chapter 4, which covers the corrosion of concrete reinforcement. Air pollutants also have the potential to damage concrete, principally where such pollutants are acid in nature, an aspect dealt with in Chapter 3.

Chemical incompatibility is also an important issue for concrete. The hardened cement matrix of concrete is, to a very limited extent, soluble in water, which means that some of the constituents will gradually leach from the concrete surface with time. This process is exacerbated when aggressive chemical species, particularly acidic ones, are dissolved in the water. Additionally, sulphate ions in solution can lead to the progressive deterioration of the surface of concrete. In many soils, both sulphates and acid conditions are present, creating a hostile environment for concrete foundations, and in contaminated land, such conditions can potentially reach even greater levels of aggression. These matters are also covered in Chapter 3.

Throughout Chapters 2 to 4, an approach has been adopted to examine the mechanisms that cause deterioration, what factors (both in terms of environmental conditions and material characteristics) influence deterioration, and, hence, what is required to minimise the rate of deterioration.

Chapter 5 examines how durable concrete can be specified and designed, with emphasis placed on the European standards covering this topic.

Chapter 6 examines the aspects of the construction process, which can influence durability, and technologies that can be employed at the start of a structure's service life to enhance resistance to deterioration.

Finally, Chapter 7 examines the issue of service life, serviceability, and when it ceases. It then goes on to examine how durability problems in a structure can be identified and suggests how future deterioration might be predicted and how deteriorated concrete can be repaired.

REFERENCES

1. British Standards Institution. 2002. *BS EN 1990:2002: Basis of Structural Design*. London: British Standards Institution, 118 pp.
2. British Standards Institution. 2004. *U.K. National Annex for Eurocode: Basis of Structural Design*. London: British Standards Institution, 18 pp.
3. Raupach, M. 2006. Concrete Repair According to the New European Standard *EN 1504*. In M. Alexander, H. D. Beushausen, F. Dehn, and P. Moyo (eds.), *Concrete Repair, Rehabilitation, and Retrofitting*. Abingdon, United Kingdom: Taylor and Francis.
4. British Standards Institution. 2003. *BS 7543:2003: Guide to Durability of Buildings and Building Elements, Products, and Components*. London: British Standards Institution, 34 pp.
5. Dubrovskii, V. B., S. S. Ibragimov, M. Y. Kulakovskii, A. Y. Ladygin, and B. K. Pergamenshchik. 1967. Radiation damage in ordinary concrete. *Soviet Atomic Energy*, v. 23, pp. 1053–1058.

Chapter 2

Physical mechanisms of concrete degradation

2.1 INTRODUCTION

Concrete exposed to the environment may experience conditions that can cause it to deteriorate through wholly mechanical processes. These processes usually involve either loss of material from the concrete surface or cracking, or a combination of both.

Loss of material can be problematic for two reasons. First, a loss of cross-sectional area will compromise the load-bearing capacity of a structural element. It will also reduce the depth of the cover protecting steel reinforcement from external agents likely to accelerate corrosion and, possibly, providing fire protection.

Cracking may also compromise structural performance. However, the formation of cracks will also provide routes for harmful chemical species into concrete and towards reinforcement.

Many of the physical mechanisms of deterioration are the result of volume changes. These mechanisms include various forms of shrinkage resulting from the evaporation of water in either a fresh or a hardened state, the removal of free water as a result of cement hydration reactions and thermal contraction. They also include expansion resulting from the freezing of water within concrete pores and various and sometimes complex forms of mechanical attrition of concrete, which are usually included under the categories of 'abrasion' and 'erosion'.

All of these mechanisms are examined in this chapter. The general approach taken is to describe each mechanism before then examining the factors (in terms of both material characteristics and environmental conditions) that influence the ability of concrete to offer resistance. From this, means of enhancing resistance to attack are outlined, with reference to relevant standards and guidance.

2.2 SHRINKAGE

The reduction in volume of concrete can occur through a number of mechanisms. Contraction of concrete as its temperature falls can be categorised as shrinkage, and this mechanism is discussed in Section 2.3. Furthermore, the reaction of concrete with atmospheric carbon dioxide leads to a reduction in volume, which is covered in Chapter 4. Coverage of shrinkage in this section will concentrate on shrinkage mechanisms caused by changes in the levels of moisture in concrete. These mechanisms are plastic shrinkage, drying shrinkage and autogenous shrinkage, which are discussed individually below. Strategies for avoiding or reducing the effects of these processes are then examined.

2.2.1 Plastic shrinkage

After a concrete element with an exposed horizontal surface is placed and compacted, two processes will simultaneously occur at the surface. First, the solid constituents will tend to settle, leading to the formation of a layer of water at the surface – 'bleeding'. Second, there will be evaporation of free water from the surface.

When the rate of evaporation exceeds the rate of bleeding, a point is reached when the cement particles at the concrete surface are no longer covered by water, and as a result of surface tension, water menisci form between the grains (Figure 2.1c). This causes an attractive capillary pressure (p) between the particles, which is described by the Gauss–Laplace equation:

$$p = -\gamma\left(\frac{1}{R_1} + \frac{1}{R_2}\right)$$

where γ is the surface tension at the air–water interface (N/m), and R_1 and R_2 are the principal radii of curvature of the meniscus surface (m).

As water progressively evaporates, the principal radii of curvature, which are used to describe the meniscus shape, become smaller. Eventually, a point is reached where the principal radii of some menisci are too small to bridge the gap between particles, at which point the water surface must rearrange itself further beneath the surface, with a consequent penetration of air (Figure 2.1d).

The capillary pressure at which this occurs is known as the 'air entry value', and its occurrence marks the beginning of a period when the risk of cracking is at its highest [1]. This is because the areas below the surface where menisci are present will undergo relatively large strains as a result of the capillary pressure, whereas areas with air-filled porosity will display

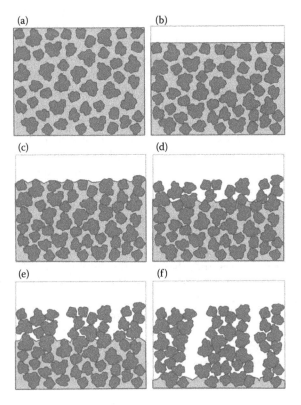

Figure 2.1 Initiation of plastic shrinkage cracks resulting from the evaporation of water from a fresh concrete surface.

much smaller strains. This discrepancy in the magnitude of localised strain has the potential to initiate cracks in the immature cement.

Plastic shrinkage cracks can take a number of forms, including map cracking (or crazing) and diagonal cracks at the edges of slabs [2]. These cracks are relatively shallow. However, further longer-term shrinkage processes – discussed in subsequent sections – will lead to the development of further stresses, and plastic shrinkage cracks will act as the initiation points for cracking extending deeper beneath the surface.

Environmental factors that influence plastic shrinkage are those that influence the rate of evaporation. These include ambient temperature, relative humidity and wind speed. The influence of these parameters on the rate of evaporation is discussed in further detail in Chapter 6 in relation to the curing of concrete.

It has been proposed that the critical rate of evaporation is generally around 1.0 kg/h/m^2, above which cracking becomes likely [3]. However, in reality the critical rate will depend not only on the rate of evaporation, but also on the rate and duration of bleeding and the rate at which the cement hydrates and develops strength. Bleeding is influenced predominantly by the cement fraction of concrete. The influence of concrete mix characteristics on bleeding and plastic shrinkage are summarised in Table 2.1. Although this table only presents an approximate means of evaluating the likelihood of cracking, the parameters most likely to cause problems with regards to plastic shrinkage are those that both reduce bleeding and increase the likelihood of shrinkage cracking. Thus, the use of finer cement, higher cement content, low water/cement (W/C) ratios and the use of lightweight aggregate are all likely to increase the risk of plastic shrinkage cracking to the greatest extent.

The influence of particle fineness on plastic shrinkage is shown in Figure 2.2, which considers a meniscus formed between just two particles. The inclusion of finer particles leads to finer pores, and the menisci that form between particles have lower principle radii of curvature. The result is higher capillary pressures and, therefore, the development of greater tensile stresses.

This effect is observed either where finer Portland cement is used or where finer materials such as fly ash (FA), ground granulated blast furnace slag (GGBS) and silica fume are used in combination with Portland cement. The risk of cracking is more pronounced when these other constituents are present. This is because their reactions are slower, resulting in a reduced rate of development of tensile strength and a greater likelihood of crack formation.

Figure 2.2 also shows the effect of reducing the W/C ratio. Again, pore diameters are reduced along with the radii of curvature of the menisci.

Plastic shrinkage is displayed by the cement paste fraction of concrete. Thus, a higher cement content (and a lower aggregate content) will lead to an increase in plastic shrinkage. The nature of aggregate used may also have an impact on plastic shrinkage. In particular, water absorption will increase the rate of shrinkage.

The use of lightweight aggregate may also reduce bleeding and increase plastic shrinkage relative to similar concrete mixes using normal weight aggregate. It is tempting to assume that this is the result of the absorption of water by the aggregate. However, the main reason appears to be that, to achieve the same strength as normal weight aggregate concrete, a higher cement content is required.

Although a low rate of bleeding may result in problems with plastic shrinkage cracking, a high rate may be just as problematic. This is because rapid bleeding indicates a high rate of settlement of solids, which may cause plastic settlement cracking.

Plastic settlement cracking occurs when settlement occurs rapidly and, therefore, before setting, in features of structural elements where different

Table 2.1 Influence of various concrete constituent characteristics on bleeding and plastic shrinkage cracking

| Parameter | Influence | | Reference(s) |
	Bleeding	Plastic shrinkage	
Cement type	Inclusion of FA and silica fume reduces bleeding.	Higher rate of evaporation when silica fume and FA are present. Higher plastic crack density when FA and silica fume are present.	[6]
	Inclusion of GGBS leads to an increased rate of bleeding.	Higher rate of evaporation when GGBS is present. Higher plastic crack density when GGBS is present.	[5]
Cement fineness	Finer particle size distribution leads to lower bleed rates.	Fine particle size leads to greater risk of cracking.	[4,7]
Cement content	High cement content reduces bleeding.	Reduced rate of evaporation with higher cement content. Increased magnitude of shrinkage with higher cement content.	[4,5,9]
W/C ratio	Direct correlation between W/C and bleed rate.	Reduced rate of evaporation at a low W/C. Greater risk of cracking at a low W/C.	[4,8]
Admixtures	Air entrainers reduce bleeding. Non–air entraining water reducers increase bleeding.	Retarding admixtures increase the risk of plastic shrinkage. Accelerating admixtures reduce risk. Air-entraining admixtures appear to reduce risk.	[2,10,11]
Aggregate	Lightweight aggregate will reduce bleeding.	Lightweight aggregate increases the rate of plastic shrinkage.	[12,13]

magnitudes of movement can occur in close proximity. Examples of element features that can display cracking are shown in Figure 2.3. In Figure 2.3a, the extent of settlement directly over reinforcement is limited relative to settlement at either side. The magnitude of settlement is greater at the surface, and this can produce cracking in elements in which there is a change in element depth (Figure 2.3b). In columns where the

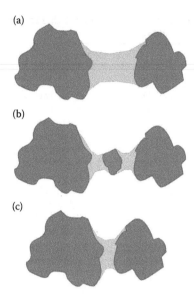

Figure 2.2 Formation of water menisci between cement particles (a) in a cement paste with a high W/C ratio; (b) with markedly different particle sizes; and (c) in a cement paste with a low W/C ratio.

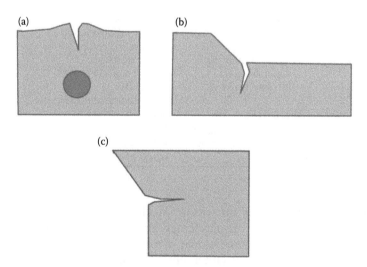

Figure 2.3 Configurations in concrete structural elements that may cause plastic settlement cracking.

cross-sectional area increases towards the top, settlement of the concrete in the outer region of the wider part will be limited relative to elsewhere (Figure 2.3c), leading to horizontal cracking.

2.2.1.1 Avoiding plastic shrinkage cracking

The most straightforward means of reducing the risk of plastic shrinkage cracking is simply to cure concrete appropriately, since this will limit the rate of evaporation. Curing is discussed in more detail in Chapter 6. An interesting approach to avoid plastic shrinkage in lightweight concrete has been proposed recently in the form of 'internal curing', – using water-saturated lightweight aggregate – which subsequently releases water into the cement matrix as moisture is lost through evaporation [14].

Although it is evident that plastic shrinkage cracking can be avoided through appropriate selection and proportioning of concrete constituents, taking a more broad view of concrete durability, this approach is likely to be limited by other requirements of the material. As will be seen in subsequent chapters, many problems relating to durability are minimised through specifying concrete with minimum cement contents and maximum W/C ratios and including materials such as FA and slag. Thus, other approaches to reducing plastic shrinkage may be necessary. Nonetheless, the selection of suitable admixtures (i.e., admixtures that retard hydration less) and the use of admixtures specifically to reduce shrinkage may be an option.

Air entrainment appears to limit the extent to which plastic shrinkage cracking occurs [2]. In addition, a number of shrinkage-reducing admixtures are commercially available. All work by the same mechanism – they reduce the surface tension of the water [15]. This has the effect of reducing the contact angle between the water and the cement surfaces, which, with reference to the Gauss–Laplace equation previously discussed, will reduce the capillary pressure. Most shrinkage-reducing admixtures are based around polyether compounds. Normally, there are side effects of using the admixtures, including reductions in strength and possible incompatibilities between other admixtures, such as air-entraining agents.

The inclusion of fibres in concrete will also control cracking. Fibres work in this instance by providing a means for stress to be transferred across cracks – 'crack bridging' – which prevents crack growth [16]. These fibres can be steel, glass, carbon, and a range of different polymers, although polypropylene is the most common. Microfibres having a diameter less than 0.3 mm (and quite often an order of magnitude less than this) and a length of less than 20 mm are most effective, with narrower and longer fibres within this range being typically superior [17]. Optimum crack control is normally achieved at volume fractions of approximately 0.2% to

0.3%, although it should be stressed that the presence of these levels of fibre will have a significant effect on workability, and mix design will need to take this into account. Moreover, it has been proposed that, to avoid problems with workability, a limit of 0.25% by volume should be observed [18].

Combinations of higher diameter (0.5 mm) steel fibres with microfibres have also been found to be beneficial [18].

2.2.1.2 Avoiding plastic settlement cracking

Microfibres can also be used to prevent plastic settlement cracking. Additional to any crack bridging effect, prevention of crack formation is also the result of reduced bleed rates. The reason for the reduction in the rate of bleeding this has been attributed to the wetting of the fibre surfaces reducing the amount of free water available. However, this explanation is not wholly adequate, since the reduction in bleeding is disproportionate to the additional surface introduced. Instead, it appears more likely that the fibres act to stiffen the cement matrix, reducing the magnitude of differential settlement that can lead to cracking [16]. Narrower fibre diameters and higher fibre volume fractions both increase the reduction in settlement.

Increasing the surface area of the solid constituents in a concrete mix is, nonetheless, effective in reducing bleeding and, thus, plastic settlement. The use of Portland cement with a finer particle size or the inclusion of finer materials, such as silica fume, will reduce bleeding. In the same way, bleed rates can be reduced through the inclusion of more fine aggregate relative to coarse.

Both the W/C ratio and the cement content influence bleed rates. Reduced bleed rates are observed at low W/C ratios. As previously discussed, because maximum W/C ratios are specified for durability, reducing plastic settlement through the manipulation of this parameter is likely to be possible.

Plastic settlement is least pronounced where high cement contents are used, because a high cement content will yield a larger wettable surface in the fresh concrete. Increasing the cement content of a concrete mix is also compatible with current approaches to specifying for durability. However, it should be noted that, for both W/C ratio and cement content, what is good for reducing plastic settlement is bad for reducing plastic shrinkage. Thus, manipulation of both parameters to control cracking in the fresh state may have an opposing effect, and, where it is essential to avoid such problems, other approaches may be wiser.

The use of air-entraining admixtures has the effect of enhancing cohesivity and reducing bleeding, and is thus a useful means of controlling plastic cracking, since it has a favourable effect on both settlement and shrinkage.

2.2.2 Drying shrinkage

The moisture content of hardened concrete has a potentially significant influence on the volume that it occupies, with an increase in the level of moisture producing an increase in volume. Moisture content is controlled by the relative humidity of the surrounding environment, as shown in Figure 2.4. Depending on the external relative humidity, different mechanisms influence the volume. The best means of understanding these mechanisms is to consider concrete that is completely saturated and the processes occurring as this water evaporates. Although there are a number of ways in which the volume of concrete can be modified by its interactions with water, the two main mechanisms derive from the development of capillary pressure and from the effects resulting from the drying of calcium silicate hydrate (CSH) gel, a reaction product formed during the setting and hardening of cement.

2.2.2.1 Capillary pressure

A pressure difference develops when two immiscible fluids are present in a capillary and only one is capable of wetting the surface. This attractive capillary pressure (p_c) is described by the Young–Laplace equation:

$$p_c = \frac{2\gamma\cos\theta}{r}$$

where γ is the surface tension (N/m), θ is the contact angle between the wetting fluid and the capillary surface and r is the capillary radius (m).

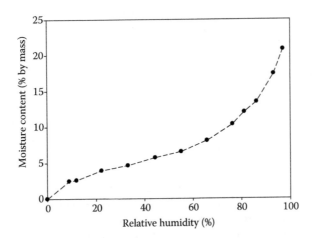

Figure 2.4 Moisture content of concrete with a W/C ratio of 0.4. (From Houst, Y. F. and F. H. Wittmann. *Cement and Concrete Research*, 24, 1994, 1165–1176.)

In a concrete pore, the two fluids are water and air. The reason why shrinkage increases as the moisture content of the concrete declines (i.e., as it dries) is that, as the relative humidity in concrete drops, water in progressively smaller pores will evaporate, causing the value of r in the previous equation to decrease.

This phenomenon is described by the Kelvin equation:

$$\ln\frac{p}{p_0} = \frac{2\gamma V_m}{rRT}$$

where p is the vapour pressure of fluid (N/m^2), p_0 is the saturated vapour pressure (N/m^2), V_m is the molar volume of fluid (m^3/mol), R is the universal gas constant (J/mol·K) and T is the temperature (K).

Relative humidity is p/p_0 in this equation, and the smallest pores in concrete will normally be approximately 2 nm in diameter. According to the equation, at 10°C these pores will be empty of all condensed water below a relative humidity of approximately 33% – it should be noted that the convention is to use a negative r value for a pore containing water vapour rather than condensed water. However, shrinkage is still observed at relative humidity levels below this point, indicating that at least one additional mechanism is effective. It is generally agreed that this mechanism is related to the manner in which CSH gel particles interact with water.

2.2.2.2 Gel particle effects

The main reaction product of Portland cement is CSH gel, which typically comprises 50% to 60% by volume of a mature Portland cement (PC) paste. The gel is an aggregation of colloidal particles (particles having at least one dimension between 1 nm and 10 μm). Each gel particle is composed of layers of calcium silicate sheets, which are distorted and arranged on top of each other in a disordered manner (Figure 2.5). The nature of this configuration means that there is much space between the layers, which can be occupied by 'interlayer' water.

Like many substances of this kind, imbibition of water by the gel leads to swelling, and evaporation of water leads to shrinkage. Several theories have been proposed to explain why this is the case for CSH gel. One theory relates to surface energy and proposes that, because the surface of the gel particles will possess a surface free energy and surface tension will act to reduce this value as much as possible, this has the effect of placing the interior of the particle into compression. These compressive forces will lead to a reduction in the particle volume. The adsorption of water molecules on the surface acts to reduce the surface energy, and so, when relative humidity increases, the volume of the particles will increase, and vice versa [20].

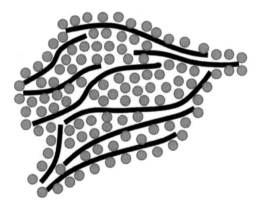

Figure 2.5 Model of the microstructure of a CSH gel particle. The distorted lines are crystallite layers, whereas the circles are water either adsorbed on the crystallite surfaces or held as 'interlayer' water between the crystallite layers.

Although this proposed mechanism is scientifically sound, it has been estimated that, if it were the main effect, it could only be responsible for a very small amount of the total shrinkage observed between 40% and 0% relative humidity [21]. An alternative explanation, which would produce appropriate magnitudes of volume change, is that shrinkage occurs as a result of the loss of interlayer water.

2.2.2.3 Aggregate shrinkage

Most aggregate materials shrink to a lesser extent relative to cement paste, and although the cement fraction comprises a much smaller volume fraction than the aggregate, it normally has the largest influence on shrinkage. As will be seen later, the higher modulus of elasticity of most aggregate relative to cement paste has a restraining effect that limits shrinkage.

There are some exceptions to this general observation. For instance, some aggregate sources in Scotland have been found to be prone to higher levels of shrinkage [22]. Contaminants such as clay and timber are also prone to shrinkage. Dolerite, basalt, mudstone, and greywacke have all been found to potentially display shrinkage, although only from certain sources.

A plot showing the typical overall influence of relative humidity on shrinkage is shown in Figure 2.6.

2.2.2.4 Shrinkage as a function of time

If the water in saturated, or partially saturated, concrete pores is not in equilibrium with the surrounding air, it will either lose water through evaporation or take in water vapour from the atmosphere. This movement of

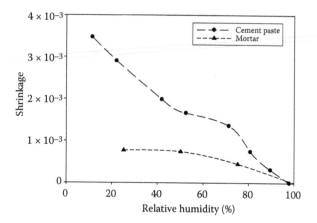

Figure 2.6 Drying shrinkage of cement paste and mortar (sand/cement ratio = 4:1)
versus relative humidity. (From Wittmann, F. H. The structure of hardened
cement paste: A basis for a better understanding of the material's properties. In Paul V. Maxwell Cook, ed., *Hydraulic Cement Pastes: Their Structure
and Properties*. Slough, United Kingdom: Cement and Concrete Association
of Great Britain, 1976, pp. 96–117; and Verbeck, G. J. Carbonation of
hydrated Portland cement. American Society for Testing Materials. *Special
Technical Publication* 205, 1958, pp. 17–36.)

water vapour is the result of diffusion, and thus shrinkage is not instanta-
neous. Figure 2.7 shows a typical plot of shrinkage versus time. The rate
of diffusion of water vapour through concrete is dependent on the amount
of water remaining and, thus, on the relative humidity in the pores. This
relationship can be described using the following equation:

$$D = D_1 \left(\alpha_0 + \frac{1-\alpha_0}{1+\left(\dfrac{1-H}{1-H_c}\right)^n} \right)$$

where D is the diffusion coefficient (m^2/s), D_1 is the maximum value of the
diffusion coefficient (m^2/s), α_0 is the ratio of the maximum and minimum
values of D, H is the relative humidity in the concrete pores, H_c is the
relative humidity around which the diffusion coefficient changes and n is a
constant [25].

The values of n and D_1 are dependent on the concrete, whereas H_c and
α_0 have values of approximately 0.75 and 0.05, respectively. The nature
of this relationship is shown in Figure 2.8 – above a relative humidity of
75%, the diffusion coefficient is high, whereas below this humidity level,
it is very low.

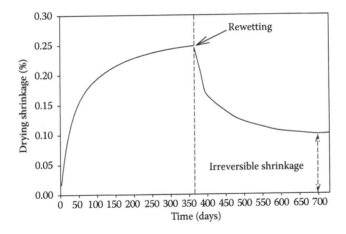

Figure 2.7 Typical plot of drying shrinkage against time.

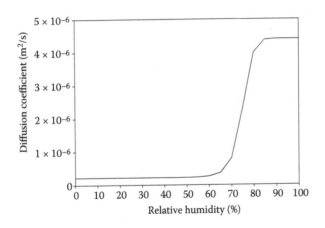

Figure 2.8 Dependence of the diffusion coefficient of water vapour through concrete as a function of relative humidity, calculated using a value of n of 10 and a maximum diffusion coefficient (D_1) of 4.40 × 10^{-10} m²/s.

The reason for the radical change in diffusion coefficient at lower relative humidities has been attributed to the manner in which water molecules interact with the surface of concrete pores. At low levels of relative humidity, where the number of water molecules will be low, the molecules will be attracted by intermolecular forces to a narrow zone at the pore surface. In this zone, these intermolecular forces will act to hinder the movement of water. As the number of water molecules increases, while a layer of molecules will still be present at the pore surface, the layer will not be able to accommodate all the water present, leaving the water molecules closer to

(a)   (b)

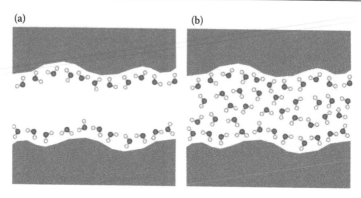

Figure 2.9 Water molecules in an idealised concrete pore. At a low relative humidity (a), the water molecules are all held close to the pore walls by intermolecular forces, resulting in a reduced mobility. At a high relative humidity (b), the same layer of water molecules is present, but the higher density of molecules means that water molecules are also present in the pore interior. These molecules are much less influenced by interactions with the pore walls and are thus more mobile.

the middle of the pore to move with relative freedom [25]. This phenomenon is illustrated in Figure 2.9.

In the situation where the surface of saturated concrete is exposed to air, the diffusion process leads to the development of a relative humidity profile, as shown in Figure 2.10.

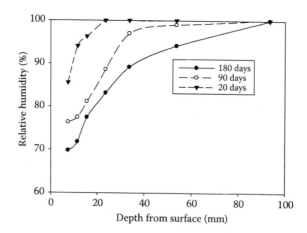

Figure 2.10 Moisture profiles in a Portland cement paste prism with a W/C ratio of 0.59 exposed to a 60% relative humidity environment. (From Parrot, L. J., *Advances in Cement Research*, 1, 1988, 164–170.)

Another feature of drying shrinkage is that a proportion of it will be irreversible – if the concrete returns to its water-saturated condition after drying, the concrete does not fully return to its original dimensions. This is also illustrated in Figure 2.7. There is still much debate regarding the mechanism behind this effect. One suggestion is that, as the CSH gel undergoes shrinkage, the gel particles reorientate themselves in a manner that cannot be recovered from fully [21].

2.2.2.5 Factors controlling drying shrinkage

A number of factors influence the drying shrinkage of concrete. These include the environmental conditions that a structure is exposed to, the composition of the concrete and the configuration of a structural element. When discussing drying shrinkage, it is important to be aware that the process can be affected both in terms of the rate at which shrinkage occurs and the ultimate magnitude of shrinkage. Ultimate shrinkage is important because, as we shall see later, the magnitude of strain caused by shrinkage determines the stresses developed within the concrete. The rate of shrinkage may, in some cases, be less significant. However, it is of great importance when drying shrinkage is occurring in immature concrete, where strength is still developing.

In terms of environmental factors that influence drying shrinkage, the main parameters are relative humidity (as previously discussed), temperature and air movement. An increase in temperature leads to a faster rate of drying shrinkage. This is due to a faster rate of evaporation and diffusion of water vapour.

Air movement around a structure will normally take the form of wind, and faster wind speeds will more rapidly remove water vapour.

The main material characteristics that influence drying shrinkage are aggregate type, aggregate/cement ratio, W/C ratio and cement type.

There is usually a good correlation between the capacity of aggregate to absorb water and the magnitude of drying shrinkage observed in concrete containing it (Figure 2.11). Instinctively, it may be tempting to think that this is related to capillary pressures, as previously discussed. However, the reason is far simpler – aggregates are normally not prone to shrinkage and have a larger modulus of elasticity in comparison with cement paste. As a result, their presence in concrete acts to internally restrain shrinkage. The reason for the correlation between water absorption and shrinkage is the result of the strong relationship between the porosity of aggregate and its stiffness.

Thus, the volume of aggregate present relative to cement paste also plays an important role, with a higher volume causing a reduction in the ultimate shrinkage of concrete, assuming an aggregate with a higher stiffness than

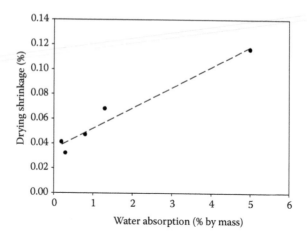

Figure 2.11 Water absorption of aggregate versus 1-year drying shrinkage in concrete. (From Carlson, R. W. Drying shrinkage of concrete as affected by many factors. *Proceedings of the 41st Annual Meeting of the ASTM*, v. 38 (Part 2), 1938, pp. 419–437.)

the cement paste. However, this relationship is not linear and is dependent on the elastic modulus of aggregate relative to cement. This is illustrated in Figure 2.12.

The relationship between shrinkage (S) and aggregate volume fraction (V_a) is described by the following equation:

$$S = S_c (1 - V_a)^\alpha$$

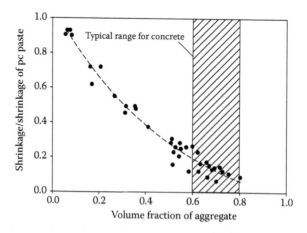

Figure 2.12 Shrinkage of mixtures of cement paste and aggregate of varying proportions. (From Hobbs, D. W. and L. J. Parrot., *Concrete*, 13, 1979, 19–24.)

where

$$\alpha = \frac{3(1-\mu_c)}{1+\mu_c+2(1-2\mu_a)\dfrac{E_c}{E_a}}$$

and S_c is the shrinkage of cement paste (%), μ is the Poisson's ratio of cement (c) or aggregate (a) and E is the Young's modulus of cement or aggregate [27].

Unhydrated cement will also act as a restraining material, and a slightly more complex equation has been developed to include this effect [28].

It is sometimes stated that shrinkage increases as water content increases. Although this is true, what is really being witnessed is, again, a change in the aggregate/cement ratio – where the W/C ratio is fixed, an increase in the water content means an increase in the volume of cement paste and hence a decrease in the volume of aggregate.

As the W/C ratio reduces, the modulus of elasticity of the hardened cement paste increases, presenting a greater resistance to shrinkage (Figure 2.13).

The chemical composition of Portland cement has some influence over drying shrinkage. In particular, a higher tricalcium aluminate and alkali content will lead to larger magnitudes of shrinkage. However, the sulphate content of the cement acts to limit this effect [32,33].

Figure 2.14 shows shrinkage curves obtained from mortars containing a combination of Portland cement and various size fractions of FA. In all cases, FA has the effect of reducing shrinkage, with the finer fractions producing the least amount of length change. The cause of this effect is that

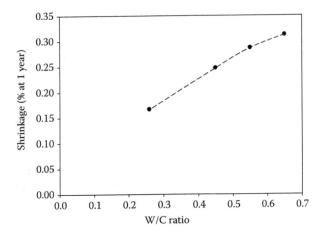

Figure 2.13 Influence of W/C ratio on 1-year drying shrinkage in cement pastes. (From Soroka, I. *Portland Cement Paste and Concrete*. London: Macmillan, 1979.)

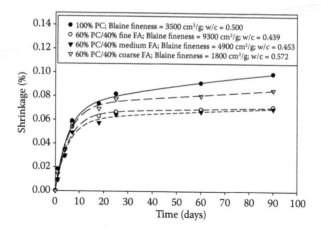

Figure 2.14 Shrinkage of Portland cement (PC)/FA mortars of equal consistency, containing FA fractions separated from the same source using air separation techniques. (From Chindaprasirta, P. et al., *Cement and Concrete Research*, 34, 2004, 1087–1092.)

the finer FA fractions permit greater reductions in water content to achieve a given consistence, which means that the mortars with finer ash fractions have lower W/C ratios. However, there may be other mechanisms effective because even the coarsest ash, which is coarser than the PC and requires a higher W/C ratio, produces lower shrinkage than the control mortar.

A review of the literature relating to research conducted into the performance of concrete containing GGBS has identified a variety of results with respect to drying shrinkage, with both increases and reductions in the shrinkage observed [35]. Where shrinkage was found to be greater than PC controls, much of the effect could be attributed to the higher paste content required when GGBS is used. The magnitude of the difference between GGBS concrete and PC control mixes is typically small.

Silica fume typically has the effect of very slightly increasing shrinkage at early ages. Later, as ultimate shrinkage is approached, shrinkage is lower at higher silica fume levels. However, as illustrated in Figure 2.15, the overall effect on shrinkage is minimal. The shrinkage of concrete has been found to reduce with increasing levels of metakaolin [36]. This shrinkage was composed of both autogenous shrinkage (see Section 2.2.3) and drying shrinkage. Once the two types of shrinkage had been resolved separately, it was evident that drying shrinkage made a relatively small contribution to shrinkage and that it was independent of metakaolin content.

Accelerating admixtures, in the form of calcium chloride, calcium formate and triethanolamine, all have the effect of increasing the rate, and possibly the magnitude, of shrinkage (Figure 2.16). Certainly in the case of calcium chloride, this is the case despite lower rates of water loss through evaporation.

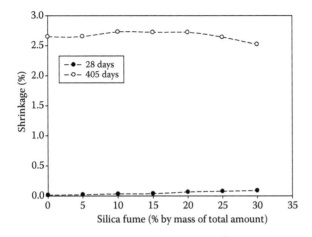

Figure 2.15 Drying shrinkage of mortar specimens containing a range of levels of silica fume. (From Rao, G. A., *Cement and Concrete Research*, 28, 1998, 1505–1509.)

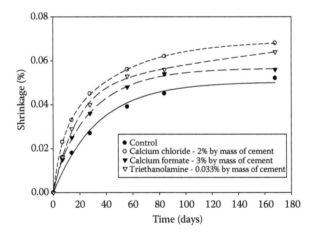

Figure 2.16 Drying shrinkage of concrete specimens containing accelerating admixtures. (From Bruere, G. M. et al., A laboratory investigation of the drying shrinkage of concrete containing various types of chemical admixtures. CSIRO. *Technical Paper 1*, 1971.)

The reason for the greater rate of shrinkage may therefore be the result of changes in microstructure. This explanation is reinforced by the observation that tricalcium silicate pastes containing quantities of $CaCl_2$ have been found to have a smaller mean pore diameter than control pastes [37]. Because pore diameter has such a significant influence over capillary pressure, higher magnitudes of shrinkage for a given quantity of moisture loss would be expected.

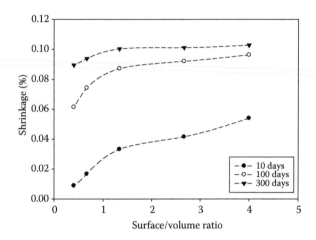

Figure 2.17 Shrinkage of mortar specimens at various ages versus the surface/volume ratio of the specimens. (From Hobbs, D. W. and A. R. Mears, *Magazine of Concrete Research*, 23, 1971, 89–98.)

The dimensions of concrete structural elements have a significant influence over the rate of drying shrinkage. A greater ratio of surface area to volume will lead to a faster rate of evaporation of water from the concrete and thus a faster rate of shrinkage. This is illustrated in Figure 2.17, which plots the shrinkage of concrete at several ages against this ratio. It should be noted that the ultimate magnitude of shrinkage is very similar regardless of the dimensions.

It is, of course, not just the dimensions of a concrete element in a structure that will define the surface/volume ratio – parts of the concrete surface may not be exposed to the atmosphere, effectively reducing the ratio.

2.2.2.6 Cracks resulting from shrinkage

Shrinkage of concrete would not necessarily be problematic in many cases were it not for the fact that it is also likely to lead to cracking. Cracking may certainly be a problem from serviceability and aesthetic perspectives, but, as we shall see in the next chapters, in terms of concrete durability, it equates to a significant compromise of the ability of the concrete to resist the ingress of harmful substances.

Concrete that can change shape freely will undergo very little cracking. However, there are very few applications where concrete is used in such a manner – in most circumstances, there will be some form of restraint acting on a concrete element. Thus, cracking resulting from drying shrinkage is likely to be a problem in the vast majority of cases where concrete is exposed to drying conditions.

In a fully restrained concrete element, actual shrinkage (in other words, a change in dimensions) will not occur. Instead, tensile strain will develop as the difference between the (theoretical) length of the element in its unrestrained state and the actual length in the restrained state increases. The development of tensile strain means that tensile stress will be developing simultaneously, and when this stress exceeds the tensile strength of the concrete, cracking will occur.

The tensile strength of concrete is substantially less than its compressive strength, usually between 5% and 15%. It develops with time in a similar manner with compressive strength, as shown in Figure 2.18. Using the shrinkage data for the control concrete mix presented in Figure 2.16, and using values for the development of the modulus of elasticity of concrete (also shown in Figure 2.18), the tensile stress in the concrete at the end of the test (if it was restrained) can be calculated to be approximately 18 N/mm². This is considerably more than the tensile strength of the concrete, and so, based on this calculation, cracking would occur. However, one feature of the manner in which concrete deforms – creep – has not yet been taken into account, which will be discussed shortly.

An alternative way of viewing cracking resulting from shrinkage is in terms of tensile strain capacity. The tensile strain capacity is the level of tensile strain beyond which concrete will crack. Because the modulus of elasticity of concrete is defined by the gradient of stress versus strain, and

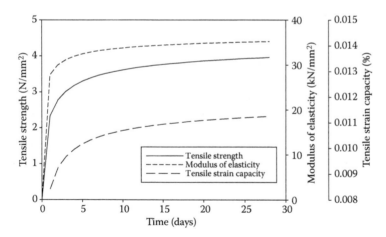

Figure 2.18 Development of tensile strength and modulus of elasticity in a 40 N/mm² concrete made with PC with a strength class of 42.5 N, calculated using the equations in Eurocode 2. Also plotted is the tensile strain capacity calculated from the strength and the modulus of elasticity values. (From British Standards Institution. *BS EN 1992-1-1:2004: Eurocode 2: Design of Concrete Structures – Part 1-1: General Rules and Rules for Buildings.* London: British Standards Institution, 2004.)

assuming that the material behaves in a wholly elastic manner, the tensile strain capacity ε_t can be calculated using the following equation:

$$\varepsilon_t = \frac{f_{ct}}{E_c}$$

where f_{ct} is the tensile strength (N/mm²), and E_c is the modulus of elasticity in compression (N/mm²) [42].

This also changes with time, as shown in Figure 2.18.

Eurocode 2 [41] includes an equation for estimating the ultimate strain resulting from drying shrinkage:

$$\varepsilon_{cd,0} = 0.85 \left[\left(220 + 110\alpha_{ds1} \right) e^{-\alpha_{ds2} \frac{f_{cm}}{10}} \right] \beta_{RH} \cdot 10^{-6}$$

where $\varepsilon_{cd,0}$ is the ultimate shrinkage strain, α_{ds1} and α_{ds2} are the constants related to the cement type (Table 2.2) and f_{cm} is the mean compressive strength.

β_{RH} is a parameter defined by the ambient relative humidity (RH) expressed as a percentage.

$$\beta_{RH} = 1.55 \left[1 - \left(\frac{RH}{100} \right)^3 \right]$$

Although this equation is potentially useful in terms of giving designers an indication of the magnitude of drying shrinkage and the likelihood of cracking, it should be stressed that it does not address some of the factors influencing drying shrinkage, in particular, the influence of aggregate type and volume.

Another aspect of shrinkage that must be borne in mind is that, in reality, most concrete structural elements will only be partly restrained. An example of this would be a concrete wall that is restrained at its foundations. The foundation will completely restrain shrinkage in the wall at its

Table 2.2 Values of the cement-related constants in the Eurocode 2 drying shrinkage equation

Cement type	Constant	
	α_{ds1}	α_{ds2}
Class S (slow early strength)	3	0.13
Class N (normal early strength)	4	0.12
Class R (high early strength)	6	0.11

base, leading to the development of tensile stresses. However, if the top of the wall is not attached to any other part of the structure, it will be able to shrink relatively freely. The extent to which a structural element is restrained is expressed in terms of a restraint factor R defined by the following equation:

$$R = \frac{\varepsilon_r}{\varepsilon_{free}}$$

where ε_r is the restrained strain, and ε_{free} is the shrinkage strain that would occur in a wholly unrestrained member.

A number of restraint factors for different structural elements are shown in Table 2.3.

The discussion of cracking so far assumes that concrete is a purely elastic material. However, in reality, this is not the case, and concrete's ability to undergo long-term deformation under load is of specific significance to shrinkage cracking. This process is creep, and in concrete undergoing shrinkage, tensile creep will act to cause a relaxation of tensile stresses, thus allowing the onset of cracking to be postponed, potentially indefinitely. The manner in which shrinkage and creep processes (as well as thermal contraction, which is discussed later in this chapter) act in combination is illustrated in Figure 2.19.

Creep is a complex process, whose rate and magnitude is dependent on a number of factors. Although a detailed discussion of the mechanisms

Table 2.3 Restraint factors for a range of structural elements

Structural element	Restraint factor (R)
Base cast onto blinding	0.2
Edge restraint in box-type deck cast in stages	0.5
Base of wall cast onto heavy preexisting base	0.6 [42]/0.5 [43]
Top of wall cast onto heavy preexisting base; L/H ratio = 1	0
Top of wall cast onto heavy preexisting base; L/H ratio = 2	0
Top of wall cast onto heavy preexisting base; L/H ratio = 3	0.05
Top of wall cast onto heavy preexisting base; L/H ratio = 4	0.3
Top of wall cast onto heavy preexisting base; L/H ratio = >8	0.5
Edge element cast onto slab	0.8
Infill bays	1.0 [42]/0.5 [43]

Source: The Highways Agency. Highway structures: Approval procedures and general design: Section 3. General design. *Early Thermal Cracking of Concrete: Design Manual for Roads and Bridges, Vol. 1.* South Ruislip, United Kingdom: Department of Environment/Department of Transport, 1987; and British Standards Institution. *BS EN 1992-3:2006: Eurocode 2 – Design of Concrete Structures: Part 3. Liquid-Retaining and Containment Structures.* London: British Standards Institution, 2006.

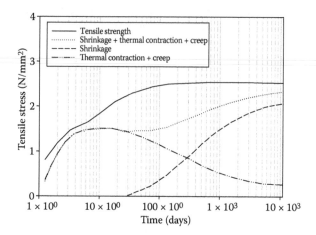

Figure 2.19 The contribution of drying shrinkage, creep, and thermal contraction on tensile stress in a restrained concrete member carrying a sustained load plotted alongside tensile strength development. (From Bamforth, P. et al., *Properties of Concrete for Use in Eurocode 2*. Camberley, United Kingdom: The Concrete Centre, 2008.)

involved in creep is beyond the scope of this book, it is worth summarising the main parameters influencing its magnitude.

The magnitude of creep is dependent on the stress acting on the concrete, with larger stresses leading to larger levels of ultimate creep. Strength also plays an important role, with a lower W/C ratio (and hence a higher stiffness) producing less creep. The influence of strength also means that less mature concrete will undergo more creep.

In addition, relative humidity influences the extent to which creep occurs, with higher magnitudes of creep at lower humidity levels. A larger ratio of volume to exposed surface area will yield a larger amount of creep.

Eurocode 2 [41] includes an equation for estimating the ultimate strain resulting from creep (ε_{cc}). Although the equation is derived from compressive creep behaviour, the standard permits the equation to be used for tensile creep. The equation, in a form notated for tensile loading, is

$$\varepsilon_{tc} = \frac{\varphi \sigma_t}{E_t}$$

where ε_{tc} is the ultimate strain resulting from tensile creep, φ is the ultimate creep coefficient, σ_t is the tensile stress resulting from restrained shrinkage (N/mm^2) and E_t is the tangent modulus (N/mm^2).

The standard also provides a means of estimating the creep coefficient at time t using the following equation:

$$\varphi = \varphi_{RH} \cdot \frac{16.8}{\sqrt{f_{cm}}\left(0.1 + t_0^{0.2}\right)} \left[\frac{t - t_0}{(\beta_H + t - t_0)}\right]^{0.3}$$

where φ_{RH} is a constant related to the ambient relative humidity, f_{cm} is the mean compressive strength (N/mm²), t_0 is the time at which the tensile stress from shrinkage and thermal contraction initiates (days) and β_H is a constant related to the relative humidity.

The manner in which φ_{RH} and β_H are calculated depends on the mean compressive strength of the concrete, as shown in Table 2.4.

The standard also includes corrections that can be used to take account of elevated temperatures and different cement types.

Although concrete is susceptible to cracking under stresses resulting from shrinkage, reinforcement is not. Reinforcement allows the transfer of tensile stresses resulting from shrinkage to the concrete in a manner that is not possible in an unreinforced member. This is illustrated in Figure 2.20, which shows the development of cracks in an unreinforced and reinforced concrete member.

Where reinforcement is absent, the increase in the tensile stress resulting from shrinkage eventually reaches the tensile strength of the concrete, leading to cracking. The formation of the crack leads to the complete relaxation of tensile stress and, due to the removal of restraint, subsequent shrinkage leads to a widening of the crack.

Table 2.4 Equations for φ_{RH} and β_H for the prediction of creep

	Equation	
	≤35 N/mm²	>35 N/mm²
φ_{RH}	$1 + \dfrac{1 - \dfrac{RH}{100}}{0.1\sqrt[3]{\dfrac{2A_c}{u}}}$	$\left[1 + \dfrac{1 - \dfrac{RH}{100}}{0.1\sqrt[3]{\dfrac{2A_c}{u}}}\left(\dfrac{35}{f_{cm}}\right)^{0.7}\right]\left(\dfrac{35}{f_{cm}}\right)^{0.2}$
β_H	$1.5[1 + (0.012RH)^{18}]\dfrac{2A_c}{u} + 250$	$1.5[1 + (0.012RH)^{18}]\dfrac{2A_c}{u} + 250\left(\dfrac{35}{f_{cm}}\right)^{0.5}$

Note: A_c = cross-sectional area of a member (m²); u = perimeter of the member exposed to the atmosphere (m); RH = relative humidity (%).

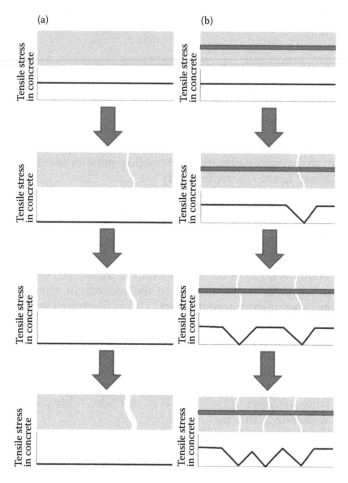

Figure 2.20 Crack formation in unreinforced concrete (a) and reinforced concrete (b) undergoing shrinkage.

When concrete is reinforced, after a crack forms, the stress it carried by reinforcement is transferred to the concrete. Thus, a tensile stress is maintained in the uncracked regions of concrete and, as shrinkage progresses, additional cracks form.

The change of crack formation behaviour caused by reinforcement is referred to as 'crack control' and is beneficial from a concrete durability perspective by producing improved resistance to the ingress of detrimental substances compared with unreinforced concrete. Instinctively, this may seem strange, because a shift from a small number of wide cracks to a large number of narrow cracks appears to be of limited value. However, as will be shown in subsequent chapters, crack width plays a much more important role than crack density (or 'crack spacing') in defining the permeation characteristics of concrete.

Eurocode 2 provides an equation for estimating the average crack width (w_k), a simplified version of which is

$$w_k = S_{r,max} \left(\frac{\sigma_s - k_t \dfrac{f_{ct}}{\rho} \left[1 + \rho \dfrac{E_s}{E_{cm}} \right]}{E_s} \right)$$

where $S_{r,max}$ is the maximum crack spacing (m), σ_s is the tensile stress in the reinforcement (N/mm²), k_t is the coefficient dependent on the duration of loading (higher for shorter periods), ρ is the reinforcement ratio of the concrete section, E_s is the modulus of elasticity of the reinforcement (N/mm²) and E_{cm} is the modulus of elasticity of the concrete (N/mm²).

The reinforcement ratio, in simple terms, is the ratio of the cross-sectional area of the reinforcement to that of the concrete. *Eurocode 2* defines the steel ratio in a more detailed manner, which takes into account the use of mixed diameters of reinforcement bars, prestressing or pretensioning, the bond between reinforcement and concrete and the matter of how much of the concrete in a section is actually in tension.

From the equation, it is evident that the crack width can be reduced by using stiffer reinforcement, a greater quantity of reinforcement or stronger concrete. Additionally, reducing the maximum crack spacing will also reduce the crack width.

The crack spacing is strongly influenced by the efficiency with which stress is transferred between the concrete and the reinforcement. A greater efficiency leads to the gradient of the decline and rise in tensile stress around a crack (as shown in Figure 2.20) to become steeper, which produces a narrower zone of stress relaxation and more opportunity for crack formation elsewhere. This efficiency is controlled by the strength of the bond and the surface area of reinforcement. *Eurocode 2* also provides an equation for estimating the average crack spacing, an abbreviated version of which is

$$S_{r,max} = k_3 c + \frac{k_1 k_4 \varnothing}{\rho}$$

where c is the concrete cover depth (m), $\varnothing$ is the bar diameter of the reinforcement (m), k_1 is the coefficient reflecting the strength of the bond between the reinforcement and the concrete (lower for a stronger bond), k_3 is a constant (3.4) and k_4 is also a constant (0.425).

Thus, the spacing is reduced by reducing the bar diameter, improving the bond (for instance, by using high bond reinforcement bars) or again increasing the quantity of steel. Reducing the depth of the cover will also reduce the spacing. As will be discussed later, the depth of the cover plays an important role in protecting concrete and its reinforcement from deterioration in aggressive chemical environments. However, here it is evident that an overspecification of cover is not necessarily beneficial in durability terms, since the cracks that may result could compromise the protection theoretically provided by the higher depth of the cover.

2.2.3 Autogenous shrinkage

Evaporation is not the only means by which water can be removed from the pores of concrete. As Portland cement undergoes hydration reactions, free water will be converted to chemically combined or chemisorbed water on hydrate surfaces. This loss of free water can also lead to shrinkage – autogenous shrinkage.

Portland cement paste that has been able to react fully will ultimately yield around 400 g of nonevaporable water (water that is chemically bound in cement hydration products or present as interlayer water in the gel) for every kilogram of original cement. The precise value is dependent on the composition of the cement.

This means that, even if evaporation from concrete is entirely prevented, there will be a considerable reduction in the volume of water present, and shrinkage will occur via the same mechanisms as for drying shrinkage. A typical autogenous shrinkage curve is shown in Figure 2.21.

Factors that influence autogenous shrinkage include the type of cement used and the W/C ratio. W/C ratio is important because of its role in defining the pore size distribution in the hardened concrete. This is significant for autogenous shrinkage because pore radius strongly influences pore pressures (see Section 2.2.2). A larger proportion of smaller pores will lead to higher pore pressures. As a result, a low W/C ratio produces higher magnitudes of autogenous shrinkage. In conventional concrete, autogenous shrinkage tends to be of a relatively small magnitude (between 0.005% and 0.01%). However, in high-strength concrete, where lower W/C ratios are necessary, autogenous shrinkage may exceed drying shrinkage.

High-strength concrete will often also contain silica fume, which also has the effect of refining pore structure and thus increasing autogenous shrinkage. The effect of both W/C ratio and silica fume content on autogenous shrinkage is shown in Figure 2.22.

Other cementitious constituents influence autogenous shrinkage. The presence of GGBS causes some increase in autogenous shrinkage [48]. FA typically appears to have the opposite effect [18]. It has been suggested

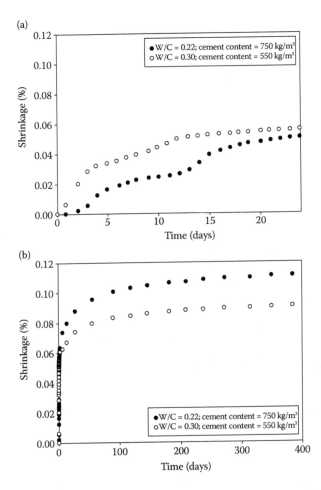

Figure 2.21 Autogenous shrinkage over different timescales – (a) 25 days, (b) 400 days – of two Portland cement concretes of low W/C ratio and high cement content. (From Termkhajornkit, P. et al., *Cement and Concrete Research*, 35, 2005, 473–482.)

that this is the result of the faster rate of the reaction of GGBS, leading to the formation of reaction products whose presence acts to reduce pore sizes toward the ranges where capillary pressure has the most influence [49]. The slower rate of the reaction of FA ensures that capillary pore sizes remain relatively large until later ages when autogenous shrinkage is of lesser importance. For the same reasons, finer cement will lead to higher levels of autogenous shrinkage [50].

As for drying shrinkage, a higher aggregate content and stiffness act to impart greater restraint against shrinkage.

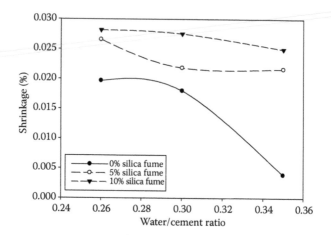

Figure 2.22 Autogenous shrinkage in concrete resulting from different W/C ratios and silica fume contents. (From Zhang, M. H. et al., *Cement and Concrete Research*, 33, 2003, 1687–1694.)

2.2.4 Reducing the problem of cracking resulting from shrinkage

From a structural design perspective, drying shrinkage cracking can be reduced through the inclusion of movement joints. By placing discontinuities of this type in a structure, restraint is lessened and cracking is reduced. However, although joints will normally be filled with a flexible and impermeable sealant, this material is likely to have a functional life significantly shorter than concrete, and between the start of its deterioration and its replacement, there exists the possibility of, for instance, chloride ions making their way into the joint and through relatively low depths of concrete to the reinforcement. For this reason, and because short spacings are required between joints to wholly avoid cracking, there has been a shift in practice toward a combination of larger joint spacings and reinforcement for crack control.

As previously seen, a very effective means of controlling cracking in concrete is to include reinforcement. When reinforcement is present solely for the control of cracking, it is used sparingly and referred to as 'minimum' or 'nominal' reinforcement. In very simple terms, the minimum reinforcement ratio (ρ_c; see Section 2.2.2) required to control cracking is equal to the ratio

$$\rho_c = \frac{f_{ct}}{f_y}$$

where f_{ct} is the tensile strength of the concrete at a given age (N/mm²), and f_y is the yield strength of the reinforcement (N/mm²) [51].

This is because the yielding of reinforcement must not occur if crack control is to be effective. If the reinforcement were to yield at around the location of a crack, it would allow the crack to widen rather than transfer stress to form more cracks.

Although this equation provides a useful rule of thumb for determining how much reinforcement is required, in reality, a more detailed approach is required for the design of many concrete elements. Guidance for several such cases is provided in *Eurocode 2*.

The option of replacing some conventional reinforcement with steel fibres or macrosynthetic fibres also exists. This is discussed in more detail in Chapter 5.

Guidance on selecting appropriate combinations of movement joint spacings and reinforcement for liquid-retaining structures is provided as part of *Eurocode 2* in *BS EN 1992-3* [43]. Similar guidance is provided in *BS 8007* [52], which has been superseded by *Eurocode 2* but which contains some useful background on the types of movement joint. The *Concrete Society Technical Reports 34* and *66* provide guidance for ground floors and external paving [53,54].

It is also clearly possible to reduce the amount of shrinkage through the modification of concrete constituents and mix proportions. However, in reality, these parameters are likely to be limited to some extent by other requirements.

A good example of this is cement content. Reducing the cement content of concrete will reduce both drying shrinkage and autogenous shrinkage. However, as shall be seen in subsequent chapters, if concrete is to be exposed to hostile conditions, using the British complementary standards to the European standard for the specification of concrete, a minimum cement content limit will be specified. This will clearly limit the extent to which shrinkage can be controlled through reducing the cement content. Nonetheless, even where the cement content is at the limit, there is a potential scope for reducing this through increasing the maximum aggregate size, which in turn permits a lower minimum cement content.

Reducing the W/C ratio may also be an option for reducing drying shrinkage. This is likely to be compatible with the specification of concrete for durability, since a maximum W/C ratio will also be required where exposure to aggressive conditions is likely. However, since the water content of the concrete will be fixed to provide the required consistence to the fresh concrete mix, a reduction in the W/C ratio will lead to an increase in the cement content. For this reason, the use of water-reducing admixtures or superplasticisers is likely to be necessary to allow a reduction in the water content. These admixtures are discussed in more detail in Chapter 5. It has already been seen that the water-reducing effect of FA can also be used in this manner.

It should also be remembered that reduction of the W/C ratio to very low levels, while reducing drying shrinkage significantly, may lead to higher levels of autogenous shrinkage.

Selection of aggregate with a high stiffness will act to control shrinkage through restraint. However, aggregate with a high modulus of elasticity will also act to reduce the tensile strain capacity of concrete, making the concrete more prone to cracking, since it will have the largest influence on the stiffness of the concrete and hence the value of f_{ct}/E_c, as discussed in Section 2.2.2. Crushed aggregate typically imparts a higher tensile strain capacity to concrete than rounded aggregate [55]. This is due to the improved cement–aggregate bond that increases the tensile strength of the concrete. However, concrete containing crushed aggregate will typically require a higher cement content.

As for plastic shrinkage cracking, shrinkage-reducing admixtures can be used to modify the surface tension of the pore water [56].

2.3 THERMAL CRACKING

The hydration reactions that Portland cement undergoes with water during setting and hardening are exothermic – heat is evolved. This has the effect of increasing the temperature of concrete, which leads to expansion. Subsequent cooling leads to shrinkage and, if restrained, will lead to cracking in a similar manner with drying shrinkage.

2.3.1 Thermal expansion and contraction

Figure 2.23 shows a typical plot of heat evolution from a Portland cement paste. After an initial period of rapid heat evolution when the cement is initially wet, the cement undergoes a dormant period during which little heat is evolved. At the end of the dormant period, the rate of heat evolution increases to a peak, as a result of the hydration of the tricalcium silicate and tricalcium aluminate phases, after which it gradually declines.

Several models have been developed to predict temperature profiles within volumes of concrete. To help illustrate the influence of key parameters on thermal cracking, a finite-difference model has been used in this book [58]. The model considers a concrete section that is divided into nodes, each with a distance of Δx between them (Figure 2.24). The model progresses in time steps of duration Δt, with temperatures at each node calculated in terms of the heat evolved by the cement, the heat conducted from adjacent nodes and the heat lost via convection, using the Fourier law.

Taking the example of node 1 in Figure 2.24, the temperature at a time step m is given by the equation

$$T_1^m = T_1^{m-1} - \frac{2\Delta t}{\Delta x \rho c}\left[\frac{k_c\left(T_1^{m-1} = T_2^{m-1}\right)}{\Delta x} + h\left(T_1^{m-1} = T_e^{m-1}\right) - \frac{\Delta x}{2}\Delta QC\right]$$

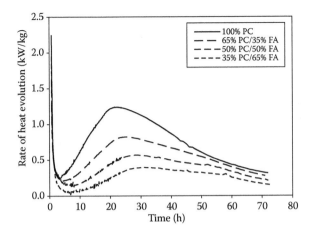

Figure 2.23 Rate of heat evolution carried out on a range of Portland cement (PC)/FA pastes at 5°C. (From Paine, K. A. et al., *Advances in Cement Research*, 17, 2005, 121–132.)

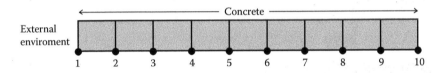

Figure 2.24 The form of the one-dimensional concrete section used in the finite difference model.

where T_y^z is the temperature at node y at time step z (K), ρ is the density of the concrete (g/m³), c is the specific heat capacity of the concrete (J/gK), k_c is the thermal conductivity of the concrete (W/mK), h is the convection heat transfer coefficient (W/m²K), T_e^{m-1} is the ambient temperature at time step $m-1$ (K), ΔQ is the heat evolved by the cement over the period Δt (W/g) and C is the cement content of the concrete (g/m³).

The model used also includes a simulation of the evolution of heat from cement and the inclusion, and possible subsequent removal, of a layer of formwork at the concrete surface.

The evolved heat is lost from the surface of concrete to the atmosphere primarily as a result of convection. However, it is usually the case that the rate of heat evolution exceeds the rate of heat loss, causing the temperature of the concrete to increase. Heat loss leads to the material close to the exterior having a lower temperature than the interior. Figure 2.25 shows the type of temperature profile that can arise.

Most materials expand as their temperature increases, and this is true both for the individual components of concrete in its fresh state and for the

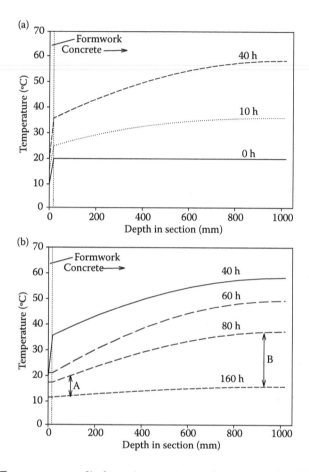

Figure 2.25 Temperature profile from the exterior to the interior of a 1000-mm concrete section exposed to an ambient temperature of 10°C during the evolution of heat from cement hydration (a) and during cooling (b), calculated using a finite-difference model. The concrete contains 350 kg/m³ of Portland cement (PC) in the cement fraction and is placed with a temperature of 20°C against an 18-mm plywood formwork, which is stripped after 48 h. (From Dhir, R. K. et al., *Magazine of Concrete Resesarch*, 60, 2008, 109–118.)

hardened mass. Thus, the increase in temperature leads to an increase in volume or, where a concrete member is restrained, a development of compressive stress once the concrete has started to set and harden.

Once the rate of heat evolution starts to decline, a point is reached where the rate of heat loss exceeds the rate of heat evolution, leading to a net decrease in temperature. This leads to a contraction of the concrete, which, where restrained, leads to the development of tensile stresses and possible cracking.

Cracking occurs during contraction rather than expansion for two reasons. First, the low tensile strength of concrete means that the tensile stresses that develop during contraction (rather than the compressive stresses) are most likely to cause cracking. Second, the modulus of elasticity of concrete during expansion will be lower than that of the more mature concrete during contraction. Because stress is equal to the modulus of elasticity multiplied by strain, the stress for a given strain will be of a greater magnitude during contraction compared with during expansion.

The differences in temperature between the surface and the interior can also lead to tensile stresses developing, which may also lead to cracking.

2.3.2 Cracks resulting from thermal contraction

CIRIA Report C660 [59] details a methodology for predicting the strain resulting from both shrinkage and thermal contraction. It uses the equation

$$\varepsilon_r = \varphi([\alpha_c T_1 + \varepsilon_{ca}] R_1 + \alpha_c T_2 R_2 + \varepsilon_{cd} R_3)$$

where α_c is the coefficient of the thermal expansion of concrete (°C^{-1}), ε_{ca} is the autogenous shrinkage, ε_{cd} is the drying shrinkage, R_{1-3} are the restraint factors, T_1 is the difference between the peak temperature in the concrete and the mean ambient temperature (°C), T_2 is the long-term fall in temperature and φ is the creep coefficient.

The coefficient of thermal expansion of hardened concrete is primarily defined by the coefficient of the aggregates that it contains, as discussed later.

The term in the equation containing T_2 relates to the thermal shrinkage that occurs as a result of seasonal changes in ambient temperature. In the United Kingdom, the mean monthly temperature currently oscillates between approximately 3°C in January and February to approximately 17°C in July and August. This means that concrete cast in the summer will not only undergo contraction as a result of the loss of heat evolved from cement hydration, but will also slowly contract further over the next 5 or 6 months as the ambient temperature drops. Although it is impossible to predict the precise drop in ambient temperature that concrete may experience, from a design perspective, it is only necessary to consider the potential drop in temperature. Thus, The Highways Agency *Design Manual for Roads and Bridges* recommends using a T_2 value of 20°C for concrete construction in summer and 10°C in winter [42].

R_1, R_2 and R_3 are restraint factors, as discussed in Section 2.2.2. Three different restraint factors are necessary because the equation takes into account contraction processes that occur over different time frames. The restraint placed on a concrete element will, in part, be defined by its own stiffness, and so the immature material will be restrained less than at later ages.

Cracks resulting from the cooling of concrete are referred to as 'early-age thermal contraction cracks'. Their nature depends on whether they are formed by temperature differences between the exterior and the interior (referred to as 'internal restraint') or by the overall contraction of a member under external restraint after the peak in cement heat evolution. In both cases, relaxation of strain through tensile creep will act to reduce the development of tensile stresses and possibly prevent cracking. The combined effect of drying (and autogenous) shrinkage, thermal contraction and tensile creep has already been presented in Figure 2.19.

Where cracking results from external restraint, cracks initiate at the surface and extend into the concrete. At the surface, the cracks are typically isolated from one another (they do not form interlinked networks) and are relatively straight or curve gently.

Internal restraint can lead to both the type of surface cracking described above and internal cracking [59]. Surface cracking initiates during the rise in the rate of heat evolution, due to the interior of the concrete expanding more than the surface, putting the surface into tension. However, once the concrete starts to cool, the rate of cooling in the interior exceeds the rate of cooling at the surface (compare the magnitude of lines A and B on Figure 2.25), leading to a faster rate of contraction. A fast rate of contraction is more likely to cause cracking than a slow rate, since, if contraction is fast, relaxation via creep is not allowed to happen, and the stress is more likely to exceed the tensile strength. Thus, the cooling of the interior concrete places it in tension to a greater extent than the surface, potentially leading to crack initiation in the interior and to surface crack widths decreasing as the interior contraction pulls them back together.

2.3.3 Factors influencing early-age thermal contraction

The factors that influence early-age thermal contraction can be subdivided into environmental conditions, construction practices and concrete composition.

2.3.3.1 Environmental conditions

The main environmental condition that influences the temperature rise of a concrete element is the ambient temperature or more accurately the ambient temperature relative to the temperature of the concrete as it is placed. Generally, a lower ambient temperature leads to both a greater temperature difference between the interior and the exterior, and a larger cooling range from the peak temperature to the ambient temperature (Figure 2.26). In both cases, this equates to larger tensile strains.

As for drying shrinkage, higher wind speeds will lead to a greater rate of thermal contraction [59]. However, in this case, it is the result of the wind

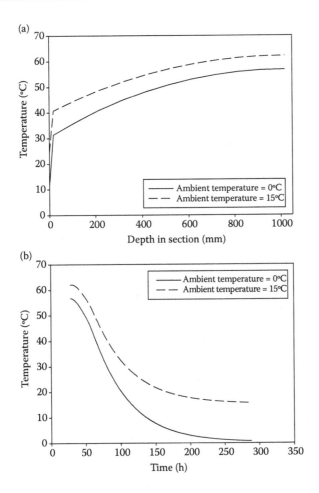

Figure 2.26 Results of calculations for two 1000-mm concrete sections under ambient conditions of 0°C and 15°C. (a) Temperature profile through the sections at the point where the concrete reaches its peak temperature. (b) Change in temperature with time during cooling. Other than ambient temperature, conditions are identical with those in Figure 2.21.

removing warmer air from the immediate exterior of the concrete. The incidence of sunlight on a concrete surface will provide heat to the surface and may reduce the rate of thermal contraction, but may increase the maximum temperature reached within its volume.

2.3.3.2 Construction practices

Formwork plays a role in influencing temperatures and temperature differences in two ways. First, the thermal conductivity of the material used will influence the rate of heat loss. Second, the time of striking of the formwork will influence heat loss, since the removal of the formwork will, in most cases, allow heat to escape more easily.

Materials used as formwork include plywood, steel and fibre-reinforced polymer composites. These materials have very different thermal conductivities and, thus, have different influences on the manner in which heat loss occurs from concrete contained within. Steel has a high thermal conductivity (see Table 2.5), whereas plywood and polymer composites typically have low conductivities. Also provided in the table is a range of thermal conductivities for concrete, since concrete may also be placed against itself. The conductivity of concrete is primarily a function of its density and the degree of water saturation and, therefore, can occupy a relatively wide range.

A lower thermal conductivity (and a greater thickness) will cause the formwork to act as an insulator. Where insulating formwork remains in place for a significant period, this acts to reduce heat loss and causes the maximum temperature reached by the concrete to reach a higher temperature. Thus, the use of a steel formwork will reduce the maximum temperature reached.

However, it is unlikely that the formwork will remain in place for a very long period – it is more likely that it will be stripped relatively early in the construction process. The time at which the formwork is stripped is also an important factor, since after stripping, heat loss will be greater. Thus, if stripping occurs prior to the maximum rate of the cement heat evolution being reached, the maximum temperature reached is typically slightly less than that reached if the formwork were to remain in place. However, it should be remembered that rapid rates of heat loss should be avoided and that stripping either shortly before or after the maximum rate of heat evolution is reached is likely to lead to rapid rates of contraction. Figure 2.27 illustrates how formwork materials and stripping times can influence the change in temperature with time of a point within a concrete section.

Rapid rates of contraction are also possible when curing is achieved through the spaying of water (or at least cold water) onto the concrete surface.

Table 2.5 Thermal conductivities of formwork materials

Formwork material	Thermal conductivity (W/mK)
Steel	43
Plywood	0.13
Fibre-reinforced polymer composite	0.30
Concrete	0.4–1.8

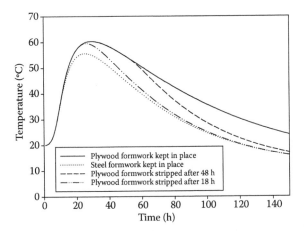

Figure 2.27 Results of calculations of the temperature of the deepest point in a 1000-mm concrete section exposed to an ambient temperature of 10°C using different formwork materials and stripping times. Other than formwork parameters, conditions are identical with those in Figure 2.20.

2.3.3.3 Concrete composition

Because the increase in temperature of concrete results from the presence of PC, an increased PC content will lead to higher temperatures. Cementitious materials such as FA and GGBS undergo lesser exothermic reactions (see Figure 2.23), and so combinations of PC and these materials will give lower temperature rises than for an equivalent mass of PC alone.

From the previous discussion of the mechanism of thermal contraction cracking, it is evident that the coefficient of the thermal expansion of concrete plays a defining role. This coefficient can be estimated using the equation

$$\alpha_c = f_t(f_m f_a \beta_p \alpha_p + \beta_{fa} \alpha_{fa} + \beta_{ca} \alpha_{ca})$$

where f_t is the correction factor for ambient temperature fluctuations, f_m is the correction factor for the moisture content, f_a is the correction factor accounting for the concrete age, β_p is the volume fraction of cement paste, β_{fa} is the volume fraction of fine aggregate, β_{ca} is the volume fraction of fine aggregate, α_p is the coefficient of the thermal expansion of cement paste (°C^{-1}), α_{fa} is the coefficient of the thermal expansion of the fine aggregate (°C^{-1}) and α_{ca} is the coefficient of the thermal expansion of the coarse aggregate (°C^{-1}) [60].

f_t has a value of 1.00 for constant ambient temperature and 0.86 for a fluctuating ambient temperature. f_m takes into account the influence of the moisture content of concrete. It has a value of 1.0 when fully saturated and increases up to approximately 1.8 at intermediate moisture contents before dropping close to 1.0 at very low moisture levels. The correction factor for concrete age (f_a) is also dependent on moisture content and, at very low and very high levels of moisture, has a value of 1. At intermediate moisture levels, it has values between 1 and 0.7, with a lower value for older concrete. From the perspective of early-age thermal contraction, assuming a value of 1.0 for all correction factors is probably valid.

It is clear from this equation that aggregate plays the most important role in determining the coefficient of thermal expansion, since it makes up the majority of the volume of concrete. Table 2.6 presents the results of a study in which the coefficient of the thermal expansion of a range of rock types was measured along with that of concrete containing aggregates derived from these rocks.

It has already been shown that the thermal conductivity of the formwork influences heat loss. The composition of concrete will also influence the rate of heat loss in terms of the rate at which heat is conducted from the interior to the surface. The thermal conductivity of concrete (k_c) can be estimated using the equation

$$k_c = k_p \left[\frac{V_a^{2/3}}{V_a^{2/3} - V_a + \left(\frac{V_a}{\left(\frac{k_a V_a^{2/3}}{k_p} \right) + 1 - V_a^{2/3}} \right)} \right]$$

where k_p is the thermal conductivity of cement paste (W/mK), k_a is the thermal conductivity of aggregate (W/mK) and V_a is the volume fraction of aggregate [62].

Table 2.6 Coefficient of thermal expansion for various rock groups and concrete containing aggregate deriving from them

Rock group	Coefficient of thermal expansion ($10^{-6}/°C$)	
	Rock	Saturated concrete
Chert/flint	7.4–13.0	11.4–12.2
Quartzite	7.0–13.2	11.7–14.6
Sandstone	4.3–12.1	9.2–13.3
Granite	1.8–11.9	8.1–10.3
Basalt	4.0–9.7	7.9–10.4
Limestone	1.8–11.7	4.3–10.3
Manufactured lightweight aggregate, coarse and fine	–	5.6

Source: Browne, R. D., Concrete, 6, 1972, 51–53.

The thermal conductivity of most materials decreases with temperature, and so the previous equation is only applicable to a given temperature when using k values measured at the same temperature.

The thermal conductivity of aggregate is dependent on the minerals that comprise the rock and the proportion of porosity that it contains. Porosity is important because the thermal conductivity of air is extremely low. Since both mineralogy and porosity can vary considerably, the conductivity range for a given rock type is typically broad. This is illustrated in Table 2.7, which provides published ranges for various rock types, plus a commonly manufactured lightweight aggregate.

The thermal conductivity of cement paste is dependent on porosity, moisture content and the type of cement used. As for aggregate, an increase in porosity gives a decrease in thermal conductivity and can be described, for oven-dry cement at 20°C, by the equation

$$k_p = 0.072e^{\left(\frac{3.05}{2.39^{W/C}}\right)}$$

where W/C is the water/cement ratio [63].

Thus, as the W/C ratio increases, the thermal conductivity decreases. The presence of entrained air will act to reduce the thermal conductivity of cement paste as a result of its low thermal conductivity.

At early ages, however, water will be present, and this causes the cement to initially possess a higher thermal conductivity, which drops as the cement hydrates (Figure 2.28). This drop occurs over a very limited period

Table 2.7 Thermal conductivities and specific heat capacities of a range of aggregate materials

Rock group	Thermal conductivity (k) at 20°C [62–64] (W/mK)	Specific heat capacity (c) at 20°C [65] (J/gK)	Density (ρ) (kg/m³)	Thermal diffusity, k/ρc (m²/s)
Chert/flint (quartz)	5.8	0.74	2650	2.96 × 10⁻⁶
Quartzite	3.2–7.9	0.73–1.01	2600–2800	1.1–4.2 × 10⁻⁶
Sandstone	1.3–4.3	0.78	2200–2800	6.0 × 10⁻⁷–2.5 × 10⁻⁶
Granite	1.9–4.0	0.60–1.17	2600–2700	6.0 × 10⁻⁷–2.6 × 10⁻⁶
Basalt	1.4–3.8	0.90	2800–3000	5.2 × 10⁻⁷–1.5 × 10⁻⁶
Limestone	1.0–3.3	0.68–0.88	2300–2700	4.2 × 10⁻⁷–2.1 × 10⁻⁶
Manufactured lightweight aggregate	0.2–0.6	0.7–0.8	960–1760	1.4–8.9 × 10⁻⁷
Portland cement, W/C 0.3–0.8	0.3–0.6	0.73–0.74	1200–1800	2.3–6.8 × 10⁻⁷

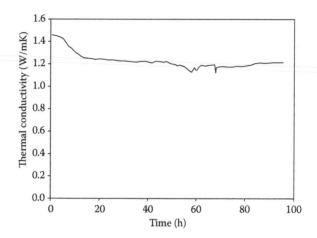

Figure 2.28 Change in the thermal conductivity of a Portland cement paste with a W/C ratio of 0.3. Spikes in the plots are instrument effects. (From Mikulić, D. et al., Analysis of thermal properties of cement paste during setting and hardening. *Proceedings of the International Symposium on Non-Destructive Testing of Materials and Structures*, Istanbul, Turkey, May 15–18, 2011, p. 62.)

at very early ages, and beyond this point, conductivity remains relatively constant. This relatively rapid drop corresponds to the incorporation of water in cement hydration products.

The calculated thermal conductivity of concrete can be corrected for moisture content using the equation

$$k_c(\text{corrected}) = k_c \left[1 + \frac{(6d_m - d_0)}{d_0} \right]$$

where d_m is the density of concrete in a moisture-containing condition (kg/m³), and d_0 is the density of concrete in an oven-dry condition (kg/m³).

The influence of cement type is illustrated in Figure 2.29, which shows the effect of including FA and GGBS in the cement fraction. Silica fume has also been found to reduce thermal conductivity [68].

Examination of the sample equation from the temperature profile model discussed in Section 2.3.1 indicates that the specific heat capacity of concrete also controls the extent to which heat transfer within concrete manifests itself as a change in temperature, with a high specific heat capacity leading to smaller temperature differences.

The specific heat capacity of concrete can be estimated using an equation based on a law of mixtures approach [70]:

$$c_{\text{concrete}} = c_{\text{water}} m_{\text{water}} + c_{\text{cement}} m_{\text{cement}} + c_{\text{aggregate}} m_{\text{aggregate}}$$

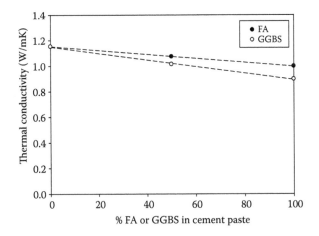

Figure 2.29 Influence of FA and GGBS on the thermal conductivity of cement pastes. (From Kim, K.-H. et al., *Cement and Concrete Research*, 33, 2003, 363–371.)

where c is the specific heat capacity, and m is the mass fraction of each constituent.

Where different coarse and fine aggregates are used or where cement combinations (for instance, PC and FA) are present, this equation needs to be subdivided further to take into account the different specific heat capacities.

Like thermal conductivity, specific heat capacity is dependent on temperature. At 20°C, the specific heat capacity of free water is 4.18 J/gK. This value is much higher than that of the other constituents of concrete, and as a result, free water makes the largest contribution. Water, which has been incorporated into cement hydration products, makes a much smaller contribution – estimated to be approximately 2.2 J/gK [71]. This is the reason for the drop in specific heat capacity with increasing cement maturity, as shown in Figure 2.30. The higher specific heat capacity of cement paste with a higher water content (and therefore a higher W/C ratio) is also illustrated in Figure 2.30.

The specific heat capacity of rock types commonly used as aggregate varies less than thermal conductivity, as shown in Table 2.7.

The specific heat capacities of cementitious materials are very similar, as shown in Table 2.8, with the exception of metakaolin.

Aside from changes in moisture content, there appears to be little change in the specific heat capacity of cement pastes before and after hydration [72]. The exception to this is cement paste containing silica fume, whose specific heat capacity increases with time [68]. It has been proposed that the manner in which atomic lattice vibrations are transferred through the cement matrix is a possible explanation for this, although this has not been proven through experimental measurement.

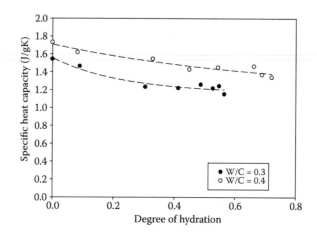

Figure 2.30 Change in specific heat with degree of cement hydration for two PC pastes with different W/C ratios. (From Bentz, D. P., *Materials and Structures*, 40, 2007, 1073–1080.)

Table 2.8 Specific heat capacities of some cement constituents

Material	Specific heat capacity at 20°C (J/gK) [66,70,73,74]
Portland cement	0.73–0.74
Siliceous FA	0.72
Calcareous FA	0.73
GGBS	0.67–0.74
Silica fume (SiO$_2$)	0.74
Metakaolin	1.01

It can be seen from the temperature profile equation in Section 2.3.1 that the rate of heat transfer is dependent on the term $k/\rho c$. Thus, the rate of heat transfer is high when the thermal conductivity is high and the specific heat capacity and density of the material are low. $k/\rho c$ has units of square meters per second and is referred to as thermal diffusivity. Table 2.7 also includes calculated diffusivity values for aggregate materials (plus Portland cement). It is evident that aggregates (other than lightweight aggregates) permit more rapid transfer of heat.

Admixtures that alter the rate of hydration will clearly also impact the rate of heat evolution. Accelerating admixtures will increase the maximum rate of heat evolution, and this will lead to a higher maximum temperature being achieved in concrete. Retarding admixtures will have the opposite effect. It is worth stressing that admixtures with a purpose other than changing the rate of setting and hardening may also influence the rate of cement hydration.

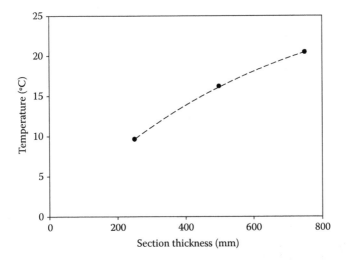

Figure 2.31 Influence of thickness on the maximum temperature reached in a concrete section. (From Dhir, R. K. et al., *Magazine of Concrete Research*, 60, 2008, 109–118.)

The dimensions of a concrete element play an important role in determining the maximum temperature reached, with a larger volume giving a higher temperature, as shown in Figure 2.31. This is the result of the heat generated in the concrete interior having to travel a larger distance by conduction to reach the surface.

2.3.4 Preventing thermal cracking

The most effective means of reducing the maximum temperature is to reduce the PC content of a concrete mix. As for drying shrinkage, although there may be a scope to reduce the cement content somewhat, mix proportion requirements for durability may limit the extent to which this can be done (see Section 2.2.3). However, by using cementitious materials that contribute less to the early heat of hydration, such as FA or GGBS, a significant reduction in the maximum temperature reached is possible. It is sometimes stated that FA and GGBS do not undergo an exothermic reaction. In reality, both reactions are exothermic, but the rate at which heat is released is much slower. The heat of reaction from these materials is dependent on their CaO content, with GGBS and calcareous FA producing almost as much heat as Portland cement, and siliceous FA producing significantly less [75].

The use of aggregates with a lower coefficient of thermal expansion will reduce the magnitude of thermal contraction. Although the selection of aggregates on the basis of their capacity to transfer heat may seem a possible means of controlling the manner in which temperature profiles develop in concrete, it is evident from Table 2.7 that, for normal weight aggregates, there is a limited variation in their thermal diffusivity characteristics. Selecting lightweight aggregates over normal weight aggregates will have a more profound influence, but because of their impact on the engineering properties of concrete and the higher cost of using lightweight aggregate, it is unlikely that such a decision would be viable solely on the grounds of avoiding early-age thermal contraction cracks.

As for drying shrinkage, the use of crushed aggregates or aggregates with a low modulus of elasticity will increase the tensile strain capacity, reducing the susceptibility of concrete to cracking.

Another way in which the maximum temperature reached within concrete can be reduced is through the cooling of the constituents. Although this can be achieved by using ice for cooling mix water, blowing refrigerated air through aggregate stockpiles and even spraying aggregates or entire fresh concrete mixes with liquid nitrogen, more basic technologies can be used to limit the temperature reached by ingredients [59]. These include running pipes and storage tanks for water used for concrete production beneath the ground, painting the surfaces of exposed pipes and tanks white, storing aggregates under shade and sprinkling with water. The temperature of concrete prior to the onset of cement hydration can be estimated using the equation

$$T = \frac{0.75(T_c M_c + T_a M_a) + 4.18 T_w M_w - 334 M_i}{0.75(M_c + M_a) + 4.18(M_w + M_i)}$$

where T is the temperature (°C), M is the proportion by mass (kg/m³) and c, a, and w are cement, aggregate, and water, respectively [76].

In terms of construction practices, much of the previous techniques used to control cracking for drying shrinkage also apply to thermal contraction, namely, the inclusion of minimum reinforcement and movement joints. Additionally, a sequence for multiple adjacent pours can be devised such that restraint is minimised. Guidance on planning such a sequence is provided in the Construction Industry Research and Information Association (CIRIA) report *Early-Age Thermal Crack Control in Concrete* [27]. To summarise, the guidance states that, wherever possible, the shape of concrete pours (whether bays for pavements and floors, or lifts) should be as close to a square either in plan or elevation and that, where non-square volumes are placed, a minimum period should be left between placing adjacent elements whose longer edges are in contact. By minimising the time between pours, adjacent bays will shrink due to thermal contraction almost in unison, thus reducing strains resulting from restraint. In contrast, when placing adjacent elements whose shorter edges are in contact, the period between pours should be maximised (Figure 2.32). This is because this avoids stresses resulting from the thermal contraction of the first pour acting perpendicular to the joint between the two elements, causing the joint to crack.

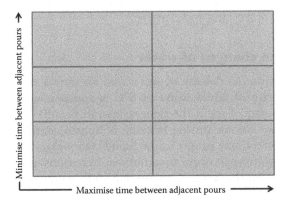

Figure 2.32 Appropriate sequence of placement for concrete elements (either bays viewed in plan or lifts viewed in elevation) to minimise restraint. (From Bamforth, P. B. Early-age thermal crack control in concrete. *CIRIA Report C660*, 2007, 9 pp.)

The manner in which formwork is used has the potential to reduce the maximum temperature reached by concrete. However, the nature of the concrete section being placed determines what approach should be taken. Where sections are relatively narrow (<500 mm), heat is likely to be able to travel from the interior to the surface relatively quickly, and so allowing heat to escape rapidly from the surface is advisable, since this will minimise the maximum temperature reached. Rapid heat loss is achieved through the use of formwork with a high thermal conductivity, for example, steel.

For deeper sections, rapid heat loss at the surface will lead to a significant temperature difference developing, which is more likely to lead to cracking. Thus, in such cases, insulation from surface heat loss is the more favourable option, and this is achieved through the use of formwork with a lower thermal conductivity (such as plywood) and long periods prior to striking to ensure that the drop in temperature is slow. The provision of additional insulation in the form of thermal blankets further reduces the rate of heat loss in such cases.

Placing concrete at lower ambient temperatures will also reduce the maximum temperature reached. Whilst ambient conditions are beyond the control of the engineer, lower night-time temperatures can potentially be exploited.

2.4 FREEZE–THAW ATTACK

Water is one of a very limited number of substances that undergoes an increase in volume when it transforms from a liquid to a solid. Within the confines of the pores of concrete, this expansion can lead to damage over repeated cycles of freezing and thawing.

2.4.1 Volume changes of water

Figure 2.33 shows the density of water and ice over a range of temperatures. The density of liquid water at 0°C is approximately 1000 kg/m³, whereas the density of ice at the same temperature is 917 kg/m³, resulting in an increase in volume during freezing of approximately 9%. It should be noted that water may persist in the liquid state below 0°C if it is supercooled. Supercooling can occur for a number of reasons, the most common of which being that the water is not mechanically agitated during cooling.

2.4.2 Action of ice formation in concrete

Free water in concrete is confined within pores. Freezing of this water leads to expansion, which, within the confined space of the pores, causes the development of stresses that can lead to cracking. However, cracking may occur in concrete whose pore system is not fully saturated. To understand why this is, it is necessary to look at the process of freezing in concrete in more detail.

Water in concrete pores does not freeze at 0°C. There are two main reasons for this. First, the freezing point of water is depressed by the presence of dissolved ions. Second, the adsorption of water on pore surfaces has the effect of permitting the supercooling of water.

The main dissolved ions in the pore solutions of uncontaminated Portland cement paste are sodium, potassium and hydroxide. Previous studies of

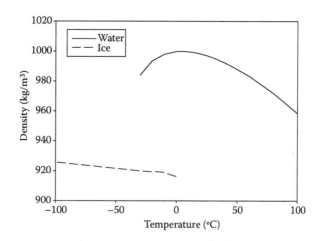

Figure 2.33 Density of water and ice over a range of temperatures.

pore solution composition in reasonably mature mortars have measured K^+ and Na^+ concentrations of almost 0.9 and 0.3 mol/L, respectively [77]. The depression of freezing point of a solvent (K_F) for dilute solutions can be calculated using the equation

$$\Delta T_F = K_F m i$$

where K_F is the cryoscopic constant of the solvent (Kkg/mol), m is the molality of the solute (mol/kg) and i is the van't Hoff factor.

The van't Hoff factor is the ratio of the concentration of solute particles present once dissolved to the concentration of solute. For ionic compounds, such as KOH and NaOH, the van't Hoff factor is equal to the number of discrete ions in the formula unit – 2. The cryoscopic constant for water is 1.853 Kkg/mol. Thus, using the concentrations above, the total depression of freezing point will be approximately 4°C.

However, water in porous media will remain partly liquid to temperatures lower than –40°C, even when there is a negligible dissolution of the solid [78]. The reason for this is that interactions between adsorbed water molecules and the pore walls prevent their arrangement into the configurations required for crystallisation. This effectively maintains a thin layer of water molecules at the pore wall in the liquid state at temperatures below 0°C – in other words, it becomes supercooled.

The overall effect of this is that freezing of free water in concrete does not occur at a single temperature. Instead, solidification occurs progressively over a temperature range, with water in large pores crystallising first. In larger pores, where the ratio of the pore wall surface to the pore volume is relatively small, freezing occurs at relatively high temperatures. However, in small pores, where the adsorbed layer of water is almost as thick as the pore radius, freezing will not occur potentially until temperatures lower than those experienced under terrestrial ambient conditions are reached.

The combination of the formation of a less dense solid and the persistence of a layer of liquid leads to the liquid exerting a hydraulic pressure. Although several mechanisms for how the concrete is damaged have been proposed, the most widely accepted explanation is that the flow of water away from areas of freezing leads to viscous resistance against water movement. This produces hydraulic pressures within the pores, which may be enough to cause tensile cracking. This resistance to flow is proportional to the length of the flow path of water in the pores, and the pressure developed (P, in N/m^2) is described through a rearrangement of the conventional equation for Darcy's law:

$$P = \frac{QL\mu}{-k}$$

where Q is the flow rate of water (m/s), L is the length of the flow path (m), μ is the viscosity of water (Ns/m^2) and k is the permeability of the flow path (m^2) (see Chapter 5) [79].

As well as damage to the concrete, the pore will undergo permanent deformation. This deformation creates additional pore space, which is subsequently filled with water when the concrete next becomes saturated. Thus, multiple freeze–thaw cycles lead to progressive damage to the concrete, as illustrated in Figure 2.34. An example of expansion (or 'dilation') resulting from cyclic freezing and thawing is shown in Figure 2.35. The formation of cracks within the concrete leads to a progressive loss in the stiffness of the concrete with each freeze–thaw cycle (Figure 2.36).

When water freezes, the resulting ice is of relatively high purity, which means that the remaining liquid contains higher concentrations of dissolved species. This results in water of higher purity being drawn into the pore through the process of osmosis, leading to hydraulic pressures of even higher magnitudes than if pure water were present.

The mechanism discussed above only applies to concrete that is saturated and that is in contact with adequate water to fill the space created by pore expansion. Below a certain level of saturation, the flow path of liquid water

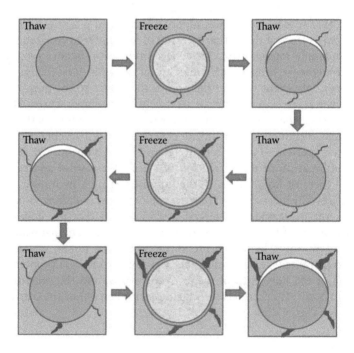

Figure 2.34 Progressive damage around a water-filled concrete pore as a result of cyclic freezing and thawing.

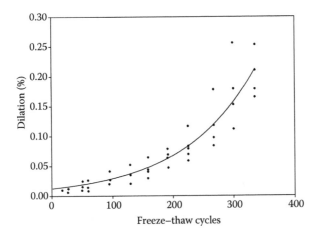

Figure 2.35 Example of the dilation behaviour of concrete specimens exposed to cyclic freezing and thawing. (From Janssen, D. J. and M. B. Snyder, *Strategic Highway Research Programme Report SHRP-C-391: Resistance of Concrete to Freezing and Thawing.* Washington, D.C.: National Research Council, 1994.)

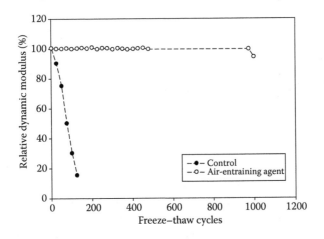

Figure 2.36 Decline in the dynamic modulus of a conventional concrete mix and a similar mix containing entrained air exposed to cyclic freezing and thawing. (From Harrison, T. A. et al., *Freeze–Thaw-Resisting Concrete: Its Achievement in the U.K.* London: Construction Information Research and Information Association, 2001.)

will be interrupted by the presence of air, thus lessening the magnitude of damage (Figure 2.37). This level of water content is referred to as the critical degree of saturation and is typically between 80% and 90%, but can be as low as 50%, and is dependent on constituents, mix proportions and the degree of cement hydration [82].

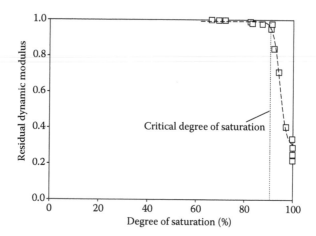

Figure 2.37 Residual dynamic modulus (obtained from measurements of the natural frequency of transverse vibration) versus degree of saturation for concrete specimens exposed to cyclic freezing and thawing. (From Fagerlund, G., *Materiaux et Constructions*, 10, 1977, 231–253.)

Both the cement matrix and the aggregate can potentially undergo damage as a result of freeze–thaw attack, with the behaviour of these different components defining the manner in which deterioration occurs. Thus, the damage produced as a result of cyclic freezing and thawing can vary and can include scaling, D-cracking and pop-outs.

2.4.2.1 Scaling

Scaling is the loss of material from the concrete surface (Figure 2.38). The process of scaling can be measured in the laboratory as a loss in mass, as shown in Figure 2.39. Since scaling involves the loss of surface, it results in a loss in the cross-sectional area and a loss of cover. Scaling is often particularly problematic where other types of mechanical action are effective on the concrete surface.

The nature of the surface of the concrete will influence its susceptibility to scaling. The surface (as will be discussed in more detail in Chapter 6) contains a higher proportion of cement paste, which tends to be prone to delamination from the aggregate-rich material beneath it once the tensile strength of the bond between the two materials is exceeded by the action of ice formation.

The surface finishing technique used can potentially render concrete more vulnerable to scaling. The use of power floats, in particular, is likely to increase the thickness of the paste-rich layer at the surface. Moreover,

Figure 2.38 Scaling of concrete steps.

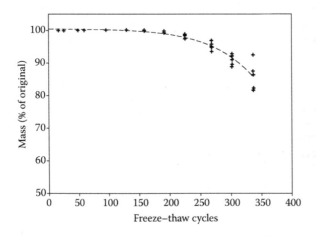

Figure 2.39 Loss of mass from a concrete specimen as a result of scaling. (From Janssen, D. J. and M. B. Snyder, *Strategic Highway Research Programme Report SHRP-C-391: Resistance of Concrete to Freezing and Thawing*. Washington, D.C.: National Research Council, 1994.)

excessive power floating may lead to a weakened interface between the surface and the bulk concrete (see Chapter 6).

The presence of de-icing compounds can have the effect of increasing the rate of scaling. This is largely the result of the osmotic pressures discussed above, because the presence of de-icing salts will considerably increase the amount of dissolved species in the pore solution. Several compounds can be used for de-icing, and it has been proposed that the extent to which scaling is exacerbated by their presence follows the approximate sequence ammonium nitrate = ammonium sulphate → potassium chloride → urea → magnesium chloride → calcium chloride → sodium chloride [84].

In addition to these compounds, ethylene glycol, propylene glycol, potassium acetate and calcium–magnesium acetate are also used as de-icers.

For scaling to occur, it is necessary for a pool of solution to be present at the concrete surface and for the minimum temperature reached to be less than –10°C. Additionally, the rate of scaling is dependent on the concentration of the de-icer present in the solution, with the fastest rate of deterioration occurring around a 'pessimum' concentration of around 3% by mass, seemingly regardless of what de-icer is used [85].

In many cases, interactions between salt and cement hydration products is also thought to contribute towards deterioration. In the case of ammonium nitrate, sodium chloride, potassium chloride, calcium chloride and acetate compounds, this occurs through the solubilisation of calcium hydroxide by substitution of hydroxide for either chloride, nitrate or acetate ions. In the case of magnesium chloride, this mechanism is effective in parallel with the conversion of calcium silicate hydrate to magnesium silicate hydrate (see Chapter 3) and calcium hydroxide to brucite (see Chapter 4). Although brucite formation can limit the ingress of de-icing salts, magnesium silicate hydrate does not contribute towards concrete strength, and the cement matrix is significantly weakened.

Glycol compounds are also known to increase the solubility of calcium hydroxide through the formation of complexes with calcium ions.

De-icer salts can potentially contribute towards the further deterioration of the concrete without freezing through a process known as 'salt hydration distress'. This occurs when concrete is subject to cyclic wetting and drying, which can cause certain salts to alternate between their hydrated and unhydrated (or less-hydrated) forms. The most common examples of this occur with thenardite (Na_2SO_4), thermonatrite (Na_2CO_3) and kieserite ($MgSO_4 \cdot H_2O$) [84]. This is damaging to concrete as a result of the significant volume changes involved in the change from unhydrated to hydrated salt. Although none of these compounds are used as de-icing salts, there exists the potential for their formation via substitution reactions between cement constituents and de-icers.

Where de-icers are applied to frozen concrete surfaces with unfrozen water below the surface, heat is required to thaw the ice at the surface, which is obtained from the water below the surface. It has been postulated that this may lead to rapid freezing, which could also potentially cause spalling [81].

2.4.2.2 D-cracking

D-cracking results from the fragmentation of aggregates rather than of cement paste. It frequently occurs in slabs and is observed at the surface in

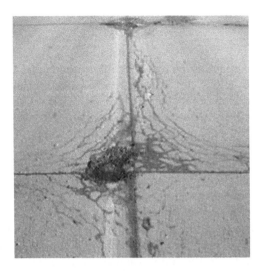

Figure 2.40 D-cracking in a concrete pavement.

the form of cracks close to edges and joints (Figure 2.40). The process of D-cracking usually initiates where the concrete (and more specifically the aggregate in the concrete) is saturated above a critical level. In pavement slabs, this typically occurs at the base, where the concrete is potentially in contact with water-bearing road-base and subbase layers.

Only certain aggregate sources are prone to freeze–thaw deterioration, and test methods exist for evaluating susceptibility (see Chapter 5). In general terms, susceptibility is dependent on the pore structure of aggregates. Specifically, freeze–thaw resistance is determined by the volume of porosity within the aggregate particles, the pore size, and the length of flow paths running through the particles. The influence of the first two of these parameters has been characterised in the form of the equation

$$\text{EDF} = \frac{K_1}{P} + K_2 D_{\text{median}} + K_3$$

where EDF is the expected durability factor, P is the pore volume down to a pore diameter of 4.5 μm (cm^3/g), D_{median} is the median pore diameter down to a pore diameter of 4.5 μm (μm) and K_1, K_2, and K_3 are constants [86].

The 'durability factor' referred to in the equation is a factor that can be calculated from fundamental frequency measurements made on concrete specimens exposed to freeze–thaw cycles, as defined in *ASTM C-666* [87]. It is defined as

$$DF = \frac{100\left(n_1^2/n^2\right)N}{M}$$

where DF is the durability factor, n_1 is the fundamental frequency of the concrete specimen at 0 cycles, n is the fundamental frequency at N cycles, N is the number of cycles at which $\left(n_1^2/n^2\right)$ reaches a specified minimum value or the specified number of cycles at which exposure is to be terminated, whichever is less and M is the specified number of cycles at which exposure is to be terminated.

A high durability factor represents a high resistance to freeze–thaw attack. Thus, a low volume of total porosity and coarser pore sizes will give a higher resistance to freeze–thaw attack.

It has been proposed that values for K_1, K_2, and K_3 of 0.579, 6.12, and 3.04, respectively, are appropriate and that an EDF below 40 indicates susceptibility to D-cracking. The reason for the exclusion of pore diameters below 4.5 μm relates to the suppression of freezing points in pores, since water in pores smaller than this is unlikely to freeze under terrestrial ambient temperatures.

The reason for the flow path length playing a role in influencing the freeze–thaw durability of aggregate is the same as discussed previously for cement paste. However, for aggregate, it has a different significance: since the flow path length will be limited by aggregate size, coarser aggregate is more susceptible to freeze–thaw damage. This effect is illustrated in Figure 2.41.

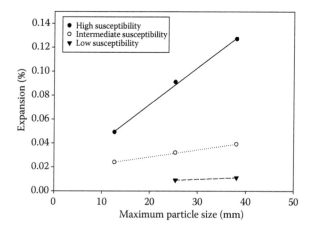

Figure 2.41 Expansion of concrete mixes containing aggregates with service records of low, intermediate and high susceptibility to D-cracking after 350 freeze–thaw cycles versus maximum aggregate size. (From Stark, D. and P. Klieger, *Highway Research Record 441*, 1973, pp. 33–43.)

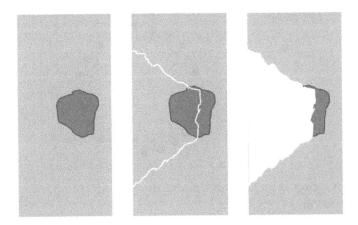

Figure 2.42 Pop-out resulting from the expansion and cracking of an aggregate particle in concrete subject to freeze–thaw attack.

Lightweight aggregates can potentially be susceptible to freeze–thaw attack, since they can accommodate large volumes of water. Generally speaking, the lower the density of lightweight aggregate, the more susceptible it will be [89]. However, as for normal-weight aggregate, pore structure is an important factor, with smaller pore size distributions giving greater resistance.

2.4.2.3 Pop-outs

Pop-outs are also the result of aggregate susceptibility to freeze–thaw attack. In this case, damage occurs as a result of the expansion of aggregate particles close to the surface, leading to surface cracking and the loss of small conical volumes of mortar at the surface around the aggregate particle, as shown in Figure 2.42.

2.4.3 Avoiding damage from freeze–thaw attack

Where freeze–thaw attack is likely to result from ice formation in the cement matrix of the concrete, rather than through aggregate susceptibility, the most effective means of avoiding freeze–thaw damage is through the entrainment of air. Air entrainment is achieved with the use of chemical admixtures known as 'air-entraining agents'.

2.4.3.1 Air-entraining admixtures

Air-entraining admixtures are substances added during mixing to promote the formation and stability of air bubbles within the cement matrix. They

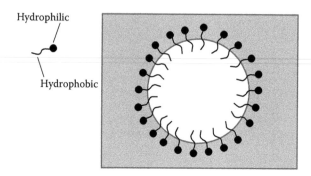

Figure 2.43 Location of air-entraining agent molecules at the air–water interface of a bubble.

are usually anionic surface active agents ('surfactants'), including salts of fatty acids (e.g., sodium oleate), alkyl sulphates (such as sodium dodecyl sulphate [SDS]), alkyl–aryl sulfonates, products derived from the neutralisation of wood resins and alkylphenol ethoxylates (APEs) [11].

Surfactants are molecules that possess hydrophobic and hydrophilic groups, usually at the opposite ends of a molecular chain. When added to a concrete mix – normally dissolved in the mix water – the vast majority of the molecules are adsorbed onto cement and aggregate surfaces, with the hydrophobic group aligned with the surface. However, a smaller number of the molecules (now almost certainly present as calcium salts) will remain in the solution. The adsorption of surfactant molecules on solid surfaces is significant, since it limits the number of molecules that can take part in air entrainment. The vast majority of solid surface in a fresh concrete mix will be cement, and as we will see later, the fineness of cement influences the amount of air that can be entrained for a given dosage of admixture.

The dissolved molecules assist in the formation of bubbles produced by the agitation of mixing by reducing the surface tension of the water, reducing the amount of energy required to form each bubble and, therefore, increasing the number of bubbles formed.

Bubbles formed in this way are typically small (<0.25 mm in diameter) and, without the surfactant, would either coalesce to form larger bubbles or dissolve into the water. The surfactant molecules arrange themselves at the air–water interface (Figure 2.43), preventing both processes and, thus, stabilising the bubbles.

2.4.3.2 Effects of air entrainment

Air entrainment is effective in protecting from frost attack because the bubbles provide space into which water can flow during freezing. With reference to the theory behind the development of hydraulic pressures discussed in the previous section, the presence of air bubbles effectively acts to shorten flow paths within the cement matrix, thus reducing hydraulic pressures.

The effectiveness of air entrainment is very much dependent on the total volume of air and the size of bubbles formed. Initially, an increase in air content leads to an improvement of freeze–thaw durability, but the benefit of air content levels off, as shown in Figure 2.44. The figure shows that a larger air content is required for higher strength concrete mixes. The reason for this is that, for the mixes used to generate these data, those with higher strength contained a greater volume of cement paste, and since it is the paste that contains the bubbles, a higher volume of air is required to achieve an equal level of protection. The same effect is seen when comparing mixes containing aggregate with a range of maximum aggregate sizes – concrete with a larger maximum aggregate size will typically require a smaller air content to achieve maximum protection from freeze–thaw attack, because these mixes will require a lower cement paste content to fill the space between aggregate particles.

Bubble size is significant because it determines the distance between individual bubbles at a given air content. This distance is referred to as the 'bubble spacing factor', and theoretical calculations indicate that a spacing factor of less than 0.25 mm provides maximum protection from freeze–thaw damage [91]. In reality, this value has been observed to be between 0.20 and 0.80 mm, depending on cement type, W/C ratio and other factors.

Although the spacing factor is a superior means of defining the manner in which air is present in concrete, it cannot be determined for fresh concrete, since it requires a microscopic analysis of the hardened cement matrix. The total air content of fresh concrete, on the other hand, is measured with

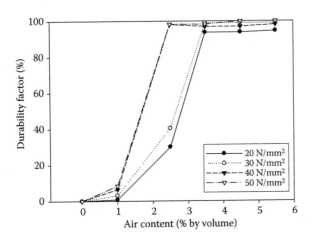

Figure 2.44 Durability factor versus air content for concrete mixes of various strengths. (From Dhir, R. K. et al., *Magazine of Concrete Research*, 51, 1999, 53–64.)

relative ease using the pressure methods described in *EN 12350-7* [92]. Therefore, it is conventional to specify concrete in terms of air content. The European standard for concrete, *BS EN 206* [93], requires that air content be specified in terms of a minimum air content, with the upper limit being the minimum air content plus 4% by volume.

The magnitude of freeze–thaw exposure is defined within the standard in terms of exposure classes for freeze–thaw attack, which are XF1 to XF4:

- XF1: moderate water saturation without de-icing agent
- XF2: moderate water saturation with de-icing agent
- XF3: high water saturation without de-icing agent
- XF4: high water saturation with de-icing agent or sea water

where XF1 is the least aggressive of the conditions.

The standard provides limiting values for a minimum air content of 4.0% by volume for exposure classes XF2 to XF4. For all exposure classes, aggregate is required to be sufficiently resistant to freeze–thaw attack, as defined by *BS EN 12620* [94] (see Chapter 5). It also provides maximum W/C ratios – an approach to reducing the damaging effects of freezing and thawing, which will be discussed later.

The complementary British standard to *BS EN 206*, *BS 8500-1* [95], provides more detailed requirements for air content. Although the air contents required for different XF exposure classes do not change, it addresses the fact that, as maximum aggregate size increases, the cement content decreases. Thus, because entrained air is exclusively present in the cement paste, as the maximum aggregate size increases, the minimum required air content reduces. The standard also defines the minimum cement contents for different maximum particle sizes.

A less desirable side effect of air entrainment is that the inclusion of bubbles in the cement matrix leads to a reduction in the strength and the stiffness of the hardened concrete. This reduction is largely dependent on the volume of air entrained (Figure 2.45). The effect on stiffness is similar, as shown in Figure 2.46.

Although the presence of entrained air often has the effect of making cement pastes more viscous, the presence of air within concrete has the effect of reducing viscosity and thus improving workability [96]. It has been proposed that the reason for this is that the presence of air increases the volume of cement paste [97]. The resulting increase in the volume ratio of paste to aggregate produces the improved workability. It has also been suggested that the presence of air bubbles acts to lubricate the movement of aggregate particles. Entrained air will also act to reduce bleeding.

This improvement in workability is clearly welcome, since it permits a reduction in water content while maintaining the desired workability. The

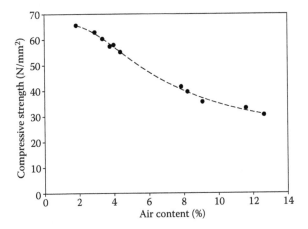

Figure 2.45 28-day compressive strengths of Portland cement-based concrete mixes with the same mix proportions but varying entrained air contents. (From Zhang, Z. and F. Ansari, *Engineering Fracture Mechanics*, 73, 2006, 1913–1924.)

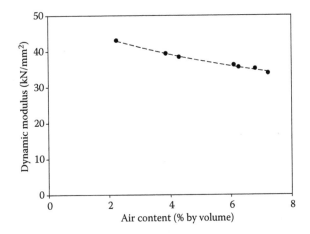

Figure 2.46 Effect of air content on the dynamic modulus of a concrete mix. (From Mayfield, B. and A. J. Moreton, *Civil Engineering and Public Works Review*, 64, 1969, 37–41.)

resulting lower W/C ratios can therefore be exploited to partly offset the loss of strength resulting from the presence of air.

The introduction of air into concrete might be expected to render the resulting material more vulnerable to ingress of substances harmful to either the concrete itself (Chapter 3) or steel reinforcement (Chapter 4). However, many researchers report either no change or a decrease in permeability [98], sulphate deterioration rates [99], chloride diffusion coefficients [100] and rates of carbonation [101] in concrete containing entrained air relative to control specimens containing no air. The reason for this is largely the result of the reduction in W/C ratio afforded by the improvement in workability. Where W/C ratios are kept constant, the presence of air does indeed increase gas permeability, along with gas diffusivity (of oxygen) [102]. Water sorptivity is generally reduced by the presence of air.

The increase in permeability is not the direct result of the presence of air voids – these are not interconnected and so can only contribute in a minor way to the movement of fluid through the cement matrix. Instead, the region around each air bubble is similar in character to the region around aggregate particles (see Chapter 5), which means that they are more porous and act as a path for fluid to flow through with less resistance than the bulk cement paste.

It has also been established that air voids at the surface of steel reinforcement act as initiation points for corrosion, leading to a reduction in the threshold chloride concentration required to initiate corrosion [103]. Thus, where protection of reinforcement is one of the objectives, it is likely that the increased resistance to chloride ingress will be offset by a greater vulnerability to corrosion. Although there are currently insufficient data to provide clear guidance on this aspect of durability, it is probably reasonable to conclude that no benefit with respect to corrosion protection is achieved through air entrainment.

2.4.3.3 Loss of entrained air

BS EN 206 makes the point that the specifier must take into account the possibility that the air content may be reduced during mixing, pumping, placing and compacting. This is a very real possibility, since any violent force acting on fresh concrete will have the effect of driving some of the bubbles to the surface. This is shown in Figure 2.47, which shows the effect of compaction by vibration on air content.

The loss of air is not necessarily as serious as it may appear at first because larger bubbles are preferentially lost from the surface of the concrete. This is the result of the upthrust force and viscous drag acting on a bubble in a fluid being equal and opposite. The upthrust force is defined by Archimedes' law as the weight of fluid displaced by the bubble, whereas viscous drag is defined by Stoke's law. Because these forces are equal, the equation below can be written:

$$6\pi\mu r\upsilon = \frac{4\pi r^3 g\rho}{3}$$

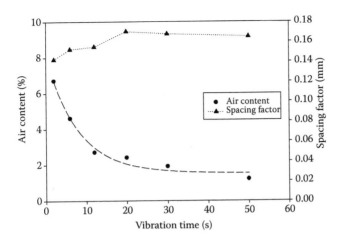

Figure 2.47 Influence of vibration time on the air content and spacing factor of fresh air-entrained concrete. (From Backstrom, J. E. et al., *Journal of the American Concrete Institute*, 30, 1958, 359–375.)

where μ is the viscosity of the fluid (Ns/m^2), r is the bubble radius (m), v is the velocity of the bubble (m/s), g is the acceleration caused by gravity (m/s^2) and ρ is the density of the fluid (kg/m^3).

The velocity at which a bubble rises to the surface is therefore

$$v = \frac{2\rho r^2 g}{9\mu}$$

which means that a bubble of 20-μm diameter will rise at a velocity four times that of a bubble of 10 μm. The overall effect of this phenomenon is shown in Figure 2.47 in the form of the spacing factors observed after different periods of vibration. Although there is an increase in spacing factor, it is very slight, indicating a retention of finer bubbles. The overall effect of this is that large bubbles that do not contribute to freeze–thaw resistance, but do contribute towards the loss in strength, are lost, whereas the more beneficial small bubbles are retained [81].

Mixing may lead to the loss of entrained air, particularly if it is carried out over prolonged periods, such as during transit. Figure 2.48 shows the manner in which mixing time affects the air content in mixes containing different air-entraining admixtures. It is evident that, in most cases, the loss in air is relatively small, even over long mixing periods. However, the figure illustrates that this is dependent on the type of admixture used, with one of the agents showing much larger losses, albeit from a higher initial air content.

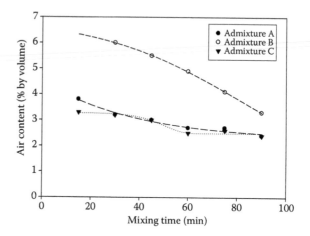

Figure 2.48 Changes in air content, with the mixing time of concrete mixes containing three different air-entraining admixtures. (From Scripture, E. W. and F. J. Litwinowicz, *ACI Journal*, 45, 1949, 653–662.)

The pumping of concrete will change air-void characteristics. This is partly dependent on the manner in which pumping is carried out, with vertical pumping producing the most profound changes. Figure 2.49 shows changes in the air content and the spacing factors of concrete mixes before and after pumping. For both horizontal and vertical pumping, there is an

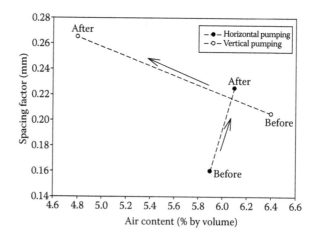

Figure 2.49 Effect of horizontal and vertical pumping over distances of approximately 20 m on the air content of fresh concrete and the bubble spacing factor in the hardened state. (From Pleau, R. et al., *Transportation Research Record* 1478, 30–36.)

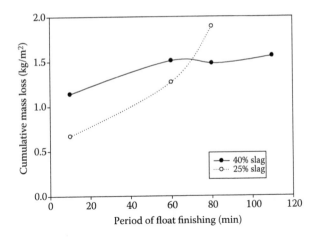

Figure 2.50 Magnitude of mass loss resulting from scaling after 50 freeze–thaw cycles from concrete specimens containing different proportions of slag. (From Hooton, R. D. and A. Boyd. Effect of finishing, forming, and curing on de-icer salt scaling resistance of concretes. In M. J. Setzer and R. Auberg (eds.), *RILEM Proceedings 34: Frost Resistance of Concrete*. London: Spon, 1997.)

increase in the bubble spacing factor. Additionally, where vertical pumping is conducted, there is also a significant decline in the total volume of entrained air.

Surface finishing activities – particularly power floating – can also remove significant quantities of air from the surface layer. This can have much more significant implications, since this zone of a concrete element is likely to be exposed to the greatest threat from freeze–thaw damage. This is shown in Figure 2.50, where prolonged periods of finishing by power float lead to more significant scaling when exposed to cyclic freezing and thawing.

2.4.3.4 Factors affecting air content and air-void parameters

Concrete with a higher workability will tend to undergo the 'deentrainment' of air at a greater rate. However, this is offset by the fact that more workable concrete will tend to contain a higher total volume of air for a given dosage of air-entraining agent.

The temperature of fresh concrete during mixing influences both the total volume of air entrained and the air-void characteristics. Figure 2.51 plots the total air and spacing factors of air-entrained concrete manufactured over a range of temperatures. The total volume of entrained air declines with increasing temperature, whereas the spacing factor increases.

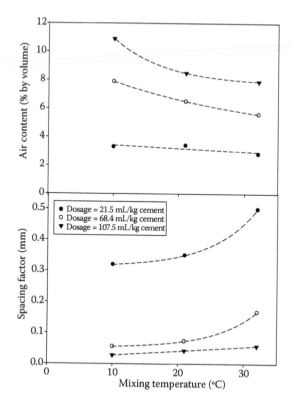

Figure 2.51 Total air content and spacing factors for concrete mixes containing three different dosages of air-entraining agent and mixed at a range of temperatures. (From The Concrete Society. *Technical Report 34: Concrete Industrial Ground Floors, 3rd ed.* Surrey, United Kingdom: The Concrete Society, 2003.)

The reason for this is related to workability, as previously discussed. Fresh concrete at low temperatures has a lower viscosity because such conditions limit the extent to which the early hydration products that increase viscosity are formed. Thus, the concrete is more workable at low temperatures and permits larger volumes of air to be entrained.

A number of characteristics of the constituents of a concrete mix will play a role in determining the volume of air entrained for a given dosage of air-entraining agent. These factors are summarised in Table 2.9.

The amount of air entrained for a given dosage of air-entraining agent is significantly influenced by the W/C ratio of a concrete mix. This is illustrated in Figure 2.52, which plots air content obtained for the same air-entraining agent dosage for a series of concrete mixes of different W/C ratios. As the W/C ratio reduces, so does the air content achieved.

Table 2.9 Concrete constituent characteristics influencing the amount of air entrained by a given dosage of air-entraining agent

Constituent	Characteristic	Details	Reference(s)
Portland cement	Fineness	Finer cements display lower levels of air entrainment for a given admixture dosage.	[110]
	Alkali content	Higher alkali levels have the effect of increasing the volume of air entrained at a given dosage level, as well as decreasing the spacing factor for a given air content. Higher alkali levels also improve the stability of bubbles. Alkalis can come from other sources, such as FA and aggregates.	[111–113]
FA	Carbon content	Unburnt carbon in FA acts to reduce the air content achieved with a given dosage of admixture. Thus, ash with high loss-on-ignition values can be problematic.	[114]
		See Portland cement.	
Slag	Fineness	The presence of slag will normally reduce the volume of air entrained and increase the spacing factor for a given dosage of air-entraining agent.	[115]
		See Portland cement.	
Silica fume	Fineness	Silica fume has little influence on air content or spacing factor.	[116]
Aggregate	Fine aggregate content	Fine aggregate >300 μm will enhance the air-entraining capabilities of a mix.	[117]
	Fine content	Material <300 μm will typically reduce the air content.	[117]
Superplasticising admixtures		Superplasticisers used alongside air-entraining agents will typically reduce the total volume of air. However, they preferentially reduce the number of larger bubbles, with the overall effect of reducing the spacing factor.	[118]

2.4.3.5 Other approaches

BS 8500-1 also provides an alternative route to freeze–thaw resistance by including the option of increasing the minimum strength class and, for more aggressive conditions, a reduction in the maximum W/C ratio. This approach is likely to reduce the extent to which water penetrates beneath

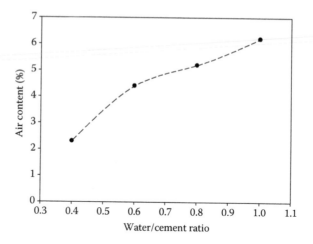

Figure 2.52 Air content of concrete mixes with different W/C ratios and an equal dosage of air-entraining admixture. (From Backstrom, J. E. et al., *Journal of the American Concrete Institute*, 30, 1958, 359–375.)

the concrete surface while also increasing its resistance to the stresses produced during freezing. The validity of this approach is borne out by the results of laboratory testing, such as those shown in Figure 2.53.

There has been much discussion of whether low W/C ratios are an adequate measure against freeze–thaw attack. Although there are certainly examples in the literature of low W/C mixes that perform well under freeze–thaw exposure [119], there are many examples of those that do not [120].

BS 8500-1 indicates that the aforementioned approach is likely to be inferior to air entrainment, but includes it as an option on the grounds that, at very low W/C ratios, achieving appropriate air contents may prove difficult.

The penetration of water into concrete can be further limited by the use of surface treatments, such as surface coatings and hydrophobic penetrants (Chapter 6). It is important that, where surface coatings are used, the formulation will permit water vapour to pass through. Where this does not happen, the concrete will not be able to dry out, and water can become trapped behind the coating, potentially making the material more vulnerable.

The inclusion of fibres in concrete has, in some cases, been demonstrated to improve freeze–thaw resistance. Fibres that have been shown to be in some way beneficial include steel, polypropylene, polyvinyl alcohol, carbon and processed cellulose [122–126]. The magnitude to which resistance is modified by the presence of fibres varies considerably across studies in the literature, with some apparent dependency on the cement used. This

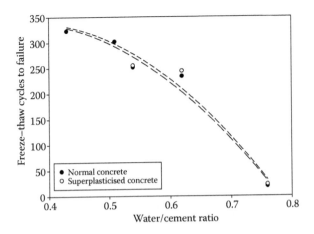

Figure 2.53 Number of freeze–thaw cycles required to cause failure (using a linear expansion of 0.3% as the failure criteria) of concrete prisms with varying W/C ratios. (From Dhir, R. K. et al., Durability of concrete with a superplasticising admixture. In J. M. Scanlon, ed., *ACI Special Publication SP-100: Concrete Durability – Proceedings of Katharine and Bryant Mather International Symposium, Volume I*. Detroit: American Concrete Institute, 1987, pp. 741–764.)

variation may partly be the result of the wide and varied array of freeze–thaw tests in operation internationally.

At least one study has found that the presence of steel fibres reduces the volume of air and increases the spacing factor when an air-entraining agent is also used [127].

It has been proposed that the improvement in performance is the result not of an increase in tensile strength, but of the ability of fibres to provide crack control (see Section 2.2.2), leading to reduced crack widths resulting from freeze–thaw damage. This would result in a reduction in the rate of ingress of additional water and an improved potential for the autogenous healing of cracks (Chapter 5) [81]. It has also been suggested that the loss of mass from test specimens is not an appropriate measure of damage resulting from freeze–thaw attack for concrete containing fibres [128]. This is because fragments of material can remain attached to specimens via fibres despite substantial damage to the concrete. Thus, largely as a result of the uncertainty surrounding the use of fibres as a means of controlling freeze–thaw damage, this option is currently not covered by UK standards.

Some success in imparting enhanced freeze–thaw durability has been achieved through the use of polymer-modified concrete (which is discussed under the subject of screeds in Chapter 6) [129,130]. However, an improvement in performance is usually observed at relatively large levels of polymer use. Furthermore, this approach is currently not supported by UK standards.

Where aggregates susceptible to freeze–thaw damage are to be used, there is a potential benefit in limiting the maximum aggregate size. Additionally, limiting water ingress through the various means previously discussed may also offer some additional protection.

2.5 ABRASION AND EROSION

Abrasion (or 'wear') of concrete usually refers to the action of traffic (in the form of wheeled vehicles, foot traffic, etc.) on concrete surfaces, which can lead to the gradual loss of material, leading to the loss of a level surface and to a potentially heightened vulnerability to other durability-compromising processes.

Erosion refers to the loss of surface material as a result of either the action of solid particles being carried by moving water or a process known as 'cavitation'. The former of these forms of erosion is very similar to abrasion. However, cavitation is related to the formation and implosion of gas cavities in water experiencing rapid changes in pressure. These mechanisms are discussed in more detail below, along with the characteristics of concrete which enhance resistance to this mode of attack, and how such characteristics are best achieved.

2.5.1 Mechanisms of abrasion and erosion

The term 'abrasion' covers a wide range of processes acting on a concrete surface to cause it to progressively lose material. In general terms, we normally think of abrasion as being the action of sandpaper on a surface – one solid surface slides over another, with one of the surfaces (the sandpaper) being harder and consequently removing a proportion of the softer surface. From experience, we know that we must also apply some pressure to the sandpaper – there must be a force acting to push the surfaces together at right angles to the direction of movement.

The aforementioned process is indeed one of the mechanisms that cause abrasion of concrete surfaces (Figure 2.54). It is referred to as 'abrasion by friction' and is one of the main processes occurring during erosion, caused by particles being carried over a concrete surface by water. However, the same effect can be caused by skidding of a wheel or foot on a pavement or floor surface. Moreover, debris, such as dust and grit, between the wheel or foot and the concrete surface will also have a similar and potentially more damaging effect.

Moving particles may also abrade a surface by simply impacting with it head-on (Figure 2.55). Impacts of this type can cause fragments to break

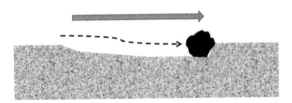

Figure 2.54 Abrasion by friction.

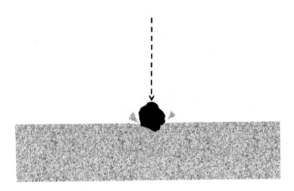

Figure 2.55 Abrasion by impact.

free from the surface or initiate surface cracks that may grow to produce fragmentation after a period of time. Of course, interactions between moving particles and concrete surfaces need not be at either of these extremes, and impacts can have angles of incidence between 90° and very low angles, with a combination of friction and impact abrasion occurring with each particle strike. Figure 2.56 shows the effect of two different angles of incidence (15° and 90°) on the rate of abrasion of a concrete surface with the higher angle being more damaging. The figure also illustrates the influence of another factor that influences erosion rates – fluid velocity – with higher velocities producing greater rates of erosion.

Other factors that influence the rate of erosion are the mass of particles carried in a suspension, particle size, shape, and hardness. Figure 2.57 illustrates the effect of the quantity of particles, with an increase in the concentration of particles producing faster erosion. Typically, larger particles travelling at a given velocity will produce a larger rate of abrasion [131]. Particle size is related to fluid velocity, since a higher velocity will allow moving water to carry larger particles [132].

The influence of particle shape on erosion rates appears dependent on the angle of incidence, with angular particles causing erosion at slightly lower rates compared with rounded particles at low angles, but causing

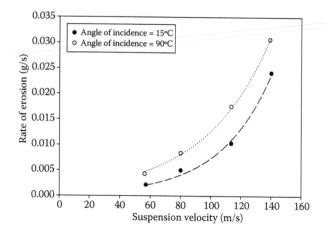

Figure 2.56 Effect of the angle of incidence of suspensions (containing 25% by mass of garnet particles with an average particle size of 0.212 mm) travelling at different velocities on the rate of erosion of concrete specimens made with Portland cement with a compressive cylinder strength of 30 MPa. (From Goretta, K. C. et al., *Wear*, 224, 1999, 106–112.)

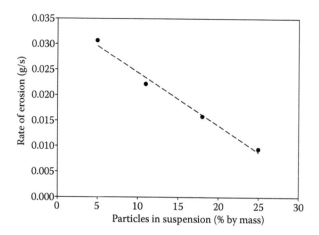

Figure 2.57 Influence of the mass of particles in a suspension travelling at a rate of 140 m/s on the rate of erosion of concrete specimens made with Portland cement with a compressive cylinder strength of 30 MPa. Data points are interpolated from plot lines between data points on a graphical figure. (From Hu, X. G. et al., *Wear*, 253, 2002, 848–854.)

erosion at much greater rates at 90° [131]. It has been proposed that this is because at 90°, angular particles striking the surface will be more effective at inducing cracking in particles of aggregate in the concrete, since their sharp corners will apply higher stresses during impact. At low angles, the manner in which the angular particles strike the concrete is such that this concentration of stress is less likely to occur.

The rate of erosion increases with the hardness of the particles carried by a fluid.

Another process that can contribute to erosion in an environment where moving water is in contact with the concrete surface is cavitation. Cavitation is a phenomenon that occurs when bubbles of water vapour are formed in moving water as it undergoes an abrupt drop in pressure.

In moving water, a drop in static pressure is caused by an increase in velocity of the fluid, as described by Bernoulli's equation for an incompressible fluid:

$$\frac{v^2}{2} + gz + \frac{p}{\rho} = c$$

where v is the velocity of the flow of the fluid at a point on a streamline (m/s), g is the acceleration caused by gravity (m/s²), z is the elevation of the point above a reference plane (m), p is the static pressure at the point (N/m²), ρ is the density of the fluid (N/m³) and c is a constant.

An example of a feature that would cause such an increase in velocity is shown in Figure 2.58, where a constricting channel will cause water to increase in velocity as it moves from the broader to the narrower part. Features such as vortices, slopes away from the direction of the flow, and voids in channels can all produce similar effects.

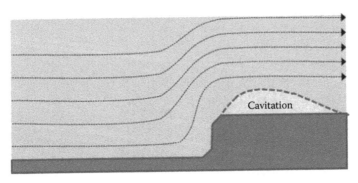

Figure 2.58 Cavitation as a result of the narrowing of a channel.

As a general rule of thumb, cavitation may occur wherever there is a curve in the streamlines followed by the flow of water, on the inside of the curve. Whether cavitation will occur can be estimated using the cavitation index (σ_c), which is calculated using the equation

$$\sigma_c = \frac{2\Delta p}{\rho v_0}$$

where Δp is the drop in static pressure moving from one point in the flow to another (N/m^2), and v_0 is the original velocity of the water (m/s).

The bubbles of water vapour violently collapse in a manner similar to that shown in Figure 2.59. This collapse produces a fast-moving jet of water that leads to the generation of significant local stresses. It is these stresses that, when they occur against a concrete surface, produce damage. Generally speaking, damage can start where the cavitation index exceeds a value of 0.2. ACI 210R-93 [132] provides approximate threshold values (ranging from 0.19 to 0.30) for a range of different features capable of causing the phenomenon.

Damage from individual cavitation events takes the form of small pits in the surface, but since the formation of bubbles will typically occur with great frequency, the cumulative effect can be significant. Generally, accumulation of damage is initially slow, but increases with time until the rate of loss of material peaks and declines.

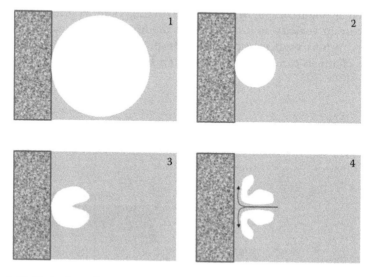

Figure 2.59 Cavitation of a water vapour bubble adjacent to a concrete surface. The collapse leads to the generation of a jet of water (image 4) that applies significant localised stress to the surface.

The action of traffic over a concrete surface will also generate stresses that can cause microcracking, ultimately leading to abrasion. Figure 2.60 shows the different stresses that take the form of vertical compressive stresses, horizontal tensile and compressive stresses and shear stresses. The relatively low strength of concrete in tension and shear makes these types of stress the largest threat to the integrity of a pavement or floor, although large vehicles will induce significant compressive strengths. Although the diagram shows a wheel, it is clear that the action of a human foot will produce similar stress distributions, albeit of a smaller magnitude.

Other processes may influence the rate of abrasion. Chemical attack processes will weaken the surface of concrete, making it more prone to abrasion. However, it is still usually necessary for mechanical abrasive action to be acting on the surface of abrasion to occur, and so chemical attack should

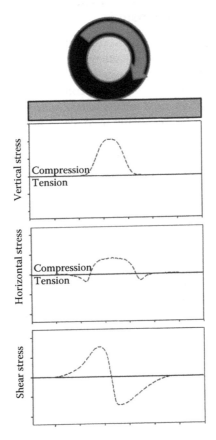

Figure 2.60 Vertical, horizontal and shear stresses in the wearing layer of a concrete pavement as a result of wheeled traffic. Curve shapes are based on References 135 and 136.

not itself be viewed as an erosion mechanism. Chemical attack is discussed in detail in Chapter 3.

Processes that cause surface cracking (for instance, freeze–thaw attack) are also likely to exacerbate abrasion damage. Furthermore, the action of living organisms has been shown to have the capacity to damage concrete surfaces in a manner that could worsen the abrasion process. A study in which concrete blocks were stored for several months in a coastal zone found their surfaces to be highly populated with microorganisms that appeared to be responsible for features in the surface, including boreholes, etched pits, surface microcracks and features characteristic of dissolution [133]. These features affected a relatively shallow depth below the surface (<30 μm) but could potentially present another cause of accelerated abrasion.

2.5.2 Factors influencing resistance to abrasion and erosion

When considering the results of research into the variables that influence abrasion and erosion resistance of concrete, it is important to note that this can be measured using a wide variety of tests. Specifically, the manner in which abrasive action is achieved varies considerably, and it is important to stress that, although in most instances a similar behaviour is likely to be observed regardless of the test being used, cases have been reported where this is not the case.

In terms of British standards, the tests available can be divided into those appropriate for floors and for pavements. Pavement abrasion is dealt with by *BS EN 13863-4* [137]. The method involves exposing concrete surfaces to a road testing machine mounted with wheels with studded car or truck tyres. The standard specifies the contact pressure, speed and tyre type for three variants of the same method. For each method variant, a testing schedule is defined, which involves a series of runs where the number of revolutions and whether testing is carried out wet or dry are specified. After testing, the depth of rutting of the surface is measured.

Abrasion of flooring screeds is covered by several parts of *BS EN 13892*. *BS EN 13892-3* [138] is a method known as the 'Bohme test'. The test involves a test specimen being applied to a revolving plate with a fixed load. A granular abrasive of standardised composition is distributed across the plate surface and abrasion measured in terms of loss of volume from the test specimen.

BS EN 13892-4 [139] is the Bristish Cement Association (BCA) method, which determines abrasion resistance by measuring the depth of abrasion produced by exposure to a specified number of revolutions (taking ~15 min) of a revolving plate mounted with three steel wheels through which a fixed load is applied.

Abrasion resulting from the action of a loaded steel swivel castor on a screed or concrete surface is measured in the method described in *BS EN*

13892-5 [140]. The castor is moved over the surface through the mechanical action of the table on which the specimen is mounted, which oscillates in both longitudinal and transverse directions at different rates, leading ultimately to the castor uniformly covering the test area. The test is required to be carried out for 10,000 cycles, which takes approximately 24 h. Resistance is determined in terms of abrasion depth after a given period of abrasion.

ASTM C779 [141] addresses abrasion of horizontal concrete surfaces. The standard describes three different techniques. The first two techniques involve a test machine consisting of a revolving carousel that is able to apply a constant load to the surface that is being tested. In the first test, the load is applied through three revolving steel discs. The rate of the revolution of both the carousel and the discs is defined in the standard. The carousel is run for a period of 30 or 60 min depending on whether long-term performance is a key concern. The machine feeds abrasive silicon carbide grit, which is fed onto the test surface at a constant defined rate. Abrasion resistance is evaluated by measuring the depth of wear at the end of the test.

The second of the horizontal surface tests does not use grit, and the revolving discs are replaced by three dressing wheels—spiked steel rollers that travel over the concrete surface. Again, after the machine is run for a period of either 30 or 60 min, depth of wear measurements are taken.

The third test uses a different type of machine that consists of a revolving hollow drive shaft that applies a constant load to a ring containing a series of captive, but freely moving, ball bearings held against the concrete surface. The drive shaft is revolved at a defined rate, and the depth of wear is measured using a micrometer gauge that is part of the apparatus. The abrasion resistance is expressed in terms of the time required to reach a specified depth of wear.

Another abrasion test for horizontal surfaces is *ASTM C944* [142]. The test is, in some ways, similar to the previous ASTM test using dressing wheels, albeit on a smaller scale. The test apparatus consists of a downward-facing drill press whose chuck holds a 'rotating cutter'. This consists of a shaft equipped with an array of dressing wheels. The cutter head is pressed against the concrete specimen surface under a specified load, and the cutter is rotated at a specified rate. At 2-min intervals, the cutter is stopped, and the specimen is removed for weighing to determine the loss of mass. The procedure is carried out at least three times on the same spot.

There are additionally two ASTM tests for concrete resistance to erosion-type processes. The first of these, *ASTM C418* [143], uses sandblasting in a cabinet using a standardised gun nozzle and a standardised abrasive delivered onto the surface of a concrete specimen at a specified rate under a specified pressure for 1 min. The rate of abrasion is estimated by measuring the mean volume of the cavities produced on a number of spots by determining the mass of plastic clay of known density required to fill the cavity.

The second method, *ASTM 1138M* [144], is the 'underwater method' and attempts to more closely mimic the conditions encountered when concrete is exposed to sediment carried by moving water. In the method, the cylindrical concrete specimen is placed in a cylindrical tank, with the surface to be tested facing upward. Steel balls of various sizes are placed on the surface, and the tank is filled with water. A paddle of specified dimensions and geometry is submerged in the water at a fixed distance above the concrete surface and rotated at a speed of 1200 revolutions per minute. The concrete specimen is periodically removed and weighed to determine the loss of mass resulting from abrasion up to a test age of 72 h.

Because abrasion is a purely mechanical process acting on the surface of the concrete, it is not surprising that strength plays a very important role in defining the resistance of concrete to this mode of deterioration. The strength of both the aggregate and the cement matrix plays a role, and the importance of these properties is very much dependent on whether the cement matrix or the aggregate is the stronger material.

Figure 2.61 illustrates the influence of the W/C ratio of concrete on abrasion resistance. The influence of cement content on abrasion resistance is dependent on the strength of the cement matrix. With very low W/C ratios, where the strength of the paste is higher than that of the aggregate, increases in resistance with an increase in cement content are observed (for a fixed W/C ratio) [145]. However, the opposite is true where the strength of the concrete is lower than the aggregate.

Stronger aggregate will typically produce greater resistance to abrasion (Figure 2.62). However, this effect is more pronounced where the strength of the concrete is lower [146].

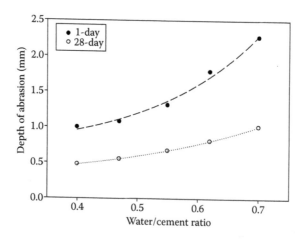

Figure 2.61 Influence of W/C ratio on abrasion depth for Portland cement concrete tested at two ages. (From Dhir, R. K. et al., *Materials and Structures*, 24, 1991, 122–128.)

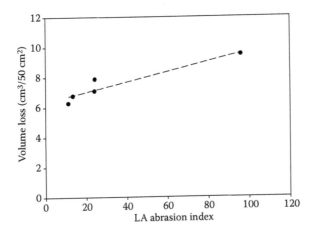

Figure 2.62 Relationship between the strength of aggregate (measured using the Los Angeles (LA) test) and abrasion of concrete containing the aggregate (measured using the Bohme abrasion test). (From Kılıç, A. et al., *Cement and Concrete Composites*, 30, 2008, 290–296.)

Maximum aggregate size appears to be important, with a larger value producing greater resistance [145]. It should be stressed that since normal mix design techniques reduce the cement content as maximum aggregate size increases, this factor may have an influence on experimental results. Nonetheless, when the proportion of coarse aggregate is reduced, there is an increase in abrasion resistance, indicating that aggregate size has a genuine influence [149].

As with most durability characteristics, adequate curing of concrete is of great importance. This is discussed in greater detail in Chapter 6.

2.5.3 Achieving abrasion resistance

BS 8204-2: Screeds, bases, and in situ floorings – Part 2. Concrete-wearing surfaces [150] is a code of practice covering the measures required to achieve abrasion resistance of a directly finished concrete surface. 'Directly finished' means that the concrete is designed and finished such that no additional layer or layers of material are required to be laid over this surface to achieve the appropriate level of abrasion resistance.

The standard categorises abrasion resistance into four classes, whose level of resistance is measured in terms of performance of the concrete surface in the wear test described in *BS EN 13892-4* [139].

The four abrasion resistance classes defined by the standard are AR0.5 (special)/DF, AR1/DF, AR2/DF and AR4/DF, which correspond to wear test depths of 0.05, 0.10, 0.20 and 0.40 mm, respectively, and DF stands for 'directly finished'. The characteristics of concrete mixes satisfying the two highest performing classes are not explicitly defined by the standard, presumably because this level of performance presents something of a technical challenge. Rather, the standard requires that a proprietary concrete is used, possibly containing special aggregate and finished using special methods.

In the case of the lower wear resistance classes, a number of mix characteristics are explicitly specified. First, a minimum compressive strength class is defined – C40/50 and C32/40 for AR2/DF and AR4/DF, respectively. Furthermore, requirements for minimum properties for coarse and fine aggregates are set. Coarse aggregate must have a Los Angeles coefficient of less than LA_{40} (Chapter 5), along with a grading in line with the requirements defined in *BS EN 12620* [94]. However, the standard also states that aggregate for which there is no British standard, but whose suitability for abrasion-resistant floors has been established, may also be used.

Fine aggregate is required to be natural aggregate deriving from the disintegration of rock and conforming to one of the three gradings defined in *BS EN 12620*. Additionally, the floor is required to be finished using power floating, followed by repeated power trowelling (Chapter 6).

The option also exists to lay a screed over a concrete surface, making design for abrasion resistance of the underlying material unnecessary. However, this approach is not without complications, particularly with regards to mismatches in the volume change of overlying and underlying layers. The performance of screeds can be enhanced through the use of dry shake/sprinkle finishes. These are powders comprising mineral particles and/or Portland cement, which are dispersed onto the screed surface prior to finishing. The high wear resistance of the mineral particles and the reduced W/C ratio obtained in the surface layer through the introduction of more cement has the effect of improving abrasion resistance. The use of screeds is discussed in greater detail in Chapter 6.

Some coating and impregnating surface treatments for concrete will enhance abrasion resistance. Such products are discussed in further detail in Chapter 6. The wide variety of products on the market and the dependence of performance on concrete quality make it unwise to make sweeping judgments with regards to the relative merits of surface treatments; comparative studies have generally found polymeric coatings and impregnants to be superior to inorganic hardeners in imparting improved abrasion resistance [151].

The specification of concrete for use as a pavement surface is covered in the United Kingdom by the *Design Manual for Roads and Bridges* [152]. The manual includes requirements for two types of concrete pavement surface: transverse textured concrete and exposed aggregate concrete. Transverse textured concrete is concrete whose surface has been brushed or tined in a direction perpendicular to the direction of traffic movement to improve skid resistance. Exposed aggregate concrete is concrete where mortar at the surface has been removed, leaving the coarse aggregate exposed. This is normally achieved through the application of a retarding agent to the surface, followed by the removal of mortar through brushing once the interior concrete has developed sufficient strength to resist this action. Again, the aim is to enhance skid resistance.

From an abrasion resistance perspective, the manual places most importance on the role of aggregate. This is partly because, in most cases, aggregate will have the most important influence. However, it is also because the long-term skid resistance of both types of surface is dependent on limiting the wear (or 'polishing') of aggregates at the surface. The crucial parameter in defining the skid resistance of a pavement surface is the 'microtexture' of the aggregate at the surface – its roughness on a micrometer scale. This roughness can be reduced by the action of friction on the exposed aggregate surfaces, caused by events such as the braking of vehicles.

The performance of aggregates in resisting abrasion is quantified in the manual in terms of the results of two tests – the Polished Stone Value (PSV) and the Aggregate Abrasion Value (AAV) tests. The first of these tests is a measure of the extent to which abrasion reduces the skid resistance of an aggregate. It involves the polishing of aggregate samples embedded in resin in a polishing machine. The polished specimens are then tested using a standardized friction test for skid resistance, and the result is compared against the value obtained from a standard material through the calculation of the PSV. The AAV is a measure of resistance to surface abrasion and is determined from mass loss after exposing similar resin-embedded aggregate specimens to a lapping wheel on which are dispersed grains of abrasive material. The mass loss is converted to an AAV.

The minimum PSV requirement for aggregate is determined in the manual based on the number of commercial vehicles (CV) travelling on a lane of a road per day (CV/lane/day) and the type of site. The type of site is defined in terms of how frequently events that might lead to the wear of aggregates might occur. Thus, a motorway on which traffic is free flowing is likely to require aggregate with less wear resistance than an approach to a junction. The type of site is further subdivided in terms of the 'investigatory level' of the site. This is a value assigned to a site to quantify the level of skid resistance required based on the type and layout of a section of a road. Procedures for determining the investigatory level are provided elsewhere in the manual [153].

REFERENCES

1. Slowik, V., M. Schmidt, and R. Fritzsch. Capillary pressure in fresh cement–based materials and identification of the air entry value. *Cement and Concrete Composites*, v. 30, 2008, pp. 557–565.
2. The Concrete Society. *Technical Report 22: Nonstructural Cracks in Concrete*, *4th ed*. Surrey, United Kingdom: The Concrete Society, 2010, 63 pp.

3. ACI Committee 305. *ACI 305.1-06: Specification for Hot Weather Concreting*. Farmington Hills, Michigan: American Concrete Institute, 2007, 8 pp.
4. Topçu I. B. and V. B. Elgün. Influence of concrete properties on bleeding and evaporation. *Cement and Concrete Research*, v. 34, 2004, pp. 275–281.
5. Wainwright, P. J. and H. Ait-Aider, Influence of cement source and slag additions on the bleeding of concrete. *Cement and Concrete Research*, v. 25, 1995, pp. 1445–1456.
6. Almussalam, A. A., M. Maslehuddin, M. Abdul-Waris, F. H. Dakhil, and O. S. B. Al-Amoudi. Plastic shrinkage cracking of blended cement concretes in hot environments. *Magazine of Concrete Research*, v. 51, 1999, pp. 241–246.
7. Cohen, M. D., J. Olek, and W. L. Dolch. Mechanism of plastic shrinkage cracking in Portland cement and Portland cement–silica fume paste and mortar. *Cement and Concrete Research*, v. 20, 1990, pp. 103–119.
8. Wittman, F. H. On the action of capillary pressure in fresh concrete. *Cement and Concrete Research*, v. 6, 1975, pp. 49–56.
9. L'Hermite, R. Volume changes of concrete. *Proceedings of the 4th International Symposium on the Chemistry of Cement, Vol. 2*, Washington, D.C.: National Bureau of Standards, 1960, pp. 659–694.
10. Vollick, C. A. Effect of water-reducing admixtures and set-retarding admixtures on the properties of plastic concrete. *Symposium on the Effect of Water-Reducing Admixtures and Set-Retarding Admixtures on Properties of Concrete*, San Francisco, California, 1959. ASTM Special Publication 266. West Conshohocken, Pennsylvania: American Society for Testing Materials, 1960, pp. 180–200.
11. Rixom, R. and N. Mailvaganam. *Chemical Admixtures for Concrete, 3rd ed.* London: Spon, 1999, 456 pp.
12. Chandra, S. and L. Berntsson. *Lightweight Aggregate Concrete: Science, Technology, and Applications*. Norwich, United Kingdom: William Andrew, 2002, 450 pp.
13. Yeğinobali, A. Shrinkage of high-strength natural lightweight aggregate concretes. In A. Bentur and K. Kovler, eds., *PRO 23: International RILEM Conference on Early Age Cracking in Cementitious Systems (EAC '01)*. Bagneux, France: RILEM, 2002, pp. 355–362.
14. Henkensiefken, R., P. Briatka, D. Bentz, T. Nantung, and J. Weiss. Plastic shrinkage cracking in internally cured mixtures made with prewetted lightweight aggregate. *Concrete International*, v. 32, 2010, pp. 49–54.
15. Mora-Ruacho, J., R. Gettu, and A. Aguado, A. Influence of shrinkage-reducing admixtures on the reduction of plastic shrinkage cracking in concrete. *Cement and Concrete Research*, v. 39, 2009, pp. 141–146.
16. Qi, C., J. Weiss, and J. Olek. Characterisation of plastic shrinkage cracking in fibre-reinforced concrete using image analysis and a modified Weibull function. *Materials and Structures*, v. 36, 2003, pp. 386–395.
17. Banthia, N. and R. Gupta. Influence of polypropylene fibre geometry on plastic shrinkage cracking in concrete. *Cement and Concrete Research*, v. 36, 2006, pp. 1263–1267.
18. Sivakumar, A. and M. Santhanam. A quantitative study on the plastic shrinkage cracking in high-strength hybrid fibre-reinforced concrete. *Cement and Concrete Composites*, v. 29, 2007, pp. 575–581.

19. Houst, Y. F. and F. H. Wittmann. Influence of porosity and water content on the diffusivity of CO_2 and O_2 through hydrated cement paste. *Cement and Concrete Research*, v. 24, 1994, pp. 1165–1176.
20. Wittmann, F. H. Creep and shrinkage mechanisms. In Z. P. Bažant and F. H. Wittmann, *Creep and Shrinkage in Concrete Structures*. Chichester, United Kingdom: Wiley, 1982, pp. 129–161.
21. Jennings, H. M. Refinements to colloid model of C-S-H in cement: CM-II. *Cement and Concrete Research*, v. 38, 2008, pp. 275–289.
22. Building Research Establishment. *BRE Digest 357: Shrinkage of Natural Aggregates in Concrete*. Watford, United Kingdom: Building Research Establishment, 1991, 4 pp.
23. Wittmann, F. H. The structure of hardened cement paste: A basis for a better understanding of the material's properties. In P. V. Maxwell Cook, ed., *Hydraulic Cement Pastes: Their Structure and Properties*. Slough, United Kingdom: Cement and Concrete Association of Great Britain, 1976, pp. 96–117.
24. Verbeck, G. J. Carbonation of hydrated Portland cement. *Special Technical Publication 205*. West Conshohocken, Pennsylvania: American Society for Testing Materials, 1958, pp. 17–36.
25. Bažant, Z. P. and L. J. Najjar. Drying of concrete as a nonlinear diffusion problem. *Cement and Concrete Research*, v. 1, 1971, pp. 461–473.
26. Parrot, L. J. Moisture profiles in drying concrete. *Advances in Cement Research*, v. 1, 1988, pp. 164–170.
27. Pickett, G. Effect of aggregate on shrinkage of concrete and a hypothesis concerning shrinkage. *Proceedings of the American Concrete Institute*, v. 52, 1956, pp. 581–590.
28. Hansen, W. and J. A. Almudaiheem. Ultimate drying shrinkage of concrete: Influence of major parameters. *ACI Materials Journal*, v. 84, 1987, pp. 217–223.
29. Carlson, R. W. Drying shrinkage of concrete as affected by many factors. *Proceedings of the 41st Annual Meeting of the ASTM*, v. 38 (Part 2), 1938, pp. 419–437.
30. Hobbs, D. W. and L. J. Parrot. Prediction of drying shrinkage. *Concrete*, v. 13, 1979, pp. 19–24.
31. Soroka, I. *Portland Cement Paste and Concrete*. London: Macmillan, 1979, 362 pp.
32. Lawrence, C. D. Physicochemical and mechanical properties of Portland cements. In P. C. Hewlett, ed., *Lea's Chemistry of Cement and Concrete*, 4th ed. Oxford, United Kingdom: Butterworth-Heinemann, 1998, pp. 343–420.
33. Lerch, W. The influence of gypsum on the hydration and properties of Portland cement pastes. *Portland Cement Association Bulletin 12*. Chicago, Illinois: Portland Cement Association, 1946, 41 pp.
34. Chindaprasirta, P., S. Homwuttiwongb, and V. Sirivivatnanonc. Influence of fly ash fineness on strength, drying shrinkage, and sulphate resistance of blended cement mortar. *Cement and Concrete Research*, v. 34, 2004, pp. 1087–1092.
35. Hooton, R. D., K. Stanish, and J. Prusinski. The effect of ground, granulated blast furnace slag (slag cement) on the drying shrinkage of concrete: A critical review of the literature. *8th CANMET/ACI International Conference on Fly Ash, Silica Fume, Slag, and Natural Pozzolans in Concrete*, Supplementary Papers Volume. American Concrete Institute, 2004, 22 pp.

36. Brooks, J. J. and M. A. Megat Johari. Effect of metakaolin on creep and shrinkage in concrete. *Cement and Concrete Composites*, v. 23, 2001, pp. 495–502.

37. Skalny, J., I. Odler, and J. Hagymassy. Pore structure of hydrated calcium silicates: I. Influence of calcium chloride on the pore structure of hydrated tricalcium silicate. *Journal of Colloid and Interface Science*, v. 35, 1971, pp. 434–440.

38. Rao, G. A. Influence of silica fume replacement of cement on expansion and drying shrinkage. *Cement and Concrete Research*, v. 28, 1998, pp. 1505–1509.

39. Bruere, G. M., J. D. Newbegin, and L. M. Wilson. A laboratory investigation of the drying shrinkage of concrete containing various types of chemical admixtures. *Technical Paper 1*. Floreat Park, Australia: Division of Applied Mineralogy, CSIRO, 1971, 26 pp.

40. Hobbs, D. W. and A. R. Mears. The influence of specimen geometry upon weight change and shrinkage or air-dried mortar specimens. *Magazine of Concrete Research*, v. 23, 1971, pp. 89–98.

41. British Standards Institution. *BS EN 1992-1-1:2004: Eurocode 2: Design of Concrete Structures – Part 1-1: General Rules and Rules for Buildings*. London: British Standards Institution, 2004, 230 pp.

42. Tasdemir, M. A., F. D. Lydon, and B. I. G. Barr. The tensile strain capacity of concrete. *Magazine of Concrete Research*, v. 48, 1996, pp. 211–218.

43. The Highways Agency. Highway structures: Approval procedures and general design: Section 3. General design. *Early Thermal Cracking of Concrete: Design Manual for Roads and Bridges, Vol. 1*. South Ruislip, United Kingdom: Department of the Environment/Department of Transport, 1987, 8 pp.

44. British Standards Institution. BS EN 1992-3:2006: Eurocode 2: Design of Concrete Structures – Part 3. Liquid-Retaining and Containment Structures. London: British Standards Institution, 2006, 28 pp.

45. Bamforth, P., D. Chisholm, J. Gibbs, and T. Harrison. *Properties of Concrete for Use in Eurocode 2*. Camberley, United Kingdom: The Concrete Centre, 2008, 53 pp.

46. Termkhajornkit, P., T. Nawaa, M. Nakai, and T. Saito. Effect of fly ash on autogenous shrinkage. *Cement and Concrete Research*, v. 35, 2005, pp. 473–482.

47. Zhang, M. H., C. T. Tam, and M. P. Leow. Effect of water-cementitious materials ratio and silica fume on the autogenous shrinkage of concrete. *Cement and Concrete Research*, v. 33, 2003, pp. 1687–1694.

48. Lee, K. M., H. K. Lee, S. H. Lee, and G. Y. Kim. Autogenous shrinkage of concrete containing granulated blast-furnace slag. *Cement and Concrete Research*, v. 36, 2006, pp. 1279–1285.

49. Li, Y., J. Bao, and Y. Guo. The relationship between autogenous shrinkage and pore structure of cement paste with mineral admixtures. *Construction and Building Materials*, v. 24, 2010, pp. 1855–1860.

50. Tazawa, E-. I. and S. Miyazawa. Influence of cement and admixture on autogenous shrinkage of cement paste. *Cement and Concrete Research*, v. 25, 1995, pp. 281–287.

51. Evans, E. P. and B. P. Hughes. Shrinkage and thermal cracking in a reinforced concrete-retaining wall. *Proceedings of the Institution of Civil Engineers*, v. 39, 1968, pp. 111–125.

52. British Standards Institution. *BS 8007:1987: Code of Practice for Design of Concrete Structures for Retaining Aqueous Liquids*. London: British Standards Institution, 1987, 32 pp.

53. The Concrete Society. *Technical Report 34: Concrete Industrial Ground Floors*, 3rd ed. Surrey, United Kingdom: The Concrete Society, 2013, 76 pp.

54. The Concrete Society. *Technical Report 66: External In Situ Concrete Paving*, 3rd ed. Surrey, United Kingdom: The Concrete Society, 2007, 83 pp.

55. Houghton, D. L. Determining tensile strength capacity in mass concrete. *Journal of the American Concrete Institute*, v. 67, 1976, pp. 691–700.

56. Bentz, D. P., M. R. Geiker, and K. K. Hansen. Shrinkage-reducing admixtures and early-age desiccation in cement pastes and mortars. *Cement and Concrete Research*, v. 31, 2001, pp. 1075–1085.

57. Paine, K. A., L. Zheng, and R. K. Dhir. Experimental study and modelling of heat evolution of blended cements. *Advances in Cement Research*, v. 17, 2005, pp. 121–132.

58. Dhir, R. K., L. Zheng, and K. A. Paine. Measurement of early-age temperature rises in concrete made with blended cements. *Magazine of Concrete Research*, v. 60, 2008, pp. 109–118.

59. Bamforth, P. B. Early-age thermal crack control in concrete. *CIRIA Report C660*. London: CIRIA, 2007, 9 pp.

60. Emanuel, J. H. and J. L. Hulsey. Prediction of the thermal coefficient of expansion of concrete. *ACI Journal*, v. 74, 1977, pp. 149–155.

61. Browne, R. D. Thermal movement of concrete. *Concrete*, v. 6, 1972, pp. 51–53.

62. ACI Committee 122. *ACI 122R-02: Guide to Thermal Properties of Concrete and Masonry Systems*. Farmington Hills, Michigan: American Concrete Institute, 2002, 21 pp.

63. Valore, R. C. Calculation of U-values of hollow concrete masonry. *Concrete International*, v. 2, 1980, pp. 40–63.

64. Clark, S. P. *Handbook of Physical Constants*. New York: The Geological Society of America, 1966, 587 pp.

65. Waples, D. W. and J. S. Waples. A review and evaluation of specific heat capacities of rocks, minerals, and subsurface fluids: Part 1. Minerals and nonporous rocks. *Natural Resources Research*, v. 13, 2004, pp. 98–122.

66. Matiašovský, P. and O. Koronthályová. Pore structure and thermal conductivity of porous inorganic building materials. *Proceedings of the Thermophysics Working Group of the Slovak Physical Society*. 2002, pp. 40–46.

67. Mikulić, D., B. Milovanović, and I. Gabrijel. Analysis of thermal properties of cement paste during setting and hardening. *Proceedings of the International Symposium on Non-Destructive Testing of Materials and Structures*, Istanbul, Turkey, May 15–18, 2011. New York: Springer, 2011, 62 pp.

68. Fu, X. and D. L. L. Chung. Effects of silica fume, latex, methylcellulose, and carbon fibres on the thermal conductivity and specific heat of cement paste. *Cement and Concrete Research*, v. 27, 1992, pp. 1799–1804.

69. Kim, K.-H., S.-E. Jeon, J.-K. Kim, and S. Yang. An experimental study on thermal conductivity of concrete. *Cement and Concrete Research*, v. 33, 2003, pp. 363–371.

70. Bentz, D. P., M. A. Peltz, A. Durán-Herrera, P. Valdez, and C. A. Juárez. Thermal properties of high-volume fly ash mortars and concretes. *Journal of Building Physics*, v. 34. 2011, pp. 263–275.
71. Waller, V., F. de Larrard, and P. Roussel. Modelling the temperature rise in massive HPC structures. *Proceedings of the 4th International Symposium on Utilization of High-Strength/High-Performance Concrete*, Paris, France: RILEM, 1996, pp. 415–421.
72. Bentz, D. P. Transient plane source measurements of the thermal properties of hydrating cement pastes. *Materials and Structures*, v. 40, 2007, pp. 1073–1080.
73. Steenkamp, J. D., M. Tangstad, and P. C. Pistorius. Thermal conductivity of solidified manganese-bearing slags: A preliminary investigation. In R. T. Jones and P. den Hoed, eds., *Southern African Pyrometallurgy*. Johannesburg, South Africa: Southern African Institute of Mining and Metallurgy, 2011, pp. 327–343.
74. Knacke, O., O. Kubaschewski, and K. Hesselmann. *Thermochemical Properties of Inorganic Substances*, 2nd ed. Berlin, Germany: Springer-Verlag, 1977, 861 pp.
75. Schindler, A. K. and K. J. Folliard. Influence of supplementary cementing materials on the heat of hydration of concrete. *Proceedings of the 9th Conference on Advances in Cement and Concrete*. 2003, pp. 17–26.
76. Nambiar, O. N. N. and V. Krishnamurthy. Control of temperature in mass concrete pours. *The Indian Concrete Journal*, v. 58, 1984, pp. 67–73.
77. Struble, L. and S. Diamond. Influence of cement pore solution on expansion. In M. Kawamura, ed., *Proceedings of the 8th International Conference on Alkali-Aggregate Reactions*. 1989, pp. 167–172.
78. Antoniou, A. A. Phase transition of water in porous glass. *Journal of Physical Chemistry*, v. 68, 1964, pp. 2754–2763.
79. Powers, T. C. A working hypothesis for further studies of frost resistance of concrete. *Journal of the American Concrete Institute*, v. 41, 1945, pp. 245–272.
80. Janssen, D. J. and M. B. Snyder. *Strategic Highway Research Programme Report SHRP-C-391: Resistance of Concrete to Freezing and Thawing*. Washington, D.C.: National Research Council, 1994, 201 pp.
81. Harrison, T. A., J. D. Dewar, and B. V. Brown. *Freeze–Thaw-Resisting Concrete: Its Achievement in the U.K.* London: Construction Information Research and Information Association, 2001, 15 pp.
82. Fagerlund, G. The significance of critical degrees of saturation at freezing of porous and brittle materials. *ACI Publication SP47: Durability of Concrete*. Detroit, Michigan: American Concrete Institute, 1975, pp. 13–65.
83. Fagerlund, G. The international cooperative test of the critical degree of saturation method of assessing the freeze/thaw resistance of concrete. *Materiaux et Constructions*, v. 10, 1977, pp. 231–253.
84. Jana, D. Concrete scaling: A critical review. *Proceedings of the 29th Conference on Cement Microscopy*. Quebec City, Canada, May 20–24, 2007. Quebec, Canada: International Cement Microscopy Association, 2007, pp. 91–130.
85. Valenza, J. J. and G. W. Scherer. A review of salt scaling: I. Phenomenology. *Cement and Concrete Research*, v. 37, 2007, pp. 1007–1021.

86. Kaneuji, M., D. N. Winslow, and W. L. Dolch. The relationship between an aggregate's pore size distribution and its freeze–thaw durability in concrete. *Cement and Concrete Research*, v. 10, 1980, pp. 433–441.
87. American Society for Testing and Materials. ASTM C666: Standard Test Method for Resistance of Concrete to Rapid Freezing and Thawing. West Conshohocken, Pennsylvania: American Society for Testing and Materials, 2008, 6 pp.
88. Stark, D. and P. Klieger. Effect of maximum size of coarse aggregate on D-cracking in concrete pavements. *Highway Research Record 441*, 1973, pp. 33–43.
89. Mao, J. and K. Ayuta. Freeze–thaw resistance of lightweight concrete and aggregate at different freezing rates. *Journal of Materials in Civil Engineering*, v. 20, 2008, pp. 78–84.
90. Dhir, R. K., M. J. McCarthy, M. C. Limbachiya, H. I. El Sayad, and D. S. Zhang. Pulverised fuel ash concrete: Air entrainment and freeze/thaw durability. *Magazine of Concrete Research*, v. 51, 1999, pp. 53–64.
91. Powers, T. C. The air requirements of frost-resistant concrete. *Proceedings of the Annual Meeting of the Highway Research Board*, v. 29, 1949, pp. 184–211.
92. British Standards Institution. *BS EN 12350-7:2009: Testing Fresh Concrete: Part 7. Air content – Pressure Methods*. London: British Standards Institution, 2009, 28 pp.
93. British Standards Institution. *BS EN 206: Concrete. Specification, Performance, Production, and Conformity*. London: British Standards Institution, 2013, 74 pp.
94. British Standards Institution. *BS EN 12620:2013: Aggregates for Concrete*. London: British Standards Institution, 2013, 60 pp.
95. British Standards Institution. *BS EN 8500-1:2006: Concrete: Complementary British Standard to BS EN 206-1 – Method of Specifying and Guidance for the Specifier*. London: British Standards Institution, 2006, 66 pp.
96. Cornelius, D. F. *RRL Report LR 363: Air-Entrained Concretes – A Survey of Factors Affecting Air Content and A Study of Concrete Workability*. Crowthorne, United Kingdom: Road Research Laboratory, 1970, 18 pp.
97. Golaszewski, J., J. Szwabowski, and P. Soltysik. Influence of air-entraining agents on workability of fresh high-performance concrete. In R. K. Dhir, P. C. Hewlett, and M. D. Newlands, eds., *Admixtures: Enhancing Concrete Performance*. Slough, United Kingdom: Thomas Telford, 2005, pp. 171–182.
98. Warris, B. Influence of water-reducing and air-entraining admixtures on the water requirement, air content, strength, modulus of elasticity, shrinkage, and frost resistance of concrete. *Proceedings of the International Symposium on Admixtures for Mortar and Concrete*, Brussels, Belgium, 1967, Theme IV. Brussels, Belgium: ABEM, 1967, pp. 11–15.
99. Wright, P. J. F. Entrained air in concrete. *Proceedings of the Institution of Civil Engineers*, v. 2, 1953, pp. 337–358.
100. Tang, M., Y. Tian, X. B. Mu, and M. Jiang. Air-entraining concrete bubble distribution fractal dimension and antichloride ion diffusion characteristics. *Journal of Advanced Materials Research*, v. 255–260, 2011, pp. 3217–3222.

101. Nishi, T. On the recent studies and applications in Japan concerned to admixtures in use of building concretes and mortars: Part 2. Effects of surface active agents upon hardened concretes and mortars. *Proceedings of the International Symposium on Admixtures for Mortar and Concrete*. Brussels, Belgium, 1967, Theme IV. Brussels, Belgium: ABEM, 1967, pp. 112–117.

102. Wong, H. S., A. M. Pappas, R. W. Zimmerman, and N. R. Buenfeld. Effect of entrained air voids on the microstructure and mass transport properties of concrete. *Cement and Concrete Research*, v. 41, 2011, pp. 1067–1077.

103. Viles, R. Chemical admixtures. In M. Soutsos, ed., *Concrete Durability*. London: Thomas Telford, 2010, pp. 148–163.

104. Zhang, Z. and F. Ansari. Fracture mechanics of air-entrained concrete subjected to compression. *Engineering Fracture Mechanics*, v. 73, 2006, pp. 1913–1924.

105. Mayfield, B. and A. J. Moreton. Effect of fineness of cement on the air-entraining properties of concrete. *Civil Engineering and Public Works Review*, v. 64, 1969, pp. 37–41.

106. Backstrom, J. E., R. W. Burrows, R. C. Mielenz, and V. E. Wolkodoff. Origin, evolution, and effects of the air-void system in concrete: Part 3. Influence of water–cement ratio and compaction. *Journal of the American Concrete Institute*, v. 30, 1958, pp. 359–375.

107. Scripture, E. W. and F. J. Litwinowicz. Effects of mixing time, size of batch, and brand of cement on air entrainment. *ACI Journal*, v. 45, 1949, pp. 653–662.

108. Pleau, R., M. Pigeon, A. Lamontagne, and M. Lessard. Influence of pumping on characteristics of air-void system of high-performance concrete. *Transportation Research Record 1478*, 1995, pp. 30–36.

109. Hooton, R. D. and A. Boyd. Effect of finishing, forming, and curing on de-icer salt scaling resistance of concretes. In M. J. Setzer and R. Auberg, eds., *RILEM Proceedings 34: Frost Resistance of Concrete*. London: Spon, 1997, pp. 174–183.

110. Mayfield, B. and A. J. Moreton. Effect of fineness on the air-entraining properties of concrete. *Civil Engineering and Public Works Review*, v. 64, 1969, pp. 37–41.

111. Greening, N. R. Some causes for variation in required amount of air-entraining agent in Portland cement mortars. *Journal of the PCA Research and Development Laboratories*, v. 9, 1967, pp. 22–36.

112. Pistilli, M. F. Air-void parameters developed by air-entraining admixtures, as influenced by soluble alkalis from fly ash and Portland cement. *ACI Journal*, v. 80, 1983, pp. 217–222.

113. Pigeon, M., P. Plante, R. Pleau, and N. Banthia. Influence of soluble alkalis on the production and stability of the air-void system in superplasticised and nonsuperplasticised concrete. *ACI Materials Journal*, v. 89, 1992, pp. 24–31.

114. Dhir, R. K., M. J. McCarthy, M. C. Limbachiya, H. I. El Sayad, and D. S. Zhang. Pulverised fuel ash concrete: Air entrainment and freeze–thaw durability. *Magazine of Concrete Research*, v. 51, 1999, pp. 53–64.

115. Giergiczny, Z., M. A. Glinicki, M. Sokołowski, and M. Zielinski. Air-void system and frost-salt scaling of concrete containing slag-blended cement. *Construction and Building Materials*, v. 23, 2009, pp. 2451–2456.

116. Pigeon, M., P. C. Aitcin, and P. La Plante. Comparative study of the air-void stability in a normal and a condensed silica fume field concrete. *ACI Journal*, v. 84, 1987, pp. 194–199.
117. Neville, A. M. and J. J. Brooks. *Concrete Technology*. Harlow, United Kingdom: Longman, 1987, 438 pp.
118. Litvan, G.G. Air entrainment in the presence of superplasticisers. *ACI Journal*, v. 80, 1983, pp. 326–331.
119. Graybeal, B. and J. Tanesi. Durability of an ultrahigh-performance concrete. *Journal of Materials in Civil Engineering*, v. 19, 2007, pp. 848–854.
120. Cohen, M. D., Y. Zhou, and W. L. Dolch. Non–air-entrained high-strength concrete: Is it frost resistant? *ACI Materials Journal*, v. 89, 1992, pp. 406–415.
121. Dhir, R. K., K. Tham, and J. Dransfield. Durability of concrete with a superplasticising admixture. In J. M. Scanlon, ed., *ACI Special Publication SP-100: Concrete Durability – Proceedings of Katharine and Bryant Mather International Symposium, Vol. 1*. Detroit, Michigan: American Concrete Institute, 1987, pp. 741–764.
122. Mu, R., C. Miao, X. Luo, and W. Sun. Interaction between loading, freeze–thaw cycles, and chloride salt attack of concrete with and without steel fibre reinforcement. *Cement and Concrete Research*, v. 32, 2002, pp. 1061–1066.
123. Karahan, O. and C. D. Atiş. The durability properties of polypropylene fibre-reinforced fly ash concrete. *Materials and Design*, v. 32, 2011, pp. 1044–1049.
124. Şahmaran, M., E. Özbay, H. E. Yücel, M. Lachemi, and V. C. Li. Frost resistance and microstructure of engineered cementitious composites: Influence of fly ash and micro–polyvinyl-alcohol fibre. *Cement and Concrete Composites*, v. 34, 2012, pp. 156–165.
125. Soroushian, P., M. Nagi, and A. Okwuegbu. Freeze–thaw durability of lightweight carbon fibre–reinforced cement composites. *ACI Materials Journal*, v. 89, 1992, pp. 491–494.
126. Soroushian, P. and S. Ravanbakhsh. High–early-strength concrete: Mixture proportioning with processed cellulose fibres for durability. *ACI Materials Journal*, v. 96, 1999, pp. 593–599.
127. Quanbing, Y. and Z. Beirong. Effect of steel fibre on the de-icer–scaling resistance of concrete. *Cement and Concrete Research*, v. 35, 2005, pp. 2360–2363.
128. ACI Committee 544. Measurement of properties of fibre-reinforced concrete. *ACI Materials Journal*, v. 85, 1988, pp. 583–593.
129. Balaguru, P., M. Ukadike, and E. Nawy. Freeze–thaw resistance of polymer-modified concrete. In J. M. Scanlon, ed., *ACI Special Publication SP-100: Concrete Durability – Proceedings of Katharine and Bryant Mather International Symposium, Vol. 1*. Detroit, Michigan: American Concrete Institute, 1987, pp. 863–876.
130. Bordeleau, D., M. Pigeon, and N. Banthia. Comparative study of latex-modified concretes and normal concretes subjected to freezing and thawing in the presence of a de-icer solution. *ACI Materials Journal*, v. 89, 1992, pp. 547–553.
131. Goretta, K. C., M. L. Burdt, M. M. Cuber, L. A. Perry, D. Singh, A. S. Wagh, J. L. Routbort, and W. J. Weber. Solid-particle erosion of Portland cement and concrete. *Wear*, v. 224, 1999, pp. 106–112.

132. ACI Committee 210. *ACI 210R-93: Erosion of Concrete in Hydraulic Structures.* Farmington Hills, Michigan: American Concrete Institute, 1993, 24 pp.
133. Coombes, M. A., L. A. Naylor, R. C. Thompson, S. D. Roast, L. Gómez-Pujol, and R. J. Fairhurst. Colonisation and weathering of engineering materials by marine microorganisms: An SEM study: *Earth Surface Processes and Landforms*, v. 36, 2011, pp. 582–593.
134. Hu, X. G., A. W. Momber, and Y. G. Yin. Hydroabrasive erosion of steel fibre–reinforced hydraulic concrete. *Wear*, v. 253, 2002, pp. 848–854.
135. Lekarp, F. and A. Dawson. Modelling permanent deformation behaviour of unbound granular materials. *Construction and Building Materials*, v. 12, 1998, pp. 9–18.
136. Akbulut, H. and K. Aslantas. Finite-element analysis of stress distribution on bituminous pavement and failure mechanism. *Materials and Design*, v. 26, 2005, pp. 383–387.
137. British Standards Institution. *BS EN 13863-4:2012: Concrete Pavements: Test Methods for the Determination of Wear Resistance of Concrete Pavements to Studded Tyres.* London: British Standards Institution, 2012, 14 pp.
138. British Standards Institution. *BS EN 13892-3:2004: Methods of Test for Screed Materials: Determination of Wear Resistance – Bohme.* London: British Standards Institution, 2004, 14 pp.
139. British Standards Institution. *BS EN 13892-4:2002: Methods of Test for Screed Materials: Part 4. Determination of Wear Resistance – BCA.* London: British Standards Institution, 2002, 12 pp.
140. British Standards Institution. *BS EN 13892-5:2003: Methods of Test for Screed Materials: Determination of Wear Resistance to Rolling Wheel of Screed Material for Wearing Layer.* London: British Standards Institution, 2003, 14 pp.
141. American Society for Testing and Materials. *ASTM C779/C779M-05 (2010): Standard Test Method for Abrasion Resistance of Horizontal Concrete Surfaces.* West Conshohocken, Pennsylvania: American Society for Testing and Materials, 2010, 6 pp.
142. American Society for Testing and Materials. *ASTM C944/C944M-99 (2005) e1: Standard Test Method for Abrasion Resistance of Concrete or Mortar Surfaces by the Rotating-Cutter Method.* West Conshohocken, Pennsylvania: American Society for Testing and Materials, 2005, 5 pp.
143. American Society for Testing and Materials. *ASTM C418-05: Standard Test Method for Abrasion Resistance of Concrete by Sandblasting.* West Conshohocken, Pennsylvania: American Society for Testing and Materials, 2005, 3 pp.
144. American Society for Testing and Materials. *ASTM C1138M-05 (2010) e1: Standard test method for abrasion resistance of concrete (underwater method).* West Conshohocken, Pennsylvania: American Society for Testing and Materials, 2010, 4 pp.
145. Ghafoori, N. and M. W. Tays. Resistance to wear of fast-track Portland cement concrete. *Construction and Building Materials*, v. 24, 2010, pp. 1424–1431.
146. Smith, F. Effect of aggregate quality on resistance of concrete to abrasion. *ASTM STP 205: Cement and Concrete.* Philadelphia, Pennsylvania: American Society for Testing and Materials, 1958, pp. 91–105.

147. Dhir, R. K., P. C. Hewlett, and Y. N. Chan. Near-surface characteristics of concrete: Abrasion resistance. *Materials and Structures*, v. 24, 1991, pp. 122–128.

148. Kılıç, A., C. D. Atis, A. Teymen, O. Karahan, F. Özcan, C. Bilim, and M. Özdemir. The influence of aggregate type on the strength and abrasion resistance of high-strength concrete. *Cement and Concrete Composites*, v. 30, 2008, pp. 290–296.

149. Price, W. H. Erosion of concrete by cavitation and solids in flowing water. *ACI Journal*, v. 43, 1947, pp. 1009–1024.

150. British Standards Institution. *BS 8204-2:2003: Screeds, Bases, and In Situ Floorings: Part 2. Concrete Wearing Surfaces – Code of Practice*. London: British Standards Institution, 2003, 46 pp.

151. Sadegzadeh, M. and R. J. Kettle. Abrasion resistance of surface-treated concrete. *Cement, Concrete, and Aggregates*, v. 10, 1988, pp. 20–28.

152. The Highways Agency. *Design Manual for Roads and Bridges, Vol. 7: Pavement Design and Maintenance: Section 5. Surfacing and Surfacing Materials. Part 3. Concrete Surfacing and Materials*. London: Highways Agency, 1997, 7 pp.

153. The Highways Agency. *Design Manual for Roads and Bridges, Vol. 7: Pavement Design and Maintenance: Section 3. Pavement Maintenance Assessment. Part 1. Skid Resistance*. London: Highways Agency, 2004, 38 pp.

Chapter 3

Chemical mechanisms of concrete degradation

3.1 INTRODUCTION

During the service life of concrete, a wide range of circumstances can bring it into contact with chemical species that can cause deterioration of either the cement matrix or aggregate. Furthermore, some constituents can present chemical threats from within the concrete itself. This chapter examines three of the main mechanisms of chemical degradation: sulphate attack, alkali–aggregate reaction and acid attack.

3.2 SULPHATE ATTACK

Sulphate attack occurs as a result of the ingress of dissolved sulphate ions into concrete, which subsequently undergo reactions with the hardened cement. Depending on the exposure conditions, a number of different reactions are possible, whose impact on concrete properties results either from expansion and cracking or loss in strength and integrity. These different effects will be examined subsequently. However, it is first useful to examine where sulphates arise in the environment.

3.2.1 Sulphates in the environment

Sulphates can arise from two main sources in the environment – seawater and soil and groundwater. These sources are discussed below.

3.2.1.1 Seawater

Seawater contains relatively large concentrations of sulphate ions – between 2500 and 3000 milligrammes per litre (mg/L), depending on the salinity of the water [1]. The anion is associated with sodium and magnesium cations and, to a lesser extent, potassium and calcium ions [2]. Using typical proportions of association, $MgSO_4$ can be estimated to be present at concentrations between 2400 and 2900 mg/L, with Na_2SO_4 present at concentrations between 4800 and 5800 mg/L.

As shall be discussed in subsequent sections, the high concentrations of magnesium sulphate mean that seawater can potentially be a relatively aggressive sulphate environment.

3.2.1.2 Soil and groundwater

The extent to which sulphate in soil is available for ingress into concrete depends on the solubility of the minerals present, as well as the extent to which groundwater is present and its mobility. Soil can potentially contain a number of sulphate minerals, although some of these are of low solubility. Common sulphate minerals are shown in Table 3.1, which shows that the most soluble are sodium and magnesium sulphates.

Sulphide minerals such as pyrite, marcasite and pyrrhotite may also be present. Most of these sulphide minerals are of very low solubility. In undisturbed soils rich in sulphide minerals, sulphate will only be present in significant quantities in the first few metres of soil that has undergone long-term weathering processes that have led to the gradual oxidation of sulphides. Moreover, the first metre of soil will typically contain relatively low levels of sulphate resulting from leaching to lower levels by infiltrating rainwater.

Disruption of soil during construction can potentially bring sulphide minerals in contact with air, which allows the minerals to be oxidised to sulphate minerals at a relatively high rate. In high-pH conditions, such as those produced in close proximity to hydrated Portland cement, this process can be accelerated [3,4]. Sulphur-oxidising bacteria, which may be present in soil, are also capable of converting sulphide minerals into sulphates [5].

Table 3.1 Common sulphur-bearing minerals and their solubilities

Mineral	Chemical Formula	Solubility at 25°C (mg/L)
Barite	$BaSO_4$	2
Anhydrite	$CaSO_4$	3178
Gypsum	$CaSO_4 \cdot 2H_2O$	2692
Epsomite	$MgSO_4 \cdot 7H_2O$	1,481,658
Jarosite	$KFe_3(OH)_6(SO_4)_2$	5
Mirabilite	$Na_2SO_4 \cdot 10H_2O$	340,561
Glauberite	$Na_2Ca(SO_4)_2$	78,146
Pyrite	FeS_2	0
Marcasite	FeS_2	0
Pyrrhotite	FeS	0

In both cases, sulphide is oxidised to sulphuric acid, which proceeds to react with other cations in the soil to form sulphate compounds.

Peaty soils are also rich in sulphur, although much of this is present as organic compounds rather than as inorganic minerals [6]. However, this too can be oxidised to sulphate as a result of soil disruption.

Human activities may also introduce sulphur onto sites. Previous industrial operations – particularly those involving the handling and processing of fossil fuels – may contaminate soil with sulphates and sulphides. Activities of this type include coal mining, gas and coke processing plants and iron and steel manufacture. Additionally, activities such as fertilizer manufacture and metal finishing may also introduce sulphates onto a site. In some cases, sulphate-bearing by-products from industrial activities may be introduced onto sites as granular fill. These include blast-furnace slag, colliery spoil, furnace bottom ash from coal-fired power generation and incinerator bottom ash from the combustion of municipal waste [7]. Demolition rubble can also potentially contain reasonably large quantities of soluble sulphate [8].

3.2.1.3 Other sources

Sulphate attack can also occur when concrete comes into contact with sulphuric acid (H_2SO_4). Exposure to this source of sulphate produces both sulphate attack and corrosion of the concrete. Acid attack is discussed in detail in Section 3.3.

Sulphuric acid can arise from a number of sources. Sulphur dioxide deriving from the combustion of fossil fuels will oxidise in the atmosphere to produce H_2SO_4. Sulphate-reducing and oxidising bacteria act together to convert sulphur compounds to H_2S gas, which will be converted to sulphur as it reacts with atmospheric oxygen. This sulphur can then be digested by oxidising bacteria, producing H_2SO_4.

Various industrial processes also use H_2SO_4, or produce it as a by-product, and concrete in plants where such activities are conducted may be brought into contact with this substance.

3.2.2 Conventional sulphate attack

The most commonly encountered form of sulphate attack occurs in parallel with the formation of ettringite ($3CaO \cdot Al_2O_3(CaSO_4)_3 \cdot 32H_2O$) as a reaction product. This type of attack occurs when the cations associated with the sulphate ions are sodium, potassium or calcium. Ettringite is a reaction product of Portland cement hydration and plays an important role during the early setting and hardening of cement. However, when formed within relatively mature concrete in large enough quantities, the effect can be problematic.

For ettringite to be formed by sulphate ions entering hardened concrete, the sulphate needs to come into contact with a source of aluminium and calcium. In a mature cement paste, aluminium is present as ettringite, Al_2O_3–Fe_2O_3–mono (AFm) hydration products or is substituted within the calcium silicate hydrate (CSH) gel [9]. Aluminium from AFm phases is usually more soluble than that from CSH gel, and so it is normally these phases that are involved in ettringite formation. Normally, the AFm phase present will be monosulphate ($3CaO \cdot Al_2O_3 \cdot CaSO_4 \cdot 12H_2O$), although mix constituents and environmental conditions may produce AFm phases including Friedel's salt ($3CaO \cdot Al_2O_3 \cdot CaCl_2 \cdot 10H_2O$), monocarbonate ($3CaO \cdot Al_2O_3 \cdot CaCO_3 \cdot 11H_2O$) and hemicarbonate ($3CaO \cdot Al_2O_3 \cdot 1/2CaCO_3 \cdot 1/2CaO \cdot 12H_2O$).

The reaction with monosulphate is as follows:

$$3CaO \cdot Al_2O_3 \cdot CaSO_4 \cdot 12H_2O + 2Ca^{2+}$$
$$+ 2SO_4^{2-} + 20H_2O \rightarrow 3CaO \cdot Al_2O_3(CaSO_4)_3 \cdot 32H_2O$$

The additional calcium required for this reaction is obtained from different sources, depending on the cation associated with sulphate. Where sulphate is present as $CaSO_4$, the associated ion provides all of the calcium required for the reaction. However, where sodium or potassium sulphate is involved, calcium comes either from the hydration products portlandite ($Ca(OH)_2$) or CSH gel. Where calcium is obtained from CSH gel, the Ca/Si ratio of the gel declines, and the gel is said to have become decalcified.

When no further AFm phases remain, the sulphate ions combine with calcium ions to form gypsum ($CaSO_4 \cdot 2H_2O$).

It is often stated that the expansion of concrete undergoing this type of sulphate attack is the result of the solid volume of ettringite being significantly greater than that of solid reactants. However, this is true of other cement hydration products, yet their precipitation has little impact on paste volume. It has been proposed that the expansion is the result of the decalcification of CSH gel [10], which leads to greater swelling of the gel as a result of water imbibition. Regardless of whether this is the case, decalcification will lead to loss in strength.

An alternative explanation is that the precipitation of ettringite – and possibly gypsum – crystals produces crystallisation pressures between crystals and pore walls. Different theories have been proposed regarding how crystallisation pressures arise. One theory states that they are produced when repulsive forces (electrostatic and solvation forces) exist between crystal faces and pore walls. Their existence permits a thin layer of water to occupy the space between the two surfaces, which acts as a means for new material to be supplied to the crystal face. This supply of material allows the crystal to grow to the extent that sufficient pressures develop to fracture the confining material [11]. An alternative hypothesis is that the pressures are created as a result of an increase in the concentration of dissolved species in pore solutions, which in turn are produced by the increase in solubility of ettringite or gypsum crystals caused by pressures resulting from their growth in a confined space [12].

Both CSH decalcification and crystallisation are equally feasible mechanisms, and it is possible that both are effective simultaneously.

Figures 3.1 and 3.2 show the shape of typical expansion curves resulting from conventional sulphate attack. The curves show an initially low rate of expansion followed by accelerated expansion. This has been interpreted as indicating that, during the early stage, the pores of the concrete are being filled by growing crystals of ettringite and gypsum, with accelerated expansion being observed only when the pores are filled [13].

Cracking as a result of conventional sulphate attack typically starts at the corners and edges of concrete elements, with these cracks penetrating to greater depths as the process progresses. The formation of cracks leads to a decrease in both compressive strength and stiffness. The extent to which stiffness declines is strongly dependent on compressive stresses acting on the concrete, with high levels of stress leading to an accelerated deterioration in this property [14]. However, it has been reported that moderate sulphate concentrations (~10,000 mg/L Na_2SO_4) can have a beneficial influence on strength when concrete is loaded in compression. It has been

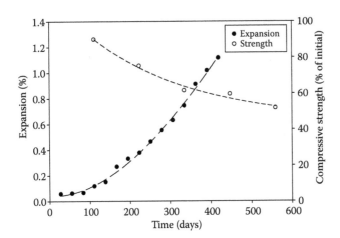

Figure 3.1 Change in specimen dimensions and compressive strength versus time for a Portland cement mortar exposed to a 25,000-mg/L sulphate (or 37,000-mg/L Na_2SO_4) solution at 23°C. (From Al-Dulaijan, S. U. et al., *Advances in Cement Research*, 19, 2007, 167–175.)

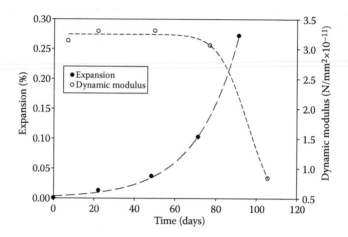

Figure 3.2 Change in specimen dimensions and dynamic modulus versus time for a Portland cement mortar exposed to a 50,000-mg/L Na_2SO_4 solution at 23°C. (From van Aardt, J. H. P. and S. Visser, *Cement and Concrete Research*, 15, 1985, 485–494.)

proposed that this effect is due to a reduced rate of ingress of sulphate into concrete whose expansion is restrained by the imposed stresses [15].

The high level of damage at corners and edges means that these features eventually become rounded as material is lost from these areas. Although one might expect a loss in mass during sulphate attack, laboratory studies show widely varying behaviour in terms of change in mass. A number of processes are occurring, which are likely to influence mass [16]. These include the following:

- An increase in mass as a result of uptake of water as a result of continued cement hydration
- An increase in mass as a result of assimilation of external sulphate ions into the cement matrix
- Mass loss as a result of disintegration
- Mass loss as a result of leaching
- Mass loss caused by movement of water out of the concrete as a result of osmosis

Thus, differences in mass change results obtained in laboratory studies can largely be attributed to differences in experimental conditions and practices that cause variation in the above processes.

3.2.3 Factors influencing the resistance of concrete to conventional sulphate attack

Three factors strongly influence the ability of concrete to resist conventional sulphate attack – the concentration of sulphate ions in the solution, the material's permeation properties and the composition of the cement constituents.

3.2.3.1 Sulphate concentration

The rate of degradation of concrete subject to conventional sulphate attack is dependent on the concentration of sulphate ions in the solution, with an increase in concentration leading to an increase in the rate of expansion [19]. This effect is illustrated in Figure 3.3, which shows the magnitude of expansion at different levels of sulphate concentration for mortars made with Portland cement.

At very high sulphate concentrations (>70,000-mg/L), where sodium is the main associated cation, a compound referred to as the U-phase is formed [20]. The U-phase is an AFm phase containing sodium, with the formula $4CaO·0.9Al_2O_3·1.1SO_3·0.5Na_2O·16H_2O$. Its formation in high quantities leads to deterioration in the form of expansion and loss of strength. Furthermore, a reduction in pore solution pH (such as that caused by carbonation) will lead to a conversion of the U-phase into ettringite, with consequent further expansion.

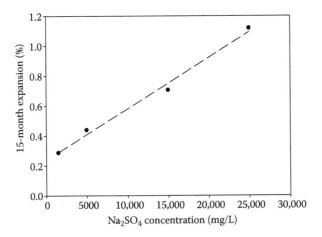

Figure 3.3 15-month expansion of Portland cement mortar exposed to solutions containing various concentrations of Na_2SO_4 at 23°C. (From Al-Dulaijan, S. U. et al., *Advances in Cement Research*, 19, 2007, 167–175.)

Where sulphate exposure derives from contact with soil, the mobility of groundwater plays an important role. Where groundwater is effectively static, the formation of ettringite in the concrete will eventually lead to the external soluble sulphates becoming depleted around the concrete surface. However, where groundwater is mobile, additional sulphate will continue to be supplied, which means that the concentration of sulphate ions in the soil surrounding the concrete surface will be maintained, and the extent of attack will be greater.

3.2.3.2 Temperature

The rate of sulphate attack is accelerated with increasing temperature (Figure 3.4), although the ultimate level of expansion is generally the same regardless of temperature. Elevated temperatures also tend to promote the formation of U-phase (see above) [21].

3.2.3.3 Cement composition

The most influential aspect of Portland cement composition with regards to sulphate resistance is tricalcium aluminate (C_3A) content. There is some correlation between C_3A content and expansion, with a level of below 8% by mass giving higher resistance to attack [22]. However, the correlation is weak, and it is evident that relating C_3A content to susceptibility to attack

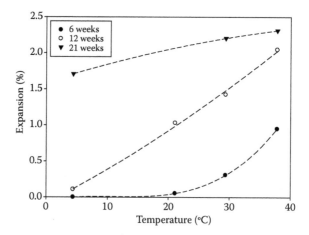

Figure 3.4 Expansion of Portland cement mortar prism exposed to a 25,000-mg/L Na_2SO_4 solution over a range of temperatures. (From Santhanam, M. et al., *Cement and Concrete Research*, 32, 2002, 585–592.)

is oversimplistic. One reason for this is that aluminate present as ettringite in hardened cement deriving from normal hydration cannot become involved in sulphate attack, and only aluminate present as AFm phases (or, possibly, still as C_3A) will contribute towards expansion [23]. This effect is clearly demonstrated by the improved sulphate resistance of concrete to which additional gypsum has been added – the higher sulphate content permits greater quantities of ettringite to persist in the mature cement matrix, rather than converting to AFm, which would be available for conversion into ettringite [24].

Increasing alkali levels in cement have been found to reduce expansion resulting from sulphate attack [16]. Taking into account the collective influence of C_3A, sulphate and alkalis, in general terms, conventional sulphate attack can be avoided when

$$\frac{C_3A\%}{SO_3\% + Na_2O_{eq}\%} < 3$$

and

$$1 < \frac{SO_3\%}{Na_2O_{eq}\%} < 3.5$$

where % denotes the percentage by mass of each constituent in the cement, and $Na_2O_{eq}\%$ is the alkali content of the cement expressed as sodium oxide equivalent (see Section 3.3).

Aluminium from the tetracalcium aluminoferrite (C_4AF) phase will also potentially contribute towards the formation of ettringite, although C_4AF has been found to sometimes reduce susceptibility to attack [25]. This has been attributed to the slower conversion of iron-containing AFm phases to ettringite, which may permit the cement matrix to better resist expansion [26].

Higher ratios of tricalcium silicate (C_3S) to dicalcium silicate (C_2S) make cement more susceptible to attack, although seemingly only when C_3A is present in its cubic structural form [22,27,28]. The increased deterioration is possibly the result of higher levels of $Ca(OH)_2$, which will provide a larger source of readily available calcium ions for the formation of ettringite.

The use of other cementitious materials in combination with Portland cement often, although not always, has the effect of enhancing resistance to conventional sulphate attack.

The incorporation of fly ash as a cement component in concrete can potentially impart greater sulphate resistance, although the chemical composition of fly ash is critical. Fly ash containing higher levels of calcium (Class C under the classification system of the United States) offers little benefit, whereas low-calcium fly ash typically yields a significant improvement in sulphate resistance [29].

Performance of concrete containing low-calcium fly ash is usually only improved after a period of adequate curing [30], indicating that the pozzolanic reaction of fly ash plays a crucial role. It is likely that the improvement in resistance is caused by a number of factors. First, concrete containing fly ash will typically display low permeability, thus reducing the rate of diffusion of sulphate ions. Second, $Ca(OH)_2$ levels within the cement matrix will be reduced, both through dilution of the Portland cement and as a result of the pozzolanic reaction, limiting the amount of calcium ions available for ettringite formation. Third, pozzolanic reactions usually have the effect of increasing the amount of aluminium incorporated in the CSH gel, limiting its availability and thus the capacity of the concrete to produce ettringite. The low pozzolanic reactivity of coarse fly ash has the effect of imparting little benefit with respect to sulphate resistance [31].

Silica fume also enhances the sulphate resistance of concrete, and some researchers have found that, on a weight-for-weight basis, it is more effective than fly ash [32]. This is presumably a result of the material's finer particle size, which yields a faster rate of pozzolanic reaction, as well as good pore-blocking properties. The low calcium and aluminium content of silica fume is also possibly beneficial, since this means that the material's presence will make little contribution towards ettringite formation. However, it should be stressed that, for economic and practical reasons, use of silica fume at levels comparable with those typical of fly ash is unlikely.

The incorporation of ground granulated blast-furnace slag (GGBS) can also have the effect of improving resistance to conventional sulphate attack. The main factor in determining whether improved resistance will be achieved is the Al_2O_3 content of the slag, with lower levels producing greater resistance [24]. This can partly be attributed to a simple dilution effect, where higher levels of GGBS with low Al_2O_3 content reduce the total amount of Al_2O_3 available for ettringite formation. However, it would also appear to be the case that the reactions of GGBS lead to significant quantities of Al_2O_3 becoming incorporated into the CSH gel structure, reducing availability still further. The influence of various other cementitious materials in combination with Portland cement is shown in Figure 3.5.

The presence of calcium carbonate in the form of limestone powder has been found to magnify the enhancement to sulphate resistance imparted by GGBS [33]. It has been proposed that the reason for this is that the calcium carbonate provides an additional source of calcium for ettringite formation, which reduces the extent to which decalcification of CSH gel occurs.

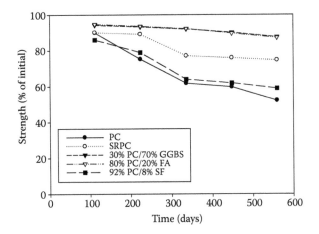

Figure 3.5 Deterioration in the strength of mortars containing Portland cement (PC), sulphate-resisting Portland cement (SRPC), ground granulated blastfurnace slag (GGBS), fly ash (FA) and silica fume (SF) exposed to a 25,000-mg/L sodium sulphate solution. (From Al-Dulaijan, S. U. et al., *Advances in Cement Research*, 19, 2007, 167–175.)

3.2.3.4 Permeation properties

The water/cement (W/C) ratio of a concrete mix has a strong influence over its subsequent sulphate resistance [34]. This is to be expected, since a lower W/C ratio will reduce the diffusion coefficient of sulphate through the material, thus reducing the volume of concrete exposed to sulphate attack after a given period of service. However, the effect of this parameter is more complex because a reduced W/C ratio will increase the strength of the cement matrix, allowing the concrete to better resist any expansive effects. As a general rule, resistance to sulphate attack is significantly enhanced through the use of concrete with a W/C ratio below 0.45 [22].

Good curing of concrete has the effect of improving the microstructure such that of concrete such that external chemical species will diffuse through it at lower rates. However, it has been found that less-than-ideal curing conditions have a beneficial effect on the resistance of concrete to conventional sulphate attack [35]. The most likely explanation for this is that inadequate curing will yield lower levels of $Ca(OH)_2$, leading to reduced ettringite formation.

In the case of sulphate attack, the coefficient of diffusion of sulphate ions through concrete is a dynamic property. This is the result of both cracking of the concrete (which will increase the diffusion coefficient) and deposition of reaction products in pores (which will reduce the coefficient). Typically, deposition of reaction products plays the more important role, leading to modification of the pore structure in terms of total porosity, pore size or both [30]. This in turn leads to a decrease in the diffusion coefficient with increasing duration of sulphate exposure.

A UK-based field study of long-term sulphate attack has found that damage resulting from sulphate attack was present in concrete to a depth of 50 mm after a period of 30 years in contact with sulphate-bearing soil [36]. It should be stressed that sulphate resistance is clearly dependent on the initial quality of the concrete and will therefore vary considerably with constituent materials and mix proportions. However, calculation of sulphate profiles using developing diffusion coefficients characterised in a laboratory-based study on PC concrete with a 0.42 W/C ratio and a cement content of 370 kg/m^3 [37] yields results that fit well with this observation.

3.2.3.5 Cement content

Resistance to sulphate attack is higher in concrete mixes with higher cement contents [38]. This is to be expected, since a mix with a larger volume of cement requires a greater quantity of sulphate ions to undergo ettringite formation and decalcification to the same extent as a mix with a smaller volume.

3.2.4 Magnesium sulphate attack

Where sulphate ions enter the concrete in association with magnesium ions, a greater damage to the integrity of concrete is often observed. This is the result of the formation of brucite:

$$Mg^{2+} + 2OH^- \rightarrow Mg(OH)_2$$

The formation of brucite is itself not necessarily problematic, since it is formed on the exterior concrete surface and has been attributed to a reduction in permeability. At higher concentrations of $MgSO_4$ (>7500-mg/L $MgSO_4$), the formation of ettringite does not occur and instead a combination of brucite and gypsum is formed. However, brucite precipitation leads to a reduction in the pH of the concrete pore solution, which produces more significant decalcification of CSH, leading to its disintegration and the subsequent replacement of calcium ions with magnesium giving a weak magnesium silicate hydrate gel [39]. Thus, where magnesium sulphate is present in higher concentrations, the main effect is a loss of strength, although expansion is still frequently observed. The visual clues for magnesium sulphate attack tend to take the form of a loss of surface and a rounding of edges and corners as material is lost. Figure 3.6 shows typical strength and mass loss behaviour.

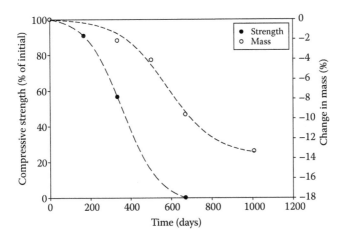

Figure 3.6 Strength and mass loss of a Portland cement mortar exposed to a 50,000-mg/L MgSO₄ solution at room temperature. (From Binici, H. and O. Aksoğan, *Cement and Concrete Composites*, 28, 2006, 39–46.)

3.2.5 Factors influencing the resistance of concrete to magnesium sulphate attack

As with conventional sulphate attack, the permeation properties of concrete have a strong influence on resistance to attack. However, other factors also play an important role, as discussed below.

3.2.5.1 Sulphate concentration

The rate of deterioration of concrete exposed to magnesium sulphate increases with concentration, as illustrated in terms of expansion in Figure 3.7. Deterioration of strength and loss of mass is particularly severe at concentrations above 10,000 mg/L SO_4^{2-} [41].

3.2.5.2 Temperature

There is a correlation between exposure temperature and the rate of expansion of concrete attacked by magnesium sulphate [12]. This is illustrated in Figure 3.8, indicating a significant increase in the rate of expansion at temperatures of approximately 30°C. However, as previously discussed, expansion is not necessarily the main cause of deterioration under this type of attack. Loss of strength, however, is also strongly influenced as a result of accelerated decalcification at elevated temperatures [42].

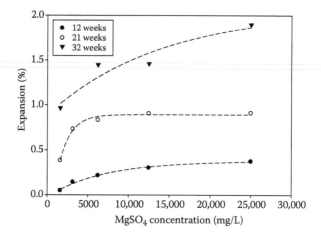

Figure 3.7 Expansion of Portland cement mortar prism versus MgSO₄ solution concentration. (From Santhanam, M. et al., *Cement and Concrete Research*, 32, 2002, 585–592.)

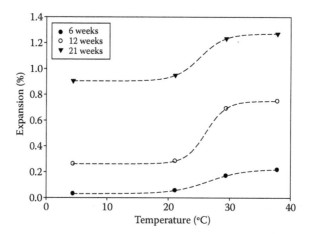

Figure 3.8 Expansion of Portland cement mortar prism exposed to a 25,000-mg/L MgSO₄ solution over a range of temperatures. (From Santhanam, M. et al., *Cement and Concrete Research*, 32, 2002, 585–592.)

strong evidence of significant concentrations of bicarbonate ions (HCO_3^-) being the crucial condition for thaumasite formation. Table 3.2 gives examples of field and laboratory studies in which thaumasite attack occurred in environments in which high bicarbonate concentrations were present in surrounding groundwater. The most significant of these is the laboratory study in which aggregates containing no carbonate minerals were found to contain thaumasite [52].

This does not exclude a role for carbonate minerals in thaumasite formation, as a result of an important feature of the chemistry of calcium bicarbonate [52]. Calcium bicarbonate is formed in soil as a result of carbon dioxide dissolving in water to form carbonic acid, which dissolves calcite:

$$CaCO_3 + CO_2 + H_2O \rightarrow Ca^{2+} + 2HCO_3^-$$

As calcium bicarbonate is formed, more CO_2, if available, dissolves in water. In soils where there is a good level of biological activity, this is usually the case. Now that bicarbonate ions are present, the newly dissolved CO_2 can play one of two roles. First, it can act as 'aggressive CO_2' and act to dissolve more $CaCO_3$ until equilibrium is reached. Second, a proportion of the CO_2 must be present to stabilise the HCO_3^- – 'stabilising CO_2'. Calcium bicarbonate becomes insoluble under higher pH conditions, which means that, as it diffuses into concrete, it will precipitate as $CaCO_3$ and thaumasite. The process of precipitation removes the requirement for stabilising CO_2, which means that aggressive CO_2 is generated, which can then dissolve calcite in the concrete.

Table 3.2 Studies whose findings suggest bicarbonate Ions as a source for thaumasite formation

Study	Aggregates	Comments
Field trial at Shipston-on-Stour [52–54]	Carbonaceous	Analysis shows groundwater to contain 482 mg/L 'CO_3 as $CaCO_3$,'. Given the limited solubility of $CaCO_3$, it can be assumed that this is bicarbonate. Thus, 482 mg/L CO_3 as $CaCO_3$ equates to 469 mg/L of bicarbonate as $CaCO_3$.
Field trial at Moreton Valence [54]	Carbonaceous	Analysis of groundwater from backfill used to cover specimens followed by ionic balance calculations indicates a bicarbonate concentration of 953 mg/L.
Laboratory study by Collet, Crammond, Swamy and Sharp [52]	Siliceous	Thaumasite detected beneath the surface of air-cured siliceous aggregate concrete prisms.

It has previously been proposed that acidic groundwater conditions promote thaumasite formation. However, research examining the effect of pH of solutions in contact with concrete has concluded that a combination of high sulphate concentration and high alkalinity (pH > 12) produced the most favourable conditions [55].

3.2.7.3 Cement composition

Although thaumasite contains no aluminium, its formation is not possible without the presence of this element. It has been demonstrated that thaumasite crystals are formed initially via nucleation onto the surface of ettringite crystals, presumably with ettringite providing a structural template [56].

The presence of $Ca(OH)_2$ appears to be required for thaumasite formation [57] and presumably acts as the necessary source of calcium. However, thermodynamic calculations have shown that the composition of CSH gel may also play a role, with a high Ca/Si ratio rendering the gel more prone to breaking down in contact with sulphate ions and thus providing Ca and Si for thaumasite formation [58]. The effect of including pozzolanic and latent hydraulic materials in concrete has the effect of reducing both $Ca(OH)_2$ levels and the Ca/Si ratio of CSH gel. GGBS, fly ash, silica fume and metakaolin have all been found to limit thaumasite formation, albeit at relatively high levels of inclusion. In the case of GGBS, levels of 70% of the cement fraction have been found to be effective [59], whereas levels exceeding 40% of fly ash are required [60]. In the case of silica fume and metakaolin, whose particle sizes usually produce a fast rate of pozzolanic reaction, a level of approximately 10% has been demonstrated to prevent thaumasite formation [61]. In general, reducing the Ca/Si ratio of the cement fraction of a concrete mix to below approximately 1 is likely to achieve good chemical resistance [58].

Regardless of whether $Ca(OH)_2$ concentrations or the composition of CSH gel (or both) controls thaumasite formation, it should also be stressed that, as with conventional sulphate attack, the improved permeation characteristics of concrete containing such materials are likely to also provide greater resistance to attack.

3.2.8 Avoiding sulphate attack

The measures required to avoid sulphate attack are dependent on the cations associated with sulphate in the environment that a structure will be operating in and other environmental factors. However, the two key objectives in protecting concrete are as follows:

1. Limit the rate of ingress of sulphate ions
2. Limit the availability of at least one chemical species required for the likely prevailing sulphate attack reaction

The first of these aims is best achieved by limiting the W/C ratio of the concrete and possibly through the inclusion of material likely to enhance the permeation characteristics of the material. However, it should be noted that this second option may be of limited benefit if magnesium sulphate is present in significant quantities.

How the second of these aims is achieved depends on which form of sulphate attack is most likely on a given site. Where it is conventional sulphate attack, limiting Ca and Al availability is the best option. In the past, sulphate-resisting cements were widely available, where sulphate resistance was achieved through low levels of C_3A. However, sulphate-resisting cement is now less common, and sulphate-resisting Portland cement is no longer manufactured in the United Kingdom. In its place are various cements that use fly ash, GGBS and other materials in combination with Portland cement clinker. Where thaumasite formation is a possibility, Ca and Si availability should be limited, an aim which can be achieved through the same means.

Guidance for providing adequate protection against sulphate attack in *Eurocode 2* refers the user to *BS EN 206: Concrete. Specification, Performance, Production, and Conformity* [62]. Sulphate attack is covered under the umbrella of 'chemical attack', which also includes acidic conditions and the presence of ammonia, magnesium and aggressive CO_2. This standard contains a means of determining the minimum requirements of a concrete mix for a given exposure class. Three exposure classes are defined (XA1-3), which are exclusive for natural soil and groundwater. No specific guidance is provided for establishing the exposure class of seawater, with a note that 'the classification valid in the place of use of the concrete applies'. The three exposure classes assume that groundwater is static and that sulphates from anthropogenic sources are not present. Using the appropriate exposure class, the maximum W/C ratio, the minimum strength class and the minimum cement content can then be established elsewhere in the standard. The two more aggressive exposure classes also require a sulphate-resisting cement to be used.

Given the limitations of the approach taken by *BS EN 206* – one of the Complementary British Standard to *BS EN 206* – *BS 8500-1: Method of Specifying and Guidance for the Specifier* [63] contains a much more detailed approach to specifying concrete for a sulphate-bearing environment. The guidance describes a specification process that initially involves assessing the nature of the site in terms of sulphate concentration (which can be assessed in various ways), the mobility of the groundwater and its pH. Contaminated sites are included in the assessment, and where a brownfield site is to be assessed, there is an additional requirement to establish magnesium concentrations.

These site conditions are then converted into an aggressive chemical environment for concrete (ACEC) class. Given the wider scope of the assessment system, 17 classes are defined. The class is then used to establish various design parameters: minimum nominal cover, minimum cement class, maximum W/C ratio, minimum cement content and a recommended cement type. Various options are available for most ACEC classes, with the general philosophy being that a higher W/C ratio demands a higher cement content and a more resistant cement type. Relatively mild sulphate exposure conditions allow cements with potentially limited sulphate resistance to be used where a low W/C ratio is used. These include conventional Portland cement (CEM I, as defined in the main British Standard for cements [64]) and CEM II cements containing up to 35% fly ash or pozzolana (either a naturally occurring pozzolana or a naturally occurring material that has been calcined to make it pozzolanic), 80% GGBS or 10% silica fume. If higher W/C ratios are used, more resistant cements are recommended. These include Portland–fly ash cements containing between 25% and 35% fly ash, blast-furnace cements (CEM III) containing 35% to 80% slag with a requirement that Al levels in the cement are limited and pozzolanic cement CEM IV/B (V). Portland–limestone cements are only recommended under relatively mild exposure conditions where very low (0.4) W/C ratios and high (360–380 kg/m³) cement contents are used. This is largely the result of concerns regarding thaumasite formation.

Where magnesium sulphate attack is likely to dominate, the use of fly ash, GGBS and other similar materials is not necessarily beneficial. For this reason, reduced W/C ratios and high cement contents are the best means of protecting concrete. However, it should be noted that, for very aggressive magnesium-bearing conditions, the complementary standard recommends blast-furnace cement with a slag content of 66% to 80% and limited Al_2O_3 levels.

Regardless of the ACEC class, a minimum nominal cover of 50 mm is required if the concrete is cast against blinding, whereas 75 mm is required when cast directly against soil.

Protection from sulphate attack deriving from seawater is also covered by *BS 8500-1*, although in this case, only minimum cement content (as a function of maximum aggregate size) and maximum W/C ratio are limited, and these limits are independent of cement type.

BS 8500-1 also recommends additional protective measures (APMs) that can be used in more aggressive environments or where long intended working lives of structures are required. These include the use of surface protection in the form of water-resistant barriers, such as polymer sheeting. Controlled permeability formwork [65] is also a possible measure – see Chapter 6. Another APM is increasing the thickness of concrete to provide a sacrificial layer that deteriorates over time, leaving an appropriate volume of unaffected material by the end of the intended service life of the structure. As previously discussed, it has been proposed that, in general, reinforced concrete components with an intended service life of 100 years would require a 50-mm sacrificial layer [7]. Finally, additional protection can be imparted simply by using concrete of a superior quality to that recommended in the standard.

The specification procedures contained within *BS 8500-1* have been adopted from the Building Research Establishment guidance *Special Digest*

1:2005: Concrete in Aggressive Ground [7], which also provides similar guidance for precast concrete.

The options available for using chemical admixtures to prevent deterioration of concrete subject to sulphate attack are limited. Air-entraining agents have, in the past, been proposed as a means of reducing expansion resulting from ettringite formation. However, there is little evidence that this approach is particularly effective, and the likely increase in permeability of the resulting material is counter to the basic requirements for sulphate resistance. Sodium citrate has been found to reduce expansion, possibly through limiting the amount of dissolved calcium in pore solutions [66]. However, this compound is also a retarder [67], which limits the applications in which it can be used.

In the case of precast concrete, the additional option of allowing surface carbonation of precast concrete prior to its use for construction in sulphate-bearing soils exists. The reason why deterioration due to sulphate attack is reduced where a carbonated surface layer is present is unclear, although in the case of conventional sulphate attack, it is almost certainly a result of the same mechanisms that are effective when limestone powder is present, as previously discussed. In environments where thaumasite formation is possible, carbonation will reduce the pH of the pore fluids in the concrete surface, making the formation of this substance much less likely. Surface carbonation can be achieved through exposing precast components to the air for a period of several days. As discussed in further detail in Chapter 4, very damp or very dry weather conditions are not conducive to the rapid formation of a carbonation layer, with optimal conditions achieved at relative humidities of between 50% and 75%. *BRE Digest 1:2005* [7] recommends at least 10 days of exposure to air.

3.2.9 Delayed ettringite formation

Another mechanism through which ettringite can cause damage to concrete is 'delayed ettringite formation'. This problem arises in concrete where no source of external sulphates is present and results exclusively as a result of their presence in the constituent materials – mainly in the form of gypsum in the cement.

Ettringite is a product of Portland cement hydration, resulting from reactions between gypsum (or other calcium sulphate minerals) added to the cement clinker during manufacture and the tricalcium aluminate (C_3A) and tetracalcium aluminoferrite (C_4AF) phases as they react with water. The formation of ettringite during the early age of concrete is normal and useful, since it acts to slow the otherwise rapid reactions of C_3A and C_4AF and contributes towards the setting process. Once the available sulphate is exhausted, ettringite progressively converts to monosulphate.

Delayed ettringite formation results from the relatively low decomposition temperature of ettringite – approximately 60°C. Above this temperature, the aluminium and sulphate from decomposed ettringite are present as poorly crystalline monosulphate, syngenite $(K_2Ca(SO_4)_2 \cdot H_2O)$ and as substitutions in the structure of CSH gel [9].

Delayed ettringite formation arises when concrete exposed to temperatures that cause decomposition of ettringite are then exposed to moist conditions (either submersion in water or exposure to air with a relative humidity close to 100%). These conditions cause ettringite to reform. However, instead of needles distributed around cement grains in the original material, the ettringite takes the form of massive crystals – which are believed to be expansive – and roughly spherical formations – which are not thought to be damaging [68]. The ettringite usually forms in pores within the cement matrix, and its progressive development leads to the formation of 'veins' of the mineral running through the matrix.

Because of the sequence of events required for delayed ettringite formation to occur, it is most commonly encountered in precast concrete components that have been cured in an autoclave or by other means at similar temperatures. However, mass concreting applications in which the volume of material is sufficient to reach temperatures in excess of 60°C may also produce the same effect.

The nature of the cement used influences the vulnerability of a concrete mix. As might be expected, an increase in the sulphate content leads to an increase in expansion, although there would appear to be a pessimum level of approximately 4% [69]. Increasing alkali levels within cement also increases the magnitude of expansion and shifts the pessimum sulphate level to higher concentrations. There is also a correlation between the magnitude of expansion and quantities of magnesium oxide in the cement. Higher levels of C_3S in Portland cement appear to reduce the stability of ettringite at elevated temperatures. Concrete made with finer cement has a tendency to be more prone to expansion resulting from delayed ettringite formation.

An equation for predicting the magnitude of expansion has been proposed:

$$e_{90} = 0.00474 \text{ SSA} + 0.0768 \text{ MgO} + 0.217 \text{ C}_3\text{A} + 0.0942 \text{ C}_3\text{S} + 1.267 \text{ Na}_2\text{O}_{eq} - 0.737|(\text{SO}_3 - 3.7 - 1.02 \text{ Na}_2\text{O}_{eq})| - 10.1$$

where e_{90} is the expansion after curing at 90°C for 12 h (%), SSA is the specific surface area of the cement (m^2/kg) and Na_2O_{eq} is the sodium oxide equivalent content (kg/m^3) – see Section 3.3.4 (kg/m^3).

3.2.9.1 Avoiding delayed ettringite formation

The previous equation clearly provides some insights as to how delayed ettringite formation might be avoided. It should be noted that, although a higher sulphate content is undesirable in this context, there may potentially be a conflict, since it is beneficial when considering resistance to conventional sulphate attack. The use of other cementitious materials in combination with Portland cement has also been shown to limit expansion resulting from delayed ettringite formation [70]. Specifically, fly ash, granulated blast-furnace slag, silica fume and metakaolin have all been shown to control expansion. It has been pointed out, however, that research into this area is relatively immature and that long-term performance is currently not fully understood [71].

3.3 ALKALI–AGGREGATE REACTIONS

Water in the pores of hardened concrete contains quantities of dissolved ions largely deriving from the cement matrix. In mature concrete, the vast majority of dissolved cations are those of the alkali metals potassium and sodium [72]. These alkali ions derive from cement. During the early periods of cement hydration, sulphate ions are removed from solution through their incorporation into hydration products such as ettringite and monosulphate, which means that the anions balancing the cations in the solution are soon exclusively hydroxide ions [9]. As a consequence, the pH value of concrete pore solutions can be as high as 13.9 [73].

Under these highly alkaline conditions, durability problems can arise as a result of reactions between aggregate minerals and hydroxide ions. Products of the reactions are capable of absorbing water, leading to expansion and cracking of concrete. These alkali–aggregate reactions can be divided into three types: (1) alkali–carbonate reaction (ACR), (2) alkali–silicate reaction, and (3) alkali–silica reaction (ASR).

3.3.1 Alkali–silica reaction

ASR involves the breaking of bonds in the framework of certain silica-bearing minerals to produce an expansive gel. Under high-pH conditions, siloxane bonds at the surface of silica minerals are attacked by hydroxide ions in the following manner [24]:

$$\equiv Si\text{-}O\text{-}Si\equiv + OH^- + R^+ \rightarrow \equiv Si\text{-}OH + R\text{-}O\text{-}Si\equiv$$

where R denotes either sodium or potassium. The reaction continues in the following manner:

$$\equiv Si\text{-}OH + OH^- + R^+ \rightarrow \equiv Si\text{-}O\text{-}R + H_2O$$

The reaction reduces the original silica network to an open, gel-like network that is more accessible to water molecules. The gel undergoes hydration:

$$\equiv Si\text{-}O\text{-}R + H_2O \rightarrow \equiv Si\text{-}O^-\text{-}\text{-}(H_2O)_n + R^+$$

As water is absorbed into the gel, it swells considerably.

It should be noted that the attack of siloxane bonds continues to an extent that the outer layer of the gel will eventually begin to break down completely [74], releasing silicate groups into solution:

$$\equiv Si\text{-}O\text{-}Si\text{-}OH + OH^- \rightarrow \equiv Si\text{-}OH + O^-\text{-}Si\text{-}OH$$

The silicate groups removed from the gel are likely to rapidly be involved in reactions between calcium ions to form CSH gel. Thus, the reactions occurring during ASR are also those that occur during the pozzolanic reactions of materials such as fly ash.

An alternative mechanism has been proposed in terms of the accumulation of osmotic cell pressure [75]. Under this mechanism, water and both alkali and hydroxide ions can move from the cement matrix into a reacting aggregate particle, but the movement of silicate ions out of the particle into the matrix is prevented by a layer of calcium alkali silica gel formed by a reaction of calcium ions with ASR gel at the cement paste–aggregate interface. Thus, the reaction product layer acts as an osmotic membrane. As water enters the ASR gel formed within the membrane, hydraulic pressure increases to a point where fracture of the aggregate and its surrounding matrix occurs, leading to expansion.

3.3.2 Alkali–carbonate reaction

The ACR is often referred to as 'dedolomitization' because it can involve the decomposition of dolomite to form brucite ($Mg(OH)_2$) and calcite ($CaCO_3$) in the following manner [76]:

$$CaMg(CO_3)_2 + 2ROH \rightarrow Mg(OH)_2 + CaCO_3$$

where R is either sodium or potassium. However, magnesite ($MgCO_3$) will also undergo a similar process:

$$MgCO_3 + 2ROH \rightarrow Mg(OH)_2 + R_2CO_3$$

In this case, the alkali carbonate products will react with the cement hydration product portlandite ($Ca(OH)_2$) to give calcite:

$$R_2CO_3 + Ca(OH)_2 \rightarrow CaCO_3 + 2ROH$$

There is a net reduction in the volume of the products of these reactions relative to the reactants, which has led to speculation that expansion is the result of expansive clay particles within the carbonate mineral matrix – exposed and unconstrained by dedolomitization – absorbing water [77].

Other proposed mechanisms include the suggestion that the source of expansion is ASR between microscopic quartz particles in the carbonate matrix [78], or the expansion of assemblages of colloidal particles present in pores within the carbonate minerals [79]. However, research carried out using carbonate rocks of high purity has led to the general conclusion that the precipitation of brucite in confined spaces is, indeed, the cause of expansion [76,80].

3.3.3 Alkali–silicate reaction

The alkali–silicate reaction involves rocks that can contain quantities of minerals with a layered phyllosilicate structure. It has been observed that, under high-pH conditions, these minerals exfoliate, permitting water to occupy the space between the layers [81]. However, gel is often also present, indicating that expansion could be the result of ASR, possibly involving strained microcrystalline quartz [82].

3.3.4 Expansion and cracking caused by alkali–aggregate reactions

The expansion characteristics of alkali–aggregate reactions are similar regardless of the reaction occurring. Examples of expansion curves are shown in Figure 3.10, showing, in some cases, a period of relative inactivity during the early part of the reaction, followed by a rapid expansion that gradually declines to an essentially constant lower rate. The nature of

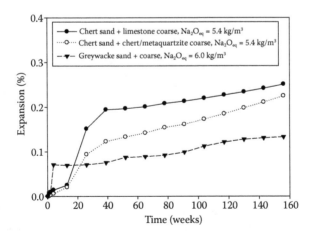

Figure 3.10 Expansion curves for concrete containing combination of coarse and fine aggregate known to display reactivity. (From Dyer, T. D. and R. K. Dhir, *Magazine of Concrete Research*, 62, 2010, 749–759.)

expansion in terms of the maximum rate and the ultimate magnitude are controlled by several factors – the alkali content of the concrete, temperature, aggregate particle characteristics, moisture levels and the W/C ratio of the mix. These factors are discussed below.

3.3.4.1 Alkali content

Before discussing the role of alkali levels in concrete in determining whether alkali–aggregate reaction will occur and the extent to which it occurs, it is important to understand the manner in which the alkali content of concrete is expressed. Because both sodium and potassium play a role in alkali–aggregate reactions, and because these elements have different atomic masses, it is useful to consider their quantities in concrete in terms of how many atoms in total are present, rather than the combined mass of potassium and sodium.

However, expressing these concentrations in terms of numbers of atoms or moles (the most convenient unit for expressing large numbers of atoms) is potentially confusing. Moreover, the manner in which the composition of concrete is expressed is in terms of kg/m^3, which means that manipulation is required of both the potassium and sodium contents of a material before the number of moles can be obtained. Thus, the convention for expressing alkali content is in terms of sodium oxide equivalent (Na_2O_{eq}). This value is obtained by expressing the sodium oxide content in kg/m^3 and adding it the number of moles of potassium oxide present, expressed in terms of the equivalent mass of sodium oxide:

$$Na_2O_{eq} = Na_2O + 0.6580.K_2O$$

where Na_2O is the mass of sodium oxide in concrete (kg/m^3), and K_2O is the mass of potassium oxide in concrete (kg/m^3).

In certain cases, alkali levels are expressed in terms of elemental alkali, in which case the sodium equivalent (Na_{eq}) is used:

$$Na_{eq} = \%Na + 0.5880.\%K$$

where %Na is the percentage mass of sodium, and %K is the percentage mass of potassium.

This approach to expressing alkali levels clearly assumes that the effects of sodium and potassium on expansion are equal. However, sodium typically produces more expansion than potassium for both ASR and ACR [84], and so some caution should be exercised in interpreting total alkali contents expressed in this manner.

The alkali content of concrete is normally established from the total alkali content of the Portland cement plus that of any other materials included in the cement fraction, such as GGBS and fly ash. This value is obtained from the chemical analysis of solutions derived from the complete digestion of each material in either hydrofluoric acid (HF) or a combination of hydrochloric acid (HCl) and other acids, or by spectrometric methods on solid materials. The alkali content of aggregate is normally discounted from the total concrete alkali content, which, in cases where aggregates are capable of releasing alkalis, can potentially underestimate the true quantity of alkali available for reaction.

Because sodium and potassium ions in the pore solution are almost entirely associated with hydroxide ions, an increase in alkali concentrations leads to an increase in pH. In the case of ACR, expansion increases with this increase in pH (Figure 3.11) [85]. However, the effect of alkalis on the expansion resulting from ASR is dependent not simply on the alkali content of the concrete but on the ratio of reactive silica to alkali. This is illustrated in Figure 3.12, in which a 'pessimum' reactive silica/alkali ratio is evident at which maximum expansion is observed [83]. The ratio is often simply interpreted as a pessimum reactive aggregate content. The ratio at which the pessimum is observed is dependent on the type of reactive aggregate present, but independent of the actual alkali levels in the concrete.

Several mechanisms have been proposed to explain this feature. These include (1) a likely greater capacity for expansion with higher levels of

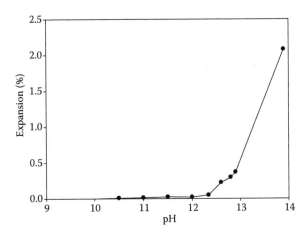

Figure 3.11 91-day expansion of reactive carbonate rock prisms exposed to solutions of increasing pH. (From Min, D. and T. Mingshu, *Cement and Concrete Research*, 23, 1993, 1397–1408.)

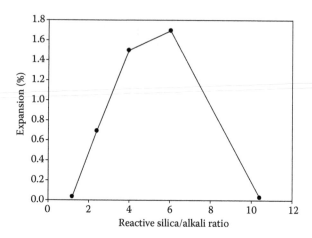

Figure 3.12 112-day expansion of concrete containing aggregate capable of undergoing ASR versus the reactive silica/alkali ratio. (From Hobbs, D. W. *Alkali–Silica Reaction in Concrete*. London: Thomas Telford, 1988.)

alkalis incorporated into the ASR gel network [86]; (2) high reactive silica/alkali ratios lead to the majority of the gel being produced while the cement paste matrix is immature and hence less susceptible to damage from expansion [87]; (3) the gel formed at increasingly high reactive silica/alkali ratios is present as thinner outer layers distributed between increasing numbers of aggregate particles, which means that a greater surface area of gel is available for reaction with calcium ions to form CSH gel [75]; and (4) using the osmotic cell pressure theory (see Section 3.3.1), the silicate-impermeable layer allows ASR gel to form and hydraulic pressure to build up within the layer, leading to fracture of the aggregate particle – at high reactive silica/alkali ratios, insufficient ASR gel is formed to lead to fragmentation [88]. All of these proposed mechanisms are credible, and the point has been made that they need not be mutually exclusive [9].

When concrete is composed in such a way that it is at the pessimum reactive silica/alkali ratio, an increase in Portland cement alkali content leads to an increase in expansion only above a certain threshold level (Figure 3.13).

3.3.4.2 Temperature

The rate of ASR increases with increasing temperature. However, the expansion of concrete containing aggregates capable of undergoing ASR typically peaks at temperatures of approximately 40°C, above which

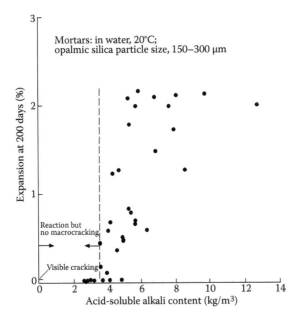

Figure 3.13 200-day expansion of mortars containing opaline silica aggregate held at 20°C versus their alkali content. (From Hobbs, D. W. *Alkali–Silica Reaction in Concrete*. London: Thomas Telford, 1988.)

expansion declines (Figure 3.14) [89]. It has been suggested that one reason for this may be the lower viscosity of ASR gel at elevated temperatures, which would lessen expansion through reduced hindrance of gel penetration of cement matrix pores.

An increase in temperature also has the effect of increasing the rate of expansion due to alkali–silicate reaction. Unlike ASR, it would appear that expansion increases steadily with increase in temperature [90].

Expansion resulting from ACR is also sensitive to reaction temperature. It would also appear that maximum expansion is observed at approximately 40°C [91], although in this case, this feature is possibly the result of magnesium ions diffusing out of the concrete [92].

3.3.4.3 Particle size and shape

In some instances, a pessimum aggregate particle size range has been identified, above and below which expansion is reduced. This reduction is most significant at smaller particle sizes (Figure 3.15). Figure 3.16 shows the

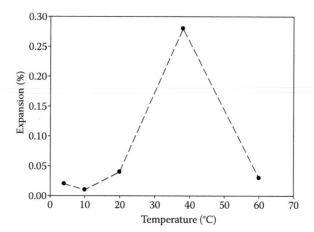

Figure 3.14 6-month expansion of mortar bars containing reactive sand tested in accordance with the ASTM C-227 method. (From Guðmundsson, G. and H. Ásgeirsson, *Cement and Concrete Research*, 5, 1975, 211–220.)

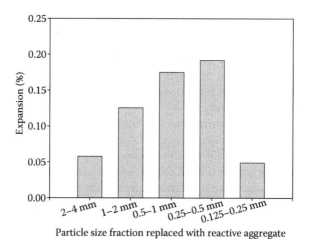

Figure 3.15 14-day expansion of mortar mixes using the ASTM accelerated mortar bar method (ASTM C1252). Each mix contains a partial replacement of low reactivity aggregate with reactive aggregate, where the reactive portion is present as a replacement of a narrow particle size fraction. (From Ramyar, K. et al., *Cement and Concrete Research*, 35, 2005, 2165–2169.)

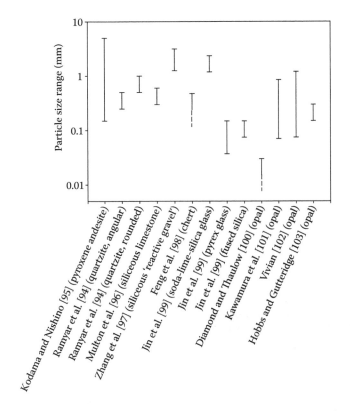

Figure 3.16 Aggregate particle size ranges containing the pessimum particle size for expansion as a result of alkali–aggregate reaction from a range of studies.

pessimum size ranges observed for a number of different reactive aggregate types. Although it should be stressed that the methods used to establish these ranges and the sieve configurations used differ between studies, it is evident that the pessimum size ranges vary considerably between aggregates. It should be noted that, in most of these examples, the likely reaction is ASR, although the example involving a siliceous limestone used a rock containing dolomite and phyllosilicates, as well as calcite and quartz, which means that both alkali–carbonate and alkali–silicate reactions could have been responsible.

It has been proposed that the reason for this pessimum effect is related to the chemical reactions outlined in Section 3.3.1 [93]. These reactions involve both the attack of silica networks to leave ASR gel and the dissolution of silica. The first of these processes acts to increase the quantity of ASR gel, whereas the second acts to reduce it. Thus, the rate at which the two processes occur defines the rate at which ASR proceeds. The mechanism

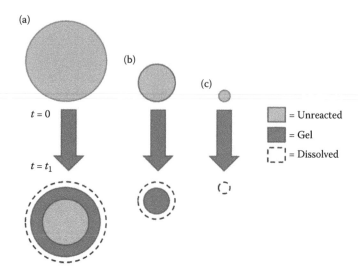

Figure 3.17 Schematic diagram showing the effect of a relatively fast rate of ASR gel formation and a slow rate of gel dissolution over a period t_1. Large particles (a) produce relatively large volumes of gel as they react. Smaller particles (b) become entirely converted to gel, which then gradually dissolves. Very small particles (c) are dissolved entirely in a relatively short period of time.

is illustrated in Figure 3.17, which shows that where the formation of gel occurs rapidly, but dissolution proceeds slowly, a large quantity of gel may be formed. In the case of small particles, a situation may be reached where the entire reactive portion of the particle is converted to ASR gel, which means that subsequent dissolution will act to reduce the particle's contribution to expansion. In the case of large particles, the high surface area and large volume of the particles means that only a small fraction of the reactive components of the total volume of aggregate will have reacted. However, at intermediate particle sizes, the capacity to produce expansive gel will be optimal. Thus, using this proposed mechanism, the pessimum particle size is defined by the rates of gel formation and dissolution, with an increasing rate of formation and slower rate of dissolution leading to a larger pessimum particle size.

It is worth pointing out that the reactions involved in ASR and the initial stages of the pozzolanic reaction of materials such as fly ash and silica fume are identical. The difference is that for the low particle sizes of the pozzolanic materials, the dissolution of ASR gel is very rapid, allowing the dissolved silicate groups to react with calcium ions to produce CSH gel, which contributes towards strength development. Moreover, as will be discussed in a subsequent section, the pozzolanic reaction offers one means of controlling alkali–aggregate reaction.

The shape of aggregates also appears to influence expansion due to ASR, with angular particles producing higher levels of expansion [94].

3.3.4.4 Moisture

Expansion resulting from ASR results from water entering the gel. Thus, the extent to which expansion occurs is dependent on the amount of water available within concrete. The humidity of air in contact with concrete undergoing ASR profoundly affects expansion, with relative humidities below approximately 75% producing little or no expansion and an increase in expansion with humidity above this level (Figure 3.18) [104,105].

Relative humidity also has a similar influence over alkali–silicate and ACRs.

3.3.4.5 W/C ratio

Another factor that can influence expansion due to alkali–aggregate reactions is the W/C ratio of a concrete mix. It has generally been observed that, as the W/C ratio increases, so does expansion [106,107]. However, it has also been found in certain cases that at higher W/C ratios, expansion declines again, which means that there would appear to be a pessimum W/C

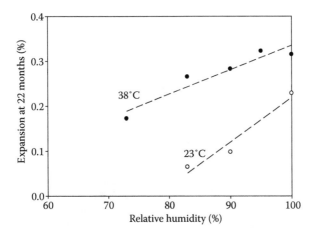

Figure 3.18 Influence of relative humidity on ASR expansion of mortar bars held for 22 months in sealed glass containers. (From Ólafsson, H. The effect of relative humidity and temperature on alkali expansion of mortar bars. In P. E. Grattan-Bellew, ed., *Proceedings of the 7th International Conference on Concrete Alkali–Aggregate Reactions,* Ottawa, Canada, August 18–22, 1986, pp. 461–465.)

ratio. The initial increase can be attributed to the reduced strength of the resulting concrete, whereas the reduced expansion at higher W/C ratios may be the result of reduced alkali concentrations in the pore fluids, as a result of the higher volume of water [106]. The pessimum W/C ratio varies depending on the constituents used. It should be stressed that, where reactive aggregate is present, expansion is substantial even at low W/C ratios, which means that adjusting this parameter during mix design will not provide a viable route to avoiding problems with alkali–aggregate reactions.

3.3.4.6 Timescale for cracking

As the pressure exerted by ASR gel increases, it eventually reaches a level that exceeds the tensile strength of concrete, leading to cracking. Typically, where the composition of the reactive aggregate is homogeneous, pressure is exerted uniformly in all directions, leading to the formation of a symmetrical configuration of three cracks located at 120° to each other (Figure 3.19), particularly when particles are angular [108]. A similar mode of cracking is also effective during ACR.

Where the composition of the particles is heterogeneous (in other words, where the rock is composed of reactive minerals in combination with non-reactive ones), the formation of cracks can be less symmetrical, particularly when reactive minerals are present as veins [108].

As expansion increases, the cracks grow and start to join together. When discussing the nature of the resulting cracks, it is necessary to consider

Figure 3.19 Typical configuration of cracks resulting from alkali–aggregate reaction expansion of a single aggregate particle. (After Figg, J. ASR: Inside phenomena and outside effects (crack origin and pattern). In P. E. Grattan-Bellew, ed., Proceedings of the 7th International Conference on Concrete Alkali–Aggregate Reactions, Ottawa, Canada, August 18–22, 1986, pp. 152–156.)

both microcracks and macrocrack structures to obtain a complete picture of the processes occurring. Where an expansive alkali–aggregate reaction has occurred, macrocracks are located at the surface and usually only penetrate to a depth of approximately one-tenth of the member thickness [109]. These cracks run perpendicular to the exposed surface and can have widths of up to 10 mm, but are more typically much less than this [82].

In contrast, microcracks are typically randomly aligned and present both at highest density and with greatest interconnectivity close to the core of a concrete member. The nature of the configuration of macrocracks and microcracks indicates that the exposed surface of concrete members undergoing expansive alkali–aggregate reactions has been placed in tension by the action of expansion, whereas the interior of the concrete is in compression. This has been reasonably interpreted as meaning that more reaction (and hence expansion) has occurred at the core than at the exposed surface [82]. This is thought to result from leaching of alkalis and evaporation of water at the surface and possibly the presence of greater porosity in the surface layer, allowing some accommodation of expansive reaction products [108].

Where concrete elements are relatively unrestrained, the crack configuration manifests itself at the exposed surface as a network of 'map cracks' (Figure 3.20). However, where there is restraint, either in the form of steel reinforcement, an applied load or adjacent structural elements, cracks tend to run parallel to the direction of the main reinforcement or the direction of the restraint.

Spalling of the surface is not observed, although 'pop-out' of reactive particles has been noted in certain cases. Pop-outs take the form of small conical plugs of concrete being forced out from the concrete surface by a reacting aggregate particle beneath (Figure 3.21) [108]. Pop-outs appear to only occur under very specific circumstances, such as where indoor concrete that has been exposed to high humidities is dried for a period of time

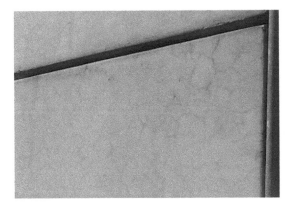

Figure 3.20 Map cracking. (Courtesy of K. Paine.)

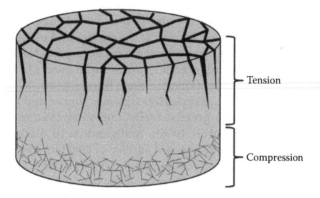

Figure 3.21 Macrocracks and microcracks resulting from the tensile and compressive forces resulting from ASR. (After Hobbs, D. W. *Alkali–Silica Reaction in Concrete.* London: Thomas Telford, 1988; Courtier, R. H., *Cement and Concrete Composites*, 12, 1990, 191–201.)

is subsequently rehumidified [110] and during steam curing of concrete (Figure 3.22) [111].

In some instances, gel may be exuded through the concrete surface. Freshly exuded, the gel is glassy in appearance and slimy to the touch [108]. However, over time, the gel reacts with carbon dioxide in the atmosphere to form an opaque white residue. An exuded gel is also often observed for ACR.

The formation of cracks in concrete undergoing ASR results in a loss of strength (Figure 3.23). Loss of tensile strength is most pronounced, with a comparable deterioration in elastic modulus. The implications of this drop

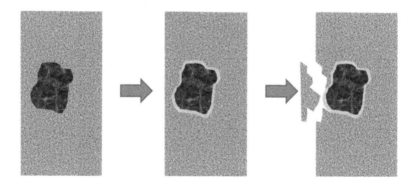

Figure 3.22 Formation of a pop-out as a result of alkali-aggregate reaction expansion.

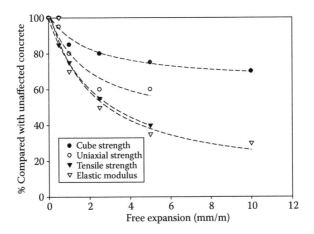

Figure 3.23 Decline in strength and elastic modulus of concrete specimens having undergone ASR expansion. (From Institution of Structural Engineers. *Structural Effects of Alkali–Silica Reaction: Technical Guidance on the Appraisal of Existing Structures.* London: Institution of Structural Engineers, 1992; Clark, L. A. *Critical Review of the Structural Implications of the Alkali–Silica Reaction in Concrete.* TRRL Contractor Report 169. Crowthorne, United Kingdom: Transport and Road Research Laboratory, 1989; Clark, L. A. and K. E. Ng. Some factors influencing expansion and strength of the SERC/BRE standard ASR concrete mix. *Proceedings of the SERC–RMO Conference,* 1989, p. 89; and Clayton, N. et al., *The Structural Engineer,* 68, 1990, 287–292.)

in strength are dependent on the type of structural element in which the reaction is occurring.

In columns, the loss of compressive strength reduces the column load-bearing capacity, although not necessarily to the same magnitude as observed in unreinforced concrete [116]. Additionally, the cover zone of the concrete can become delaminated from the reinforcement because of differences in the extent to which concrete within and outside the reinforced zone is constrained. This causes the compressive strength contribution from this part of the column section to be further compromised and removes restraint against reinforcement buckling [113].

In beams, where flexural and flexural shear strength are of key importance, there is only a relatively small decline in flexural strength with increasing free expansion resulting from ASR. Under flexure, beams appear to be stiffer than equivalent unaffected beams in the stages of elastic deformation immediately prior to yield [117]. Flexural shear strength remains unchanged or improves, with a larger deflection at failure [118]. The likely reason for this is that restrained ASR expansion 'prestresses' the beams, thus enhancing shear characteristics [109]. Similar flexural and flexural shear strength effects are observed in slabs, whereas punching shear strength remains unaffected below a free expansion level of approximately 0.06% [109,119].

Prestressed beams subject to ASR show little change in flexural strength, which has been interpreted as indicating that, along with there being no major change in compressive strength, a significant loss in pre-stress level is unlikely [120]. Flexural shear strength in beams with no shear reinforcement falls after the first cracks resulting from ASR appear, and this drop in strength persists as expansion continues. Where shear reinforcement is present, a decline in shear strength of a similar magni-tude is observed, but strength is regained on further expansion. This is most probably the result of expansion in the vertical of the beam being restrained, thus creating a degree of 'prestressing' in this direction and enhancing strength.

Although the direct effects of alkali–aggregate reactions on the proper-ties of structural elements are, in some cases, relatively minor, the forma-tion of cracks in concrete has very serious implications on performance. The nature of alkali–aggregate cracking, with potentially substantial macro-cracking in the cover zone, means that its ability to protect steel reinforce-ment can be significantly compromised. The occurrence of alkali–aggregate reaction has been demonstrated to lead to an increase in chlorides in con-crete specimens exposed to marine environments relative to specimens that had not undergone the reaction, with the extent of reinforcement corrosion increased as a consequence [121].

Concrete that has undergone alkali–aggregate reactions leading to crack-ing is also more prone to sulphate attack [121]. A study examining the resistance of concrete to freeze–thaw attack found that the onset of expan-sion resulting from cyclic freezing and thawing occurred earlier in ASR-damaged specimens [122]. However, it is interesting to note that complete disintegration of the cores occurred first in specimens containing nonreac-tive aggregate.

3.3.5 Sources of alkalis

3.3.5.1 Portland cement

The alkalis in Portland cement derive largely from feldspar, mica and clay minerals in the raw feed of cement kilns [123]. The high temperatures within the kiln cause the release of alkalis to form new compounds. In the presence of sulphate deriving from the combustion gases of the kiln, a wide range of alkali sulphate compounds are usually formed, which are vola-tilised and condensed cyclically as they move between hotter and cooler zones in the kiln. This ultimately leads to quantities of clinker being dis-charged, bearing deposits of alkali sulphates. Where there is an excess of alkali ions relative to sulphate, these ions are substituted within the structure of all the major clinker phases [124].

Although the alkali content of Portland cement can potentially be high, there has been much development of cement kiln technology aimed at reducing alkalis [125]. This has been driven not only by concerns relating to alkali–aggregate reactions, but also because alkalis have a detrimental effect on later-age strength [126].

3.3.5.2 Other cementitious constituents

Later, it will be seen that the use of other cementitious materials in combination with Portland cement can be effective in controlling alkali–aggregate expansion. However, these materials will typically contain quantities of sodium and potassium, which become available to take part in alkali–aggregate reactions. Some of the materials are discussed below.

3.3.5.2.1 Fly ash

Fly ash derived from coal-fired power generation contains alkalis whose levels depend largely on the coal source. The results of alkali analysis of fly ashes arising in the United States and the United Kingdom are shown in Figure 3.24. The US results derive from Class C (typically high in calcium) ashes largely derived from the combustion of lignite coal, whereas the UK fly ashes are all low in calcium. Alkali levels in low-calcium ashes are between 1.0 and 6.2. In the case of high-calcium ashes, although the majority of ashes contain relatively low levels of alkali, the range of alkali levels is broader and reaches higher levels.

Unlike Portland cement, where a significant amount of the total alkalis are present at the surface of cement particles, a large proportion of alkalis in fly ash are incorporated within ash particles. Low-calcium fly ash is composed of a glassy phase or phases and a variety of crystalline minerals including quartz, mullite, hematite and magnetite. Alkalis are almost exclusively incorporated into the structure of the glass, which is also the phase that undergoes pozzolanic reaction to provide cementitious products. Class C fly ashes typically possess a more varied mineralogy, with the possibility of alkalis being present in soluble crystalline phases [127]. In both cases, however, most of the alkalis are not immediately available, but are released as the pozzolanic reaction proceeds.

As will be discussed in Section 3.3.8, the presence of fly ash can reduce the extent to which expansion occurs in concrete containing reactive aggregate. However, where alkali levels in fly ash are relatively high, the contribution from the ash can be sufficient to reduce the threshold level of Portland cement alkali content above which harmful expansion occurs

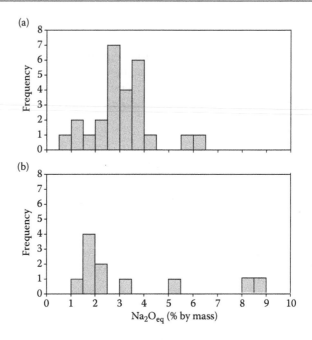

Figure 3.24 Distributions of total alkali contents of (a) 26 siliceous fly ashes from the United Kingdom and (b) 11 predominantly calcareous fly ashes from the United States. ((a) From Hubbard, F. H. et al., *Cement and Concrete Research*, 15, 1985, 185–198. (b) From Shehata, M. H. and M. D. A. Thomas, *Cement and Concrete Research*, 36, 2006, 1166–1175.)

(Figure 3.25) [128]. Moreover, the alkali availability from calcareous (>10% CaO by mass) fly ashes is generally greater than that from siliceous ash [129].

3.3.5.2.2 Ground granulated blast-furnace slag

The alkali content of GGBS can potentially be as high as 2.6% [128], although, again, the alkali ions are mainly incorporated into the glassy phase that makes up the vast majority of the material. Figure 3.26 shows a distribution of total alkali levels from a review of the literature, illustrating that levels are typically lower than those of fly ash.

Like fly ash, the alkali content of slag gradually becomes available as particles undergo latent hydraulic reaction, and this availability is again evidenced by a reduction in the Portland cement alkali content threshold above which harmful ASR expansion occurs [128].

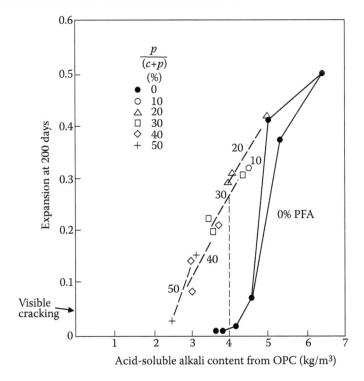

Figure 3.25 Variation of expansion of (ordinary) Portland cement (OPC)/fly ash ('pulverised fuel ash' – PFA) concretes at 200 days, with alkali contributed by Portland cement. (From Hobbs, D. W., *Magazine of Concrete Research*, 38, 1986, 191–205.)

3.3.5.2.3 Silica fume

Levels of alkali in silica fume are, on average, higher than GGBS (Figure 3.27). Moreover, although silica fume contains, on average, lower quantities of alkali than fly ash, more of this is available for release [129]. It should be stressed that silica fume is usually used in lower quantities than both fly ash and GGBS, which means that its contribution to total alkalis will typically be relatively small.

3.3.5.2.4 Water

The alkali content of mix water used in concrete production will clearly influence the concentration of available alkalis in concrete considerably, since the alkali ions will be in solubilised form and hence immediately available for reaction. The European standard for mix water for concrete requires that water used in concrete containing reactive aggregates should

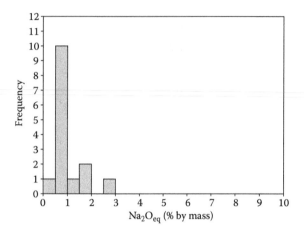

Figure 3.26 Distribution of total alkali contents of 15 blast-furnace slags from a review of the literature and the laboratory study by Hobbs. (From Hobbs, D. W., *Magazine of Concrete Research*, 38, 1986, 191–205.)

not contain alkali levels exceeding 1500 mg/L, unless other measures are taken to control alkali–aggregate reactions [132]. The drinking water standards of many countries commonly place a limit on sodium and sometimes potassium, on the grounds that they affect the taste of water [133], which means that this limit would not be exceeded when potable water is used. However, the use of water from natural underground or surface sources

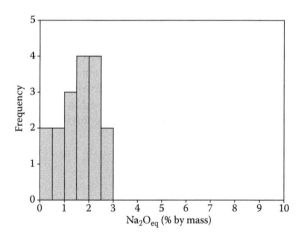

Figure 3.27 Distribution of total alkali contents of 17 silica fumes from a review of the literature and the laboratory study by de Larrard et al. (From de Larrard, F. et al., *Materials and Structures*, 25, 1992, 265–272.)

and waste water from industrial processes (including concrete manufacture) may potentially contain levels of alkali that are problematic to concrete containing reactive aggregates. Waste water recovered from concrete manufacture will typically be acceptable as long as the solids content of the water is reduced to below 1% by mass, and the water is used as a partial replacement for fresh potable or other suitable water used in the mixing process [132].

The average salinity of the planet's seawater is approximately 3.5%, with a large proportion of this being sodium chloride, which means that the use of such water in the manufacture of concrete containing reactive aggregates will almost certainly present a problem. However, where concrete for structural elements containing steel reinforcement is being manufactured, the more immediate and serious threat of chloride-induced corrosion would be of greatest concern and would hopefully rule out the use of such water in most cases.

3.3.5.2.5 Aggregates

Alkalis deriving from aggregates can be divided into two types. First, aggregates may be contaminated with highly soluble salts from the environment from which they have been obtained. The most obvious example is the contamination of aggregate from marine sources with sodium chloride [134]. Second, several rocks used as aggregates have been shown to release alkalis when exposed to high-pH conditions. These rocks include rhyolites and andesites [135,136] and certain quartzites [137]. One group of mineral constituents of many of these rocks are the alkali feldspars, and the release of alkalis from these minerals and their reaction with ASR susceptible aggregate has been clearly demonstrated [138]. Feldspar-bearing volcanic rocks, such as basalt, display a higher rate of alkali release when volcanic glasses are also present. Such glasses will contain similar levels of alkali to the feldspars present and are likely to be more soluble [139]. Release of alkalis is also possible, to a lesser extent, from clay minerals [140].

3.3.5.2.6 Chemical admixtures

Many chemical admixtures contain potassium or sodium salts of organic acids. The alkali ions introduced by these admixtures will normally contribute towards the total alkali content of a concrete mix and hence towards alkali–aggregate reactions. This has been demonstrated in the case of lignosulphonate-, sulphonated melamine formaldehyde– and sulphonated naphthalene formaldehyde–type superplasticising admixtures [106,141].

3.3.5.2.7 External sources

Alkalis may also ingress into concrete during service. The most abundant source of external alkalis is sodium chloride from seawater or de-icing salts, and the exposure of concrete to solutions of sodium chloride has been demonstrated to cause significant expansion [142]. As with any other process of ingress, the rate at which sodium ions can penetrate concrete is dependent on the age and pore structure of the material, with less mature concrete and concrete containing larger volumes of porosity displaying higher sodium diffusion coefficients [143]. Additionally, the presence of pozzolanic and latent hydraulic cement components has the effect of enhancing the extent to which sodium ions are immobilised by CSH gel (see Section 3.3.8), reducing the diffusion coefficient.

3.3.6 Aggregate reactivity

3.3.6.1 Alkali–silica reaction

Rocks and other materials capable of undergoing ASR are discussed below. Two main factors appear to favour the ASR reactivity of aggregate: the presence of strained silica, a silica morphology that presents a high surface area for reaction [144], or both. Silica can exist in strained form either from distortion of the crystal lattice through geological metamorphic processes or simply as a result of poor crystallinity.

3.3.6.1.1 Quartz (SiO₂)

The thermodynamically stable form of silica under normal ambient conditions is quartz. Quartz is commonly encountered in highly crystalline unstrained form, which is essentially unreactive in contact with alkaline solutions. However, it would appear that as the level of strain in quartz increases, so does its susceptibility to attack by alkalis [145].

Many rock types contain quartz in the form of microcrystalline or cryptocrystalline form – very fine crystalline grains discernable as such only by microscopic examination and in the case of cryptocrystalline quartz, barely so. In these rocks, the quartz grains are often separated by open grain boundaries that are potentially more accessible to water, thus permitting a fast rate of reaction [146]. The most common reactive rock types used as aggregate containing cryptocrystalline quartz are flint and chert.

Chalcedony, which was once considered a mineral, is now understood to consist of a mixture of cryptocrystalline quartz and the silica mineral moganite [147]. Although having a fibrous texture, closer inspection using electron microscope reveals that each fibre consists of spheres, hundreds of angstroms in diameter, with porosity running between them, again providing a significant surface per unit volume for reaction. It has also been proposed that moganite is more soluble and hence more reactive than quartz.

3.3.6.1.2 Opal

Opal is sometimes classified as a mineraloid rather than a mineral, on the grounds that many occurrences are noncrystalline. Furthermore, opals frequently contain microcrystalline or cryptocrystalline forms of the silica minerals cristobalite, tridymite or both [148]. The microstructure of opal consists of stacks of close-packed spheres several hundred nanometres in diameter, again potentially creating a large internal surface area for reaction.

3.3.6.1.3 Volcanic glass

Volcanic glasses are the product of the rapid cooling of magma. Although rock can consist entirely of volcanic glass (e.g., obsidian and pumice), glass is more frequently present as a matrix material in fine-grained volcanic rocks such as basalt, andesite and dacite. The amorphous nature of volcanic glass means that the bonds between atoms are strained, which leads to a higher reactivity, in the same way that quartz is more reactive in a strained condition. The potential reactivity of volcanic glasses has been shown in a study that examined the propensity for a selection of basalts to undergo ASR [149]. Basalts are typically of low reactivity, but the study found that, where petrographic analysis identified glass, ASR reactivity was higher. The chemical nature of the glass also plays a role, with acidic glasses (those richer in SiO_2) being typically more reactive.

3.3.6.1.4 Artificial glass

The occurrence of ASR in concrete containing man-made glass has been known about for some time [150]. However, in recent years, the issue has received further attention, partly as a result of research carried out into the possibility of using glass deriving from the recovery of waste container glass as aggregate in concrete [151] and partly because of growth in the recycling of rubble from the demolition of structures where glass from glazing is likely to be present. Soda-lime-silica glass, which is used in glazing and glass containers, also contains substantial amounts of sodium, making the material a potential source of alkalis.

3.3.6.2 Alkali–silicate reaction

Phyllosilicate minerals present in rocks that undergo alkali–silicate reaction are believed, at least in part, to be responsible for expansion. The rocks include greywackes, siltstones, phyllites and argillites. Phyllosilicates in these rocks include chlorite, micas, vermiculites and smectites. The process of exfoliation and swelling of vermiculite has been observed in phyllite and greywacke aggregates in concrete [81], and it has been proposed that this mineral is the main contributor to expansion. However, as discussed previously, it should be remembered that ASR may be the underlying reaction mechanism, in which case the presence of the mineral components listed for this reaction may be the source of expansion.

3.3.6.3 Alkali–carbonate reaction

The two minerals involved in ACRs are dolomite and magnesite.

3.3.6.3.1 Dolomite (CaMg(CO$_3$)$_2$)

Although dolomite is encountered in many different types of rock, the most common types encountered in the literature relating to ACR are limestones that have been exposed to magnesium-bearing solutions, partly converting calcite to dolomite. In these rocks, crystals of dolomite are typically rhombohedral in shape and are surrounded by a matrix of finer calcite crystals between which is a network of clay minerals [148]. Previous suggestions that the expansion of dolomitic limestones is the result of the expansion of this clay seem now to be unlikely. However, what is more probable is that the clay network acts as a route for water containing hydroxide ions and alkalis to migrate through the calcite to the dolomite crystals [80].

3.3.6.3.2 Magnesite (MgCO$_3$)

Magnesite is the ultimate product of the reaction between calcite and magnesium-bearing solutions and thus occurs in limestones alongside dolomite. Magnesite reacts with alkali hydroxides to form brucite, in a similar manner to dedolomitisation [76].

3.3.7 Identifying alkali–aggregate reactions

Alkali–aggregate reaction will usually first become apparent during the visual inspection of structures as 'map cracking' or similar at the surface of concrete elements. However, it should be stressed that map cracking is not unique to these types of reaction. Indeed, a wide range of processes can produce similar crack formation. These include plastic shrinkage caused by the evaporation of water from freshly placed concrete.

In the case of ASR and ACR, the presence of gel exuded from surface cracks (Section 3.3.4) may allow a firmer conclusion to be drawn with regards to whether the reaction has occurred. However, the absence of gel at the surface does not necessarily mean that an alkali–aggregate reaction is not the source of cracking.

Although nondestructive testing can assist in assessing the extent of damage, with ultrasonic techniques proving particularly useful [152], positive identification of alkali–aggregate reactions is not possible. Therefore, material must be removed from a structure in the form of cores and examined using a microscope. Examination can be conducted using either an optical or scanning electron microscope.

Optical microscopy can be carried out on either polished thin sections or simply polished specimens, although where ASR is the likely mechanism, thin sections can be beneficial. This is because, when viewed using a petrographic microscope, the presence of gels can be readily confirmed. Such microscopes feature polarising filters on either side of the specimen, permitting crystalline and amorphous materials to be distinguished. Gels are amorphous (they possess little or no order on an atomic scale). When viewed between crossed polarisers in a petrographic microscope, the configuration will usually allow some light to pass through crystalline minerals. In contrast, amorphous materials, such as ASR gel, appear black.

ASR gel will often form as both a rim around the outside of aggregate particles as well as a product in the interior of aggregates. As discussed in Section 3.3.4, the expansion of aggregate particles tends to mean that microcracks originate from the centre of particles and are symmetrically arranged. Because of the mechanism of expansion, cracks are commonly observed to narrow at greater distances from the aggregate centre.

In the case of ACR, the products of reaction form as rims around the outer parts of aggregates. Distinction between ASR products is possible using a petrographic microscope, since brucite is crystalline.

Staining may provide some additional assistance in identifying alkali–aggregate reactions under the optical microscope. An effective stain for sodium-bearing gels in concrete is uranyl acetate, which gives a yellow-green stain under ultraviolet (UV) light, although appropriate procedures are required to limit the release of this substance into the wider environment [153]. Sodium cobaltnitrite will stain gels rich in potassium, although unfortunately this also yields a yellow stain [154]. Again, the stain is most prominent using a UV light source. The use of Alizarin red has also been demonstrated to prove useful in identifying ACR reaction rims. The compound stains calcium-bearing substances red, making $Mg(OH)_2$ crystals easier to identify.

The scanning electron microscope can be used to examine either fracture surfaces of fragments broken from cores or polished sections. When viewed at fracture surfaces, ASR gel appears as a massive formation, which has been observed as having either a spongy, fine-grained or foliated surface texture [155]. In addition, the gel often contains cracks resulting from the drying of the gel. Crystalline reaction products can also often be observed

(a)   (b)

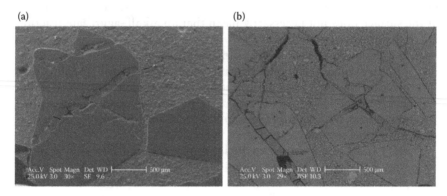

Figure 3.28 Scanning electron microscope images showing particles of glass in a Portland cement matrix that have undergone ASR. (a) Conventional image and (b) image from operating the microscope in backscattered electron imaging mode.

alongside the gel. Gel appears in polished sections as a smooth-surfaced substance, again frequently containing drying cracks. Observation of the gel in backscattered electron imaging mode is a useful method, as it permits improved discrimination between the gel and other materials (Figure 3.28).

Identification of alkali–silicate reaction using the scanning electron microscope is less likely. Although exfoliation of clay mineral particles has been observed in *in vitro* laboratory investigations [81], it is unlikely that meaningful observations could be obtained from concrete fracture surfaces.

3.3.8 Avoiding alkali–aggregate reaction

Guidance has been developed on avoiding ASR by a number of different organisations in many different countries. However, the general components in all of this guidance are similar. In summary, combinations of the following approaches are normally prescribed:

- Use an alternative source of lower reactivity aggregate
- Limit exposure to moisture
- Limit alkali levels
- Include pozzolanic or latent hydraulic materials
- Use ASR-controlling chemical admixtures

These options (with the exception of the first) are discussed below.

3.3.8.1 Limiting exposure to moisture

It has already been established that moisture plays a crucial role in alkali–aggregate reactions, and the problem of expansion can be avoided simply by limiting the extent to which concrete comes in contact with water. For structural elements in the interior of a building or in climates in which moisture is scarce, this may be a very real possibility, although it should be stressed that exposure to atmospheric moisture as well as liquid water will produce expansion, and so assessment of likely internal atmospheric conditions is required before this approach can be taken.

The size and dimensions of structural elements must also be given consideration, since where large masses of concrete are present, concrete sufficiently far beneath an element's surface will not have the opportunity to lose moisture through evaporation, which means that expansion can still occur [156].

For concrete exposed to climates where rainfall or high humidity are likely, the remaining option is to attempt to prevent water from penetrating the surface. This is best achieved either through the application of water-impermeable surface coatings or through penetrant-type treatments that impart hydrophobicity to the surface of pores at the near surface of concrete (Chapter 6). Although these approaches do reduce expansion [157], they have been demonstrated to be not particularly effective where exposure to moisture is prolonged [158]. However, in environments in which exposure to moisture is likely to be followed by prolonged periods of drying, the limited water penetration produced by such treatments is often enough to limit moisture levels to below levels likely to lead to expansion. With the additional risks associated with the reduction in the effectiveness of surface treatments with time, and the likelihood of subsequent crack formation, which is likely to produce a route for the ingress of moisture, this approach is not advisable where exposure to moisture is frequent.

3.3.8.2 Limiting alkali levels

Because soluble forms of sodium and potassium are required to be present for AAR to occur, limiting the available quantities of these ions in a concrete mix is also likely to limit expansion. In the United Kingdom, where the main concern is ASR, *BRE Digest 330* provides guidance on what alkali limits should be adhered to avoid ASR expansion [159]. The factors taken into account are the alkali content of the cement and the reactivity of the aggregate or combination of aggregates used. From these, a limit ranging from 'no limit' to less than 2.5-kg Na_2O_{eq}/m^3 is obtained. Modifications are made to this limit in terms of the amount of available alkalis coming from GGBS, fly ash and aggregates.

3.3.8.3 Admixtures

A number of chemical admixtures for the prevention of expansion deriving from ASR are commercially available. These admixtures are based around lithium salts. There is still much uncertainty regarding the precise manner in which these compounds work, with two possible mechanisms being favoured. The first of these involves lithium ions being incorporated in preference to sodium and potassium in the ASR product, leading to the formation of a crystalline product that is nonexpansive [160]. The second involves the formation of a protective crystalline lithium silicate layer at reactive aggregate surfaces, which inhibits the development of ASR gel [161].

Lithium salts have been found to be effective in controlling expansion in concrete containing greywacke aggregate [162], which may, strictly speaking, undergo alkali–silicate reaction, rather than ASR (see Section 3.3.3). Given the most probable mechanisms by which lithium admixtures work, this suggests that alkali–silicate reactions may indeed be ASR.

Lithium salts found to be effective include LiOH, LiCl, LiNO$_3$, LiF and Li$_2$CO$_3$. Adequate reduction in expansion is usually obtained with a lithium/sodium equivalent (Li/Na$_{eq}$) molar ratio of above 0.6 to 0.9, depending on the type of reactive aggregate [160,163]. Some laboratory-based studies have identified a pessimum dosage of LiOH, where ASR expansion is higher than concrete containing no admixture and above which expansion is reduced to an adequate level [163]. However, such behaviour has not been observed in other similar studies, leading to the conclusion that the appearance of a pessimum dosage may be the result of factors such as the aggregate type and grading and the W/C ratio [164].

ACR has been found to be controlled by the use of either Li$_2$CO$_3$ or FeCl$_3$ [165]. FeCl$_3$ has the disadvantage of accelerating set. Furthermore, the use of chloride compounds into reinforced concrete is likely to accelerate rates of corrosion and is thus not advisable.

The expansive nature of alkali–aggregate reactions has led several researchers to explore the possibility of using air-entraining agents to control expansion in a manner similar to their application in controlling freeze–thaw damage (Chapter 2). This research has demonstrated that the presence of air bubbles within the cement matrix reduces expansion [166]. The results of the calculations included in the UK guidance on alternative approaches to controlling ASR [167] show that, in theory, the inclusion of 3% air by volume should lead to a reduction of expansion of approximately 1%. This would be a worthwhile aim, but, in reality, the magnitude of reduction in expansion is much less than this. This has been attributed to ASR gel being unable to fully occupy air bubbles, presumably as a result of its high viscosity. As a result, the use of air entrainment is discouraged in the current UK guidance.

3.3.8.4 Pozzolanic and latent hydraulic materials

One of the most effective means of controlling alkali–aggregate reactions is through the use of pozzolanic or latent hydraulic materials as components of the cement fraction in a concrete mix. This controls ASR through two mechanisms. First, it has the effect of diluting the Portland cement, thus potentially reducing the alkali contribution from this constituent. Second, reduction in the rate and extent of ASR is achieved through the capacity of CSH gel, whose formation is contributed towards by pozzolanic and latent hydraulic materials, to incorporate alkali ions within its structure. This is possible either through sorption of the ions onto the gel surface or through their incorporation in the CSH structure.

The capacity for CSH to adsorb alkalis increases with a decrease in the Ca/Si ratio [168]. It has been suggested that this is the result of the presence of acidic silanol (Si–OH) sites in the CSH gel, which would undergo neutralisation reactions with sodium and potassium hydroxide. Because pozzolanic and latent hydraulic materials all have lower Ca/Si ratios than Portland cement, their presence in a concrete mix leads to the formation not only of more CSH but also of CSH with a greater capacity for immobilising alkalis.

Another chemical factor that strongly influences the capacity of CSH to immobilise alkali ions is the incorporation of quantities of aluminium in the CSH structure as a substitute for either calcium or silicon [169]. Aluminium substitution is likely to increase the acidity of silanol sites in the CSH gel, thus increasing the affinity for alkalis. However, it has also been proposed that aluminium substitution settles charge imbalances that would be caused by the incorporation of alkali ions in the CSH structure, allowing for an increased binding capacity [170]. Thus, alkali binding tends to be enhanced where cement components containing higher levels of aluminium are used, since this will normally increase the amount of aluminium substitution in the CSH gel.

Individual materials that have been demonstrated as being effective at controlling ASR expansion are discussed below.

3.3.8.4.1 Fly ash

Fly ash is silica-rich pozzolana (typically ~50% SiO_2) with a relatively high alumina content (~25% Al_2O_3). Although not all of these oxide constituents are available for reaction, based on the discussion above, this clearly gives the material a favourable composition for controlling alkali–aggregate reactions. However, the most significant influence on the magnitude of ASR expansion is the alkali content of the ash [171], as a result of the influence of this factor on alkalis in pore solutions [172]. Thus, if the alkali content of a fly ash is sufficiently high, the ash may have no effect, or even a detrimental effect, on expansion [128]. For this reason, the UK guidance on avoiding ASR recommends avoiding fly ash with a Na_2O_{eq} value of more than 5% [159].

Where the alkali contribution is sufficiently low to reduce expansion, reduction in expansion increases as the fly ash content of a concrete mix increases. In many instances, it has been found that lower levels of fly ash have little effect or can exacerbate expansion [173]. For this reason, the UK guidance recommends minimum levels of fly ash in the cement fraction of a concrete mix – 25% for 'low-' or 'normal-' reactivity aggregate and 40% for high-reactivity aggregate.

Another factor that influences the effectiveness of fly ash in controlling expansion is the fineness, with finer particles yielding lower expansion [171,174]. This influence clearly indicates that the reduction in expansion produced by the presence of fly ash does not derive wholly from a dilution of alkalis.

3.3.8.4.2 Ground granulated blast-furnace slag

GGBS is a latent hydraulic rather than pozzolanic material, which means that highly alkaline conditions are required to start the cementitious reactions it undergoes, but the alkaline substances causing the activation are not included in the resulting reaction products. GGBS can be used in the cement fraction of concrete at higher levels than fly ash, with less compromise in engineering properties. Thus, where GGBS has a low alkali content, the dilution effect is greater than that of fly ash. However, the effectiveness of fly ash in controlling alkali–aggregate reactions is usually greater than that of GGBS [175]. This is reflected in the UK guidance, which recommends higher minimum levels for GGBS.

3.3.8.4.3 Silica fume

The fine particle size of silica fume means that it undergoes pozzolanic reaction relatively rapidly. Possibly for this reason, it is typically even more effective than fly ash in reducing expansion. As for other materials, the inclusion of low levels of silica fume in the cement fraction of concrete containing highly reactive aggregates may worsen the magnitude of expansion. However, the highly reactive nature of silica fume means that expansion due to AAR is usually wholly eliminated above levels of approximately 10% by mass of the cement fraction [159].

3.3.8.4.4 Metakaolin

In a manner similar to silica fume, the incorporation of 10% or more by mass of metakaolin into the cement fraction of a concrete mix appears to be adequate to control expansion as a result of ASR [176]. Although little research has been done on the effectiveness of this material in controlling expansion from other forms of alkali–aggregate reaction, its ability to reduce the pH of pore solutions has been clearly demonstrated [177], and so this would appear likely.

3.4 ACID ATTACK

When concrete comes in contact with acidic solutions, constituents of the cement fraction and sometimes the aggregate are selectively dissolved. This leads to an increase in porosity, with various implications in terms of mechanical properties (principally strength) and the permeation properties of the concrete cover.

However, although contact with acids produces a deterioration of concrete at relatively rapid rates, it should be stressed that the dissolution of cement hydration products will occur in concrete exposed to water, regardless of its pH. In this section, the general mechanism involved in this process is described, after which the specifics of acid attack are detailed.

3.4.1 Leaching of concrete by water

Concrete exposed to mobile water for prolonged periods of time will eventually begin to undergo a loss in strength. Laboratory experiments have established that the loss in compressive strength of the concrete is principally the result of the leaching of portlandite ($Ca(OH)_2$) from the surface of the cement paste and a loss of calcium – 'decalcification' – from the CSH gel. As leaching progresses, a degraded zone (containing reduced levels of portlandite) develops at the concrete surface, whose depth progressively increases [178]. The extent to which deterioration of compressive strength is observed is directly proportional to a parameter referred to as the 'degraded ratio', which is the ratio of the degraded area of the cross section of the concrete (A_d) to the total cross-sectional area (A_t).

The thickness of the degraded zone is directly proportional to the loss of mass from the concrete [179]. Thus, there is usually a linear relationship between the degraded ratio and loss in strength. Furthermore, the loss in strength can be related principally to the dissolution of portlandite. This is because portlandite is present mainly as large crystals whose dissolution leaves macroporosity. Decalcification of CSH leads to the formation of microporosity, whose influence on strength is much less significant.

The increase in depth of the degraded zone is dependent on the rate at which the dissolved chemical species can diffuse through the degraded layer into the surrounding water [180] and is described by the equation

$$e = \sqrt{D_{app}t}$$

where e is the depth of the degraded zone, D_{app} is the apparent diffusion coefficient through this zone and t is time. Using this equation and the concept of the degraded ratio, the deterioration in strength of a concrete element in a structure can be described using the equation

$$\Delta s = -b\frac{l^2 - (l - 2\sqrt{D_{app}t})^2}{l^2}$$

where l is the concrete element's smallest dimension, and b is a constant dependent on the composition of the cement fraction [181].

When the water in contact with concrete possesses a pH close to 7, the process of leaching is slow, to the extent that its occurrence is usually only of concern in applications such as nuclear waste storage. However, where more acidic conditions are prevalent, the rate of leaching can be much faster and of more general concern.

3.4.2 Acidic environments

Concrete can come into contact with acidic solutions for a wide range of reasons. Industrial manufacturing processes use or produce a wide range of mineral acids. Aside from the actual manufacture of such acids, fertiliser manufacture uses nitric acid (HNO_3) and sulphuric acid (H_2SO_4) or phosphoric acid (H_3PO_4) in the production of ammonium nitrate (NH_4NO_3) and superphosphate ($Ca(H_2PO_4)_2$), respectively. HF is used in the glass industry as an etchant. Metal processing and finishing activities involve a wide range of acids and mixtures of acids. In particular, hydrochloric acid (HCl) is used in the pickling of steel to remove surface layers of rust. Organic acids also arise in industrial applications, but particularly in the food and drink industry. Acetic acid ($CH_3 \cdot COOH$) and lactic acid ($CH_3 \cdot CHOH \cdot COOH$) are commonly encountered in this type of environment, with lactic acid being present in sour milk and certain alcoholic beverages. Acetic acid is also produced by sour alcoholic beverages and is the main acidic constituent of vinegar.

Various agricultural activities also produce acidic conditions. The production of silage yields acetic and lactic acid. Milk production is likely to produce quantities of lactic acid, where spilt or discarded milk degrades under the action of bacteria. Manure is a source of a wide range of organic acids, including acetic, propanoic (CH_3CH_2COOH), butanoic ($CH_3(CH_2)_2COOH$), isobutanoic ($(CH_3)_2CHCOOH$) and pentanoic ($CH_3(CH_2)_3COOH$) acids. Superphosphate fertilizer stored on agricultural premises may contain quantities of free sulphuric and phosphoric acids.

Sewage is capable, under the correct conditions, of producing sulphuric acid. The process is initiated when sulphates in the sewage are converted to hydrogen sulphide (H_2S) by sulphate-reducing bacteria [182]. This gas reacts with oxygen to form elemental sulphur, which is subsequently oxidised by sulphur-oxidising bacteria to form sulphuric acid. The optimal temperature for this process is approximately 30°C, which means that it is more commonly encountered in hotter countries [183].

The main source of acid formation in soils and groundwater (other than pollution from industrial activity) is the oxidation of sulphides by sulphur-oxidising bacteria to form sulphuric acid, as outlined in Section 3.2.1. Groundwater may also become acidified where conditions permit the dissolution of carbon dioxide to the extent that 'aggressive CO_2' is present, as discussed in Section 3.2.6. The degradation of organic matter in soil leads to the formation of humic acid. This is not a specific compound but a complex mixture of molecules possessing phenolic and carboxylic groups, which are acidic in nature.

The release of sulphur dioxide (SO_2) and oxides of nitrogen (NO_x) deriving from combustion processes into the atmosphere leads to the precipitation of acid rain containing sulphuric and nitric acids. As will be discussed later, the extent to which this is an issue for concrete structures depends on the concentration of these acids in rainwater and the rate of acid deposition, which is dependent on both the concentration of acid and precipitation rates.

Data from precipitation chemistry networks from locations around the globe indicate pH levels in rainwater below 4.5 in many industrialised areas [184]. Moreover, modelling of the total wet deposition of acid indicates quantities as high as 0.1 $mol/m^2/year$. Acid deposition of this magnitude has the potential to affect concrete, particularly where accumulation of rainwater is possible.

3.4.3 Mechanisms of acid attack

The aggressive action of acids on concrete is partly dependent on their strength, which is measured in terms of the ease with which they lose a proton (H^+) through dissociation in a solvent. The dissociation of a monoprotonic acid, HA, is given by the equation

$$HA \rightleftharpoons H^+ + A^-$$

where A^- is referred to as the conjugate base. In an ideal case, the dissociation would be complete – the solution would contain only H^+ and A^- ions and no HA. However, in reality, the acid and its conjugate base will exist in equilibrium, and the extent to which dissociation occurs is measured using the acid dissociation constant, K_a, defined by the equation

$$K_a = \frac{[A^-][H^+]}{[HA]}$$

where the terms in square parentheses are concentrations. For convenience, acid dissociation constants are frequently expressed as pK_a, where

$$pK_a = -\log_{10}(K_a)$$

Thus, the more negative the pK_a of an acid, the greater its strength. pK_a values for a range of acids are given in Table 3.3.

In water, acids dissociate in the following manner:

$$HA + H_2O \rightleftharpoons H_3O^+ + A^-$$

The acidity or basicity of an aqueous solution is expressed as its pH value:

$$pH = -\log_{10}[H_3O^+]$$

with a low pH value denoting a high acidity.

Table 3.3 pK_a values for a range of acids

Acid	pK_a
Hydrochloric	−8
Sulphuric	−3
Nitric	−1.3
Oxalic	1.25, 4.14
Tartaric	2.99, 4.40
Citric	3.09, 4.75, 5.41
Hydrofluoric	3.17
Formic	3.77
Lactic	3.86
Acetic	4.76
Butanoic	4.83
Isobutanoic	4.86
Propanoic	4.87
Carbonic	6.35
Resorcinol	9.20, 10.90
Pyrocatechol	9.12, 12.08
Phenol	9.98
Hydroquinone	9.91, 11.44
Phloroglucinol	8.90, 9.90, 12.75
Hydroxyquinol	9.10, 11.10, 13.00
Pyrogallol	9.05, 11.19, 14.00

Note: Where multiple values are given, the acid can lose multiple protons, with each value representing the pK_a for successive protons.

Where concrete comes in contact with an acidic solution, it would be expected that the lower the pH of a solution, the faster the rate of deterioration. Thus, an aggressive solution is one that contains a strong acid at a high concentration. Although this is generally true, there are a number of other factors that influence the rate and mechanism of deterioration. These factors are discussed in further detail below and in subsequent sections.

When acid solutions come in contact with hardened concrete, various reactions involving cement hydration products and aggregate minerals will occur. In the case of calcium hydroxide, the general reaction is

$$Ca(OH)_2 + 2HA \rightarrow CaA_2 + 2H_2O$$

where HA is a monoprotonic acid. The reaction occurs at a pH value of below 12.6.

Ettringite undergoes a reaction with acid below a pH value of 10.7:

$$6CaO \cdot Al_2O_3 \cdot 3SO_3 \cdot 32H_2O + 6HA \rightarrow 3CaA_2 + 2Al(OH)_3 + 3CaSO_4 + 32H_2O$$

In the case of CSH gel, the reaction (which occurs below a pH value of ~10.5) is

$$xCaO \cdot ySiO_2 \cdot nH_2O + 2xHA \rightarrow xCaA_2 + ySi(OH)_4 + (x + n - 2y)H_2O$$

where $Si(OH)_4$ is an amorphous silica gel.

For calcium aluminate hydrates, such as monosulphate, a reaction of the type shown below occurs:

$$3CaO \cdot Al_2O_3 \cdot CaSO_3 \cdot 12H_2O + 6HA \rightarrow 3CaA_2 + 2Al(OH)_3 + CaSO_4 + 12H_2O$$

Again, an amorphous gel in the form of $Al(OH)_3$ is a product of this reaction, although, where solution pH values are less than 4, this will also react:

$$Al(OH)_3 + 3HA \rightarrow AlA_3 + 3H_2O$$

Iron-bearing hydrates behave in a similar way, although the solution pH value must be less than 2 before the second reaction can occur [185].

The overall effect of these reactions is partly dependent on the solubility of the calcium, aluminium and iron salts that are produced. Where highly soluble salts are formed, these are rapidly leached away, leaving behind a layer of weaker and more permeable gel. However, where salts of low solubility are formed, these remain within the cement matrix. The precipitation of solid calcium salts can either be beneficial or detrimental, depending on what is formed. In certain cases, the salts can form a protective layer that acts as a barrier to the ingress of the acidic species. However, it may also be the case that the formation of calcium compounds leads to crystallisation pressures that cause expansion and cracking. Table 3.4 gives solubilities of salts that may be formed as a result of the reaction of hardened cement paste with an acid.

Certain types of aggregate are also susceptible to acid attack, in particular, those containing carbonates. When limestone aggregate is brought into contact with an acid, the following reaction occurs:

$$CaCO_3 + 2HA \rightarrow CaA_2 + 2CO_2$$

Again, this can have different effects depending on the salt formed.

The initial effect of acid attack on concrete is the development of a corroded layer in which the above reactions are partially or wholly completed. Where removal of calcium, aluminium and iron occurs, the porosity of this layer is higher. Moreover, depending on the acid involved, the corroded layer can undergo shrinkage. This leads to cracking, which increases the porosity further.

Table 3.4 Solubilities of calcium and aluminium salts of various acids

Acid	Ca salt	Solubility at 20°C (g/L)	Al salt	Solubility (g/L)
Sulphuric	$CaSO_4 \cdot 2H_2O$	2.4	$Al_2(SO_4)_3$	364
Hydrochloric	$CaCl_2$	745	$AlCl_3$	458
Nitric	$Ca(NO_3)_2 \cdot 4H_2O$	1290	$Al(NO_3)_3 \cdot 9H_2O$	673
Hydrofluoric	CaF_2	0.016	AlF_3	6.7
Formic	$Ca(HCOO)_2$	166	$Al(HCOO)_3 \cdot 3H_2O$	61.9 (25°C)
Acetic	$Ca(CH_3COO)_2 \cdot H_2O$	347	$Al(CH_3COO)_3$	'Sparingly soluble'
Propanoic	$Ca(C_2H_5COO)_2$	399 (25°C)	$Al(C_2H_5COO)_3$	–
Butanoic	$Ca(C_3H_7COO)_2 \cdot H_2O$	182 (25°C)		–
Isobutanoic	$Ca(C_3H_7COO)_2 \cdot 5H_2O$	50		–
Citric	$Ca_3(C_6H_5O_7)_2$	0.95 (25°C)	$Al(C_6H_5CO_7)_3$	2.3
Tartaric	$CaC_4H_4O_6$	0.37 (0°C)	$Al_2(C_4H_4O_6)_3$	–
Oxalic	$Ca(COO)_2$	0.0067	$Al(COO)_3$	'Practically insoluble'
Lactic	$Ca(C_3H_5O_3)_2$	90	$Al(C_3H_5O_3)_3$	17
Aggressive CO_2	$Ca(HCO_3)_2$	16.6		

Beyond this layer is a volume of unaffected concrete. The rate at which the corroded layer extends into concrete is dependent on the rate at which the acidic species can diffuse through the layer to react with the unaffected concrete.

Where mechanical attrition processes that are capable of removing cement and aggregate are effective at the concrete surface, the surface may be removed at an accelerated rate. Where aggregates remain insoluble in acidic conditions, a faster rate of removal of the cement matrix relative to the aggregate is observed. Thus, a concrete surface undergoing acid attack will often be composed of exposed aggregate, left behind as the cement is preferentially removed. However, eventually this aggregate will become insufficiently bonded to the matrix and will also be lost as the process of corrosion continues.

3.4.4 Factors influencing rates of acid attack

3.4.4.1 Environmental factors

The main environmental factors that influence the rate of deterioration of concrete exposed to acidic conditions are the concentration of the acidic species, temperature and whether any mechanical attrition is also acting on the concrete surface.

Concentration is clearly an important factor, since the concentration gradient between the exterior and the interior of the concrete will control the rate of diffusion. However, it is also important because the reactions of acids with cement (and possibly aggregate) lead to their neutralisation. In situations where there is a finite quantity of acidic species in solution, and the neutralisation capacity of the concrete is sufficiently high, the acid will be wholly converted to neutral salt, leading to the deterioration process halting.

Figure 3.29 shows the loss in mass of concrete specimens exposed to acidic solutions, where attrition of the surface was periodically conducted. The effect of acidic environments on compressive strength is shown in Figure 3.30.

Because remnants of the original cement hydration products remain after acid attack, the concrete surface can potentially remain intact even after lengthy periods of reaction. However, because of the loss of strength in the corroded layer, it will become easier for the material to be worn away if any form of mechanical abrasion is acting on the surface.

Figure 3.31 shows the effect of increasingly aggressive levels of attrition on the loss of mass from concrete specimens exposed to a carbonic acid environment. As might be expected, the more aggressive the wear mechanism acting on the concrete surface, the faster the rate of surface loss. However, a more significant observation can be made from these results:

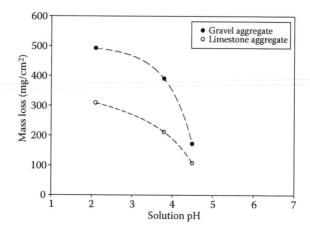

Figure 3.29 Mass loss per unit area from concrete prisms exposed to two wetting and drying cycles using lactic acid solutions of different pH. Erosion was simulated through periodic brushing of the concrete surface. (From de Belie, N. et al., *ACI Materials Journal*, 94, 1997, 546–554.)

the loss of mass where mechanical attrition is effective is approximately linear when plotted against time. Where acid attack occurs without physical damage to the surface, it should be expected that the shape of the plot would be parabolic, since the rate of attack will be limited by the rate of diffusion through the developing corroded layer. However, abrasion will act to remove the corroded layer, making diffusion unnecessary [188].

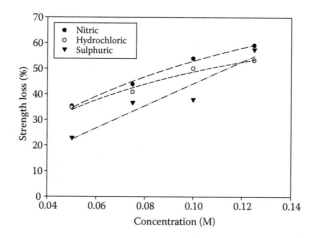

Figure 3.30 Loss of compressive strength after exposure of Portland cement mortars exposed to acid solutions of different concentrations for a period of 120 days. (From Türkel, S. et al., *Sādhanā*, 32, 2007, 683–691.)

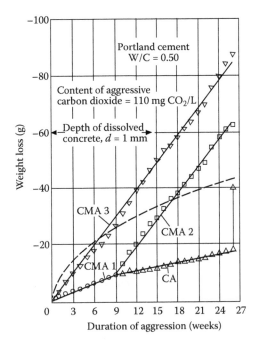

Figure 3.31 Loss of mass from Portland cement (PC) mortar prisms exposed to carbonic acid attack and different mechanical wear mechanisms. Chemical attack (CA) = no wear; chemical and mechanical attack (CMA) 1 = dabbing of prism surface prior to mass measurement; CMA 2 = surface brushed once per week; CMA 3 = surface brushed three times per day. (From Türkel, S. et al., *Sādhanā*, 32, 2007, 683–691.)

Generally, an increase in temperature will lead to an increase in the rate of deterioration as a result of acid attack. This is primarily due to an increase in the solubility of the salts formed. However, where salts possess a negative partial molar enthalpy of solution and thus become less soluble at higher temperatures, greater resistance may be observed. An example of this is calcium acetate.

3.4.4.2 Material factors

Concrete constituents can significantly influence resistance to acid attack. The main factors that play a role are calcium hydroxide content, cement content, W/C ratio and the type of aggregate used.

The influence of calcium hydroxide on acid resistance is best illustrated through examining the effect of the combination of Portland cement with other cement components, such as fly ash, since this leads to a dilution of Portland cement, resulting in lower $Ca(OH)_2$ levels. The influence of fly ash

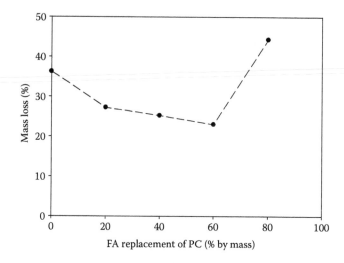

Figure 3.32 Loss of mass at 15 weeks from concrete mixes containing different levels of replacement of Portland cement (PC) with fly ash exposed to a 1% hydrochloric acid solution. Mechanical wear was achieved through removal of reaction product with a wire brush on a weekly basis. All mixes contain a 10% (by mass of total cement – PC and fly ash) addition of silica fume. (From Tamimi, A. K., *Materials and Structures*, 30, 1997, 188–191.)

content is shown in Figure 3.32, which shows that up to levels of 60%, an enhancement in acid resistance is observed. Beyond this point, resistance sharply declines. Similar improvement in acid resistance can be achieved with GGBS, silica fume and metakaolin [189–191].

The vulnerability of concrete containing higher levels of $Ca(OH)_2$ is the result of the fact that this phase is the first to dissolve as conditions move from alkaline to acidic. Thus, concrete with a high $Ca(OH)_2$ content will display a relatively rapid increase in porosity – and loss in strength – as acid attack proceeds. The increase in the rate of mass loss at very high levels of fly ash content can simply be attributed to lower strength, which makes the concrete more vulnerable to mechanical wear.

For the same reason, the Portland cement content of a concrete mix influences its resistance to acid attack. A higher Portland cement content will increase vulnerability, since the proportion of $Ca(OH)_2$ will be raised [192]. It has also been established that a higher aggregate content has the effect of preventing the formation of long cracks and reducing the crack density within the corroded layer, making diffusion through this zone slower. The first of these effects is the result of the tendency of cracks to propagate between aggregate particles, producing a network of short cracks with a more tortuous crack path. The second is the result of aggregate particles restraining shrinkage of the corroded layer, leading to less cracking [193].

It should be noted that higher Portland cement contents will impart a higher neutralisation capacity to the concrete, which may protect the concrete where acid is present in finite quantities. Also, a lower cement content may render concrete more prone to abrasion of the corroded layer [194].

Because the rate of ingress into concrete of species in solution is controlled by the capillary porosity of the cement matrix, the W/C ratio plays an important role in defining acid resistance. For the majority of acids, a lower W/C ratio yields greater resistance to corrosion, principally as a result of a reduced rate of ingress through the corroded layer and a higher strength in this layer, resulting in reduced loss of mass as a result of mechanical wear.

In the case of concrete exposed to sulphuric acid, conflicting observations have been made by researchers. In some cases, deterioration characteristics similar to other acids have been observed [196]. However, in at least one instance, measurement of loss of mass has been found to be reduced as W/C ratio increases [197]. It has been proposed that weight loss measurements at pH levels greater than 1.5 are compromised by the formation of reaction products (gypsum and probably ettringite) [197]. Although this explanation is not wholly adequate, since results where an increasing W/C ratio yielded an increased rate of mass loss have been obtained with a pH value in excess of 2 [196], it would be unwise to attempt to enhance sulphuric acid resistance through the use of high W/C ratios. A possible alternative explanation is that, depending on the mix design method used, mixes with a lower W/C ratio may contain much higher cement contents, making the concrete more vulnerable to attack [192].

Siliceous aggregates are typically highly resistant to acid attack, except in the case of highly aggressive acids such as HF. However, as indicated by the reaction outlined in Section 3.4.3, aggregates that are more readily attacked (such as limestone) neutralise acids as they dissolve. The large volume of aggregate relative to cement means that limestone aggregate will impart a high neutralising capacity to concrete. The overall effect is that concrete made with limestone aggregates is more resistant to acid attack than that made with siliceous aggregates. Evidence of this has been previously seen in Figure 3.29, which shows the effect of aggregate type on mass loss resulting from exposure to lactic acid.

3.4.5 Action of specific acids

As previously discussed, the range of acids that could potentially come into contact with concrete is wide, with the effects on concrete durability being dependent on the type of acid. This section examines attack from selected acids, illustrating the variations in acid attack resulting from the chemical nature of the acidic species.

3.4.5.1 Nitric acid (HNO₃)

The reaction of calcium hydroxide with nitric acid results in the formation of calcium nitrate:

$$2HNO_3 + Ca(OH)_2 \rightarrow Ca(NO_3)_2 + 2H_2O$$

$Ca(NO_3)_2$ is highly soluble and so is rapidly removed from the corroded layer, leaving behind porosity. Moreover, as the process of acid attack progresses, aluminium nitrate and iron (III) nitrate are formed, which are also highly soluble. There is a relatively substantial shrinkage of the corroded layer, which leads to cracking, and a further increase in the porosity of the corroded layer [193]. The overall effect is a relatively rapid rate of deterioration.

3.4.5.2 Sulphuric acid (H₂SO₄)

The threat to concrete from sulphuric acid (H_2SO_4) results partly from dissolution and also as a result of expansion. Sulphuric acid coming into contact with $Ca(OH)_2$ produces the reaction

$$H_2SO_4 + Ca(OH)_2 \rightarrow CaSO_4 \cdot 2H_2O$$

Gypsum ($CaSO_4 \cdot 2H_2O$) is precipitated as a result of its relatively low solubility (see Table 3.1), which can potentially lead to a blocking of pores at the concrete surface, leading to a slower rate of attack initially. This leads to deterioration at high sulphuric acid concentrations being slower initially than at lower concentrations, since a higher concentration yields a thicker layer of gypsum. However, this effect lasts only temporarily, with rates of deterioration ultimately greater at higher concentrations [198].

The expansion of concrete exposed to sulphuric acid is the result of gypsum and ettringite formation, as described in Section 3.2.2. As a result of ettringite's solubility at lower pH levels, ettringite is typically only formed at some distance into the interior of concrete subject to sulphuric acid attack, where pH is still relatively high [199].

Measurement of mass and length change in mortars exposed to either sodium sulphate and H_2SO_4 indicates that the acid increases the vulnerability of concrete to deterioration, even when pozzolanic materials otherwise capable of providing greater resistance are present [190]. The reason for this is most likely the result of the reduced strength and increased permeability

A range of polymeric materials can be used for polymer modification, including polymethyl methacrylate (PMMA), polystyrene (PS), polyesters, styrene acrylic esters, styrene butadiene, polyvinyl acetate (PVA), acrylic polymers, polyacrylonitrile (PAN) and epoxy resins. Investigations into the effectiveness of different polymers at imparting acid resistance have found a quite significant variation. For instance, a study of the performance of PMMA-, PS- and PAN-modified concrete exposed to H_2SO_4 and HCl found that PMMA was superior to the other polymers [205]. However, it was observed that the PMMA-modified material underwent significant expansion during exposure. Other comparative studies have identified styrene acrylic esters as superior where exposure to H_2SO_4 was used, with some polymer types giving performance inferior to polymer-free controls [206]. Further enhancements in the performance of polymer-modified concrete have been achieved through the addition of both fly ash (to a polyester-modified concrete) and Na_2SiO_3 plus a quantity of Na_2SiF_6 (to a PVA-modified concrete exposed to H_2SO_4) [207,208].

Despite some documented success with polymer-modified concrete in limiting acid attack, since modification acts only to reduce the rate of deterioration and some modifying agents produce little benefit, caution should be exercised before this approach is chosen as a means of protecting concrete. At the very least, consultation with the product manufacturer is advisable.

3.4.7.4 Protective coatings

Placing a layer of acid-resistant and water-tight material between concrete and the external environment is clearly an approach that is likely to improve durability in acidic environments. This can be achieved in a number of ways, some of which are more effective than others.

Concrete can be rendered less accessible to external acids through the application of several types of inorganic coating. The coatings of this kind most commonly used to improve resistance to acids are sodium and potassium silicates (Na_2SiO_3 and K_2SiO_3). When applied to concrete, these compounds undergo a reaction with $Ca(OH)_2$ to form CSH gel at the surface. The gel is precipitated into pores at the concrete's exterior, blocking them. Moreover, as discussed previously, CSH gel's relative stability in acidic conditions provides enhanced resistance. Zinc and magnesium fluorosilicates can be used in a similar fashion. However, it is generally agreed that the protective nature of these coatings is relatively short lived, and superior alternatives exist.

A number of specially formulated cement-bound surface layers are available, primarily for concrete floors, composed of mineral particles mixed with Portland cement [209]. These mortars are applied in relatively thick layers (2–3 cm deep). The mineral particles consist of substances such as calcium-bearing tuffs, which appear to provide acid resistance through a high neutralising capacity.

As with polymer modification, a wide range of polymer coatings are available for concrete, including epoxy and acrylic resins, bitumen, neoprene, styrene–butadiene rubber and polyesters. However, the most commonly encountered are epoxy and acrylic resin based.

It should be stressed that, where a coating is present, it should not be assumed that the surface is wholly impermeable. This is because it is highly unlikely that initial coverage will be perfect, and cracking of concrete after application may well cause the coating to rupture. Some products also contain fibre reinforcement (usually glass) to reduce the extent to which rupture is likely to occur. In experiments evaluating the acid resistance imparted by coatings, weight loss of concrete exposed to H_2SO_4 was found to be considerably higher for specimens where damaged regions had been artificially created in the coating surface, even for relatively small defects [210]. The inevitable breaches of protective coatings can be minimised by appropriate surface treatment prior to application (cleaning, use of a primer layer, etc.) and by ensuring that the coating is applied in accordance with the recommendations of the manufacturer. Generally, the thicker the coating, the more resistant the surface to acid attack.

Resin injection of cracked concrete is now an established repair technique, which is likely to provide renewed acid protection of concrete. However, in many cases, such as in concrete sewage pipes, this option is likely to be impractical.

In a comparative study examining epoxy and acrylic resin coatings, an important observation is made [211]. Two different concrete mixes were examined, with different cement contents. The extent to which loss in mass is reduced by the application of coatings was similar for both resin types, where the concrete with a low cement content was studied. However, for the higher cement content concrete, deterioration was larger, and there was a significant difference in performance between the two coatings. This highlights an important point – even when protective coatings (or polymer modification) are used, mix design for acid resistance is still an appropriate measure.

REFERENCES

1. Turekian, K. K. *Oceans, 2nd ed.* New Jersey: Prentice Hall, 1976, 152 pp.
2. Kester, D. R. and R. M. Pytkowicz. Sodium, magnesium, and calcium sulphate ion pairs in seawater at 25°C. *Limnology and Oceanography*, v. 14, 1969, pp. 686–692.

3. Casanova, I., L. Agulló, and A. Aguado. Aggregate expansivity due to sulphide oxidation: Part 1. Reaction system and rate model. *Cement and Concrete Research*, v. 26, 1996, pp. 993–998.

4. Casanova, I., A. Aguado, and L. Agulló. Aggregate expansivity due to sulphide oxidation: Part 2. Physicochemical modelling of sulphate attack. *Cement and Concrete Research*, v. 27, 1997, pp. 1627–1632.

5. Trudinger, P. A. Microbes, metals, and minerals. *Minerals Science and Engineering*, v. 3, 1971, pp. 13–25.

6. Spratt, H. G., M. D. Morgan, and R. E. Good. Sulphate reduction in peat from a New Jersey pinelands cedar swamp. *Applied and Environmental Microbiology*, v. 53, 1987, pp. 1406–1411.

7. Building Research Establishment. *BRE Special Digest 1: Concrete in Aggressive Ground, 3rd ed.* Watford, United Kingdom: Building Research Establishment, 2005, 62 pp.

8. Musson, S. E., Q. Xu, and T. G. Townsend. Measuring the gypsum content of C&D debris fines. *Waste Management*, v. 28, 2008, pp. 2091–2096.

9. Taylor, H. F. W. *Cement Chemistry, 2nd ed.* London: Thomas Telford, 1997, 480 pp.

10. Gollop, R. S. and H. F. W. Taylor. Microstructural and microanalytical studies of sulphate attack: Part 1. Ordinary Portland cement paste. *Cement and Concrete Research*, v. 22, 1992, pp. 1027–1038.

11. Scherer, G. W. Stress from crystallization of salt. *Cement and Concrete Research*, v. 34, 2004, pp. 1613–1624.

12. Ping, X. and J. J. Beaudoin. Mechanism of sulphate expansion. *Cement and Concrete Research*, v. 22, 1992, pp. 631–640.

13. Pommersheim, J. M. and J. R. Clifton. Expansion of cementitious materials exposed to sulphate solutions. *Proceedings of the Materials Research Society Symposium*, v. 333, 1994, pp. 363–368.

14. Piasta, W. G. Deformations and elastic modulus of concrete under sustained compression and sulphate attack. *Cement and Concrete Research*, v. 22, 1992, pp. 149–158.

15. Živica, V. and A. Szabo. The behaviour of cement composite under compression load at sulphate attack. *Cement and Concrete Research*, v. 24, 1994, pp. 1475–1484.

16. Lawrence, C. D. The influence of binder type on sulphate resistance. *Cement and Concrete Research*, v. 22, 1992, pp. 1047–1058.

17. Al-Dulaijan, S. U., D. E. Macphee, M. Maslehuddin, M. M. Al-Zahrani, and M. R. Ali. Performance of plain and blended cements exposed to high-sulphate concentrations. *Advances in Cement Research*, v. 19, 2007, pp. 167–175.

18. van Aardt, J. H. P. and S. Visser. Influence of alkali on the sulphate resistance of ordinary Portland cement mortars. *Cement and Concrete Research*, v. 15, 1985, pp. 485–494.

19. Santhanam, M., M. D. Cohen, and J. Olek. Modelling the effects of solution temperature and concentration during sulphate attack on cement mortars. *Cement and Concrete Research*, v. 32, 2002, pp. 585–592.

20. Li, G., P. Le Bescop, and M. Moranville. Expansion mechanism associated with the secondary formation of the U-phase in cement-based systems containing high amounts of Na_2SO_4. *Cement and Concrete Research*, v. 26, 1996, pp. 195–201.

21. Moranville, M., and G. Li. The U-phase: Formation and stability. In J. Marchand and J. Skalny, eds., *Materials Science of Concrete: Sulphate Attack Mechanism.* Westerville, Ohio: The American Ceramic Society, 1999, pp. 175–188.

22. Monteiro, P. J. M. and K. E. Kurtis. Time to failure for concrete exposed to severe sulphate attack. *Cement and Concrete Research*, v. 33, 2003, pp. 987–993.

23. Hime, W. and L. Backus. A discussion of the paper "Redefining cement characteristics of sulphate-resistant Portland cement" by Paul Tikalsky, Della Roy, Barry Scheetz, and Tara Krize. *Cement and Concrete Research*, v. 33, 2003, p. 1907.

24. Glasser, F. P. Chemistry of the alkali–aggregate reaction. In R. N. Swamy, ed., *The Alkali Silica Reaction in Concrete.* Glasgow, Scotland: Blackie, 1992, pp. 30–53.

25. Tikalsky, P. J., D. Roy, B. Scheetz, and T. Krize. Redefining cement characteristics for sulphate-resistant Portland cement. *Cement and Concrete Research*, v. 32, 2002, pp. 1239–1246.

26. Benstead, J. and J. Munn. A discussion of the paper "Redefining cement characteristics of sulphate-resistant Portland cement" by Paul Tikalsky, Della Roy, Barry Scheetz, and Tara Krize. *Cement and Concrete Research*, v. 34, 2004, pp. 355–357.

27. Shanahan, N. and A. Zayed. Cement composition and sulphate attack: Part 1. *Cement and Concrete Research*, v. 37, 2007, pp. 618–623.

28. Rasheeduzzafar, F. H. Dakhil, A. S. Al-Gahtani, S. S. Al-Saadoun, and M. A. Bader. Influence of cement composition on the corrosion of reinforcement and sulphate resistance of concrete. *ACI Materials Journal*, v. 87, 1990, pp. 114–122.

29. Bonakdar, A. and B. Mobasher. Multiparameter study of external sulphate attack in blended cement materials. *Construction and Building Materials*, v. 24, 2010, pp. 61–70.

30. Hughes, D. C. Sulphate resistance of OPC, OPC/fly ash, and SRPC pastes: Pore structure and permeability. *Cement and Concrete Research*, v. 15, 1985, pp. 1003–1012.

31. Chindaprasirt, P., S. Homwuttiwong, and V. Sirivivatnanon. Influence of fly ash fineness on strength, drying shrinkage, and sulphate resistance of blended cement mortar. *Cement and Concrete Research*, v. 34, 2004, pp. 1087–1092.

32. Torii, K. and M. Kawamura. Effects of fly ash and silica fume on the resistance of mortar to sulphuric acid and sulphate attack. *Cement and Concrete Research*, v. 24, 1994, pp. 361–370.

33. Higgins, D. D. Increased sulphate resistance of GGBS concrete in the presence of carbonate. *Cement and Concrete Composites*, v. 25, 2003, pp. 913–919.

34. Sahmaran, M., O. Kasap, K. Duru, and I. O. Yaman. Effects of mix composition and water–cement ratio on the sulphate resistance of blended cements. *Cement and Concrete Composites*, v. 29, 2007, pp. 159–167.

35. Mangat, P. S. and J. M. El-Khatib. Influence of initial curing on sulphate resistance of blended cement concrete. *Cement and Concrete Research*, v. 22, 1992, pp. 1089–1100.
36. Thaumasite Expert Group. *The Thaumasite Form of Sulphate Attack: Risks, Diagnosis, Remedial Works, and Guidance on New Construction*. London: Department of the Environment Transport and the Regions, 1999, 180 pp.
37. Tumidajski, P. J., G. W. Chan, and K. E. Philipose. An effective diffusivity for sulphate transport into concrete. *Cement and Concrete Research*, v. 25, 1995, pp. 1159–1163.
38. Bader, M. A. Performance of concrete in a coastal environment. *Cement and Concrete Composites*, v. 25, 2003, pp. 539–548.
39. Bonen, D. and M. D. Cohen. Magnesium sulphate attack on Portland cement paste: Part 2. Chemical and mineralogical analyses. *Cement and Concrete Research*, v. 22, 1992, pp. 707–718.
40. Binici, H. and O. Aksoğan. Sulphate resistance of plain and blended cement. *Cement and Concrete Composites*, v. 28, 2006, pp. 39–46.
41. Dehwah, H. A. F. Effect of sulphate concentration and associated cation type on concrete deterioration and morphological changes in cement hydrates. *Construction and Building Materials*, v. 21, 2007, pp. 29–39.
42. Gao, X., B. Ma, Y. Yang, and A. Su. Sulphate attack of cement-based material with limestone filler exposed to different environments. *Journal of Engineering and Performance*, v. 17, 2008, pp. 543–549.
43. Vuk, T., R. Gabrovšek, and V. Kaučič. The influence of mineral admixtures on sulphate resistance of limestone cement pastes aged in cold $MgSO_4$ solution. *Cement and Concrete Research*, v. 32, 2002, pp. 943–948.
44. Nehdi, M. and M. Hayek. Behaviour of blended cement mortars exposed to sulphate solutions cycling in relative humidity. *Cement and Concrete Research*, v. 35, 2005, pp. 731–742.
45. Edge, R. A. and H. F. W. Taylor. Crystal structure of thaumasite. *Acta Crystallographica*, v. B27, 1971, pp. 594–601.
46. Crammond, N. J. and M. A. Halliwell. Assessment of the conditions required for the thaumasite form of sulphate attack. In K. L. Scrivener and J. F. Young, eds., *Proceedings of the Mechanisms of Chemical Degradation of Cement-Based Systems*, 1997, pp. 193–200.
47. van Aardt, J. H. P. and S. Visser. Thaumasite formation: A cause of deterioration of Portland cement and related substances in the presence of sulphates. *Cement and Concrete Research*, v. 5, 1975, pp. 225–232.
48. Collet, G., N. J. Crammond, R. N. Swamy, and J. H. Sharp. The role of carbon dioxide in the formation of thaumasite. *Cement and Concrete Research*, v. 34, 2004, pp. 1599–1612.
49. Pipilikaki, P., D. Papageorgiou, C. Teas, E. Chaniotakis, and M. Katsioti. The effect of temperature on thaumasite formation. *Cement and Concrete Composites*, v. 30, 2008, pp. 964–969.
50. Aguilera, J., M. T. Blanco Varela, and T. Vázquez. Procedure of synthesis of thaumasite. *Cement and Concrete Research*, v. 31, 2001, pp. 1163–1168.
51. Justnes, H. Thaumasite formed by sulphate attack on mortar with limestone filler. *Cement and Concrete Composites*, v. 25, 2003, pp. 955–959.

52. Collett, G., N. J. Crammond, R. N. Swamy, and J. H. Sharp. The role of carbon dioxide in the formation of thaumasite. *Cement and Concrete Research*, v. 34, 2004, pp. 1599–1612.

53. Crammond, N. J., G. W. Collet, and T. I. Longworth. Thaumasite field trial at Shipston-on-Stour: Three-year preliminary assessment of buried concretes. *Cement and Concrete Composites*, v. 25, 2003, pp. 1035–1043.

54. Longworth, T. I. *BRE Client Report to the Department of Trade and Industry: Review of Guidance on Testing and Classification of Sulphate and Sulphide-Bearing Ground*. Report 80042, Vols. 1 and 2. Watford, United Kingdom: Building Research Establishment, 1999, 180 pp.

55. Zhou, Q., J. Hill, E. A. Byars, J. C. Cripps, C. J. Lynsdale, and J. H. Sharp. The role of pH in thaumasite sulphate attack. *Cement and Concrete Research*, v. 36, 2006, pp. 160–170.

56. Köhler, S., D. Heinz, and L. Urbonas. Effect of ettringite on thaumasite formation. *Cement and Concrete Research*, v. 36, 2006, pp. 697–706.

57. Bellman, F. and J. Stark. The role of calcium hydroxide in the formation of thaumasite. *Cement and Concrete Research*, v. 38, 2008, pp. 1154–1161.

58. Bellman, F. and J. Stark. Prevention of thaumasite formation in concrete exposed to sulphate attack. *Cement and Concrete Research*, v. 37, 2007, pp. 1215–1222.

59. Higgins, D. D. and N. J. Crammond. Resistance of concrete containing GGBS to the thaumasite form of sulphate attack. *Cement and Concrete Composites*, v. 25, 2003, pp. 921–929.

60. Mulenga, D. M., J. Stark, and P. Nobst. Thaumasite formation in concrete and mortars containing fly ash. *Cement and Concrete Composites*, v. 25, 2003, pp. 907–912.

61. Tsivilis, S., G. Kakali, A. Skaropoulou, J. H. Sharp, and R. N. Swamy. Use of mineral admixtures to prevent thaumasite formation in limestone cement mortar. *Cement and Concrete Composites*, v. 25, 2003, pp. 969–976.

62. British Standards Institution. *BS EN 206: Concrete Specification, Performance, Production, and Conformity*. London: British Standards Institution, 2013, 98 pp.

63. British Standards Institution. *BS 8500-1: Concrete – Complementary British Standard to BS EN 206: Part 1. Method of Specifying and Guidance of the Specifier*. London: British Standards Institution, 2013, 66 pp.

64. British Standards Institution. *BS EN 197-1: Cement – Part 1. Composition, Specifications, and Conformity Criteria for Common Cements*. London: British Standards Institution, 2011, 50 pp.

65. McCarthy, M. J. and A. Giannakou. In-situ performance of CPF concrete in a coastal environment. *Cement and Concrete Research*, v. 32, 2002, pp. 451–457.

66. Santhanam, M., M. D. Cohen, and J. Olek. Mechanism of sulphate attack: A fresh look – Part 1. Summary of experimental results. *Cement and Concrete Research*, v. 32, 2002, pp. 915–921.

67. Singh, N. B., A. K. Singh, and S. Prabha Singh. Effect of citric acid on the hydration of Portland cement. *Cement and Concrete Research*, v. 16, 1986, pp. 911–920.

68. Tosun, K. and B. Baradan. Effect of ettringite morphology on DEF-related expansion. *Cement and Concrete Composites*, v. 32, 2010, pp. 271–280.

69. Kelham, S. The effect of cement composition and fineness on expansion associated with delayed ettringite formation. *Cement and Concrete Composites*, v. 18, pp. 171–179.

70. Silva, A. S., D. Soares, L. Matos, M. Salta, L. Divet, A. Pavoine, A. Candeias, and J. Mirão. Influence of mineral additions in the inhibition of delayed ettringite formation in cement-based materials: A microstructural characterization. *Materials Science Forum*, v. 636–637, 2010, pp. 1272–1279.

71. Quillin, K. *BRE Information Paper IP11/01: Delayed Ettringite Formation: In situ Concrete*. Watford, United Kingdom: Building Research Establishment, 2001, 8 pp.

72. Barneyback, R. S. and S. Diamond. Expression and analysis of pore fluids from hardened cement pastes and mortars. *Cement and Concrete Research*, v. 11, 1981, pp. 279–285.

73. Taylor, H. F. W. A method for predicting alkali ion concentrations in cement pore solutions. *Advances in Cement Research*, v. 1, 1987, pp. 5–16.

74. Douglas, R. W. and T. M. El-Shamy. Reactions of glasses with aqueous solutions. *Journal of the American Ceramic Society*, v. 50, 1967, pp. 1–8.

75. Powers, T. C. and H. H. Steinour. An interpretation of some published researches on the alkali–aggregate reaction. *Journal of the American Concrete Institute*, v. 26, 1955, pp. 497–516.

76. Tong, L. and M. Tang. Expansion mechanism of alkali–dolomite and alkali–magnesite reaction. *Cement and Concrete Composites*, v. 21, 1999, pp. 361–373.

77. Gillot, J. E. and E. G. Swenson. Mechanism of the alkali–carbonate rock reaction. *Journal of Engineering Geology*, v. 2, 1970, pp. 7–23.

78. Katayama, T. The so-called alkali–carbonate reaction (ACR): Its mineralogical and geochemical details, with special reference to ASR. *Cement and Concrete Research*, v. 40, 2010, pp. 643–675.

79. Feldman, R. F. and P. J. Sereda. Characteristics of sorption and expansion isotherms of reactive limestone aggregate. *Journal of the American Concrete Institute*, v. 58, 1961, pp. 203–213.

80. Min, D. and T. Mingshu. Mechanism of dedolomitization and expansion of dolomitic rocks. *Cement and Concrete Research*, v. 23, 1993, pp. 1397–1408.

81. Gillott, J. E., M. A. G. Duncan, and E. G. Swenson. Alkali–aggregate reaction in Nova Scotia. Part 4: Character of the reaction. *Cement and Concrete Research*, v. 3, 1973, pp. 521–535.

82. Hobbs, D. W. *Alkali–Silica Reaction in Concrete*. London: Thomas Telford, 1988, 183 pp.

83. Dyer, T. D. and R. K. Dhir. Evaluation of powdered glass cullet as a means of controlling harmful alkali–silica reaction. *Magazine of Concrete Research*, v. 62, 2010, pp. 749–759.

84. Lu, D., L. Mei, Z. Xu, M. Tang, X. Mo, and B. Fournier. Alteration of alkali-reactive aggregates autoclaved in different alkali solutions and application to alkali–aggregate reaction in concrete (II) expansion and microstructure of concretence microbar. *Cement and Concrete Research*, v. 36, 2006, pp. 1191–1200.

85. Min, D. and T. Mingshu. Mechanism of dedolomitization and expansion of dolomitic rocks. *Cement and Concrete Research*, v. 23, 1993, pp. 1397–1408.
86. MacPhee, D. E., K. Luke, F. P. Glasser, and E. E. Lachowski. Solubility and ageing of calcium silicate hydrates in alkaline solutions at 25°C. *Journal of the American Ceramic Society*, v. 72, 1989, pp. 646–654.
87. Hobbs, D. W. Expansion of concrete due to alkali–silica reaction: An explanation. *Magazine of Concrete Research*, v. 30, 1978, pp. 215–220.
88. Ichikawa, T. Alkali–silica reaction, pessimum effects, and pozzolanic effect. *Cement and Concrete Research*, v. 39, 2009, pp. 716–726.
89. Guðmundsson, G. and H. Ásgeirsson. Some investigations on alkali–aggregate reaction. *Cement and Concrete Research*, v. 5, 1975, pp. 211–220.
90. Duncan, M. A. G., E. G. Swenson, J. E. Gillot, and M. R. Foran. Alkali–aggregate reaction in Nova Scotia: Part I. Summary of a 5-year study. *Cement and Concrete Research*, v. 3, 1973, pp. 55–69.
91. Mingshu, T. and L. Yinnon. Rapid method for determining the alkali reactivity of carbonate rock. In P. E. Grattan-Bellew, ed., *Proceedings of the 7th International Conference on Concrete Alkali–Aggregate Reactions*, Ottawa, Canada, August 18–22, 1986, pp. 286–287.
92. Mingshu, T., L. Yinnon, and H. Sufen. Kinetics of alkali–carbonate reaction. In M. Kawamura, *Proceedings of the 8th International Conference on Alkali–Aggregate Reactions*, 1989, pp. 147–152.
93. Dhir, R. K., T. D. Dyer, and M. C. Tang. Alkali–silica reaction in concrete-containing glass. *Materials and Structures*, v. 42, 2009, pp. 1451–1462.
94. Ramyar, K., A. Topal, and O. Andic. Effects of aggregate size and angularity on alkali–silica reaction. *Cement and Concrete Research*, v. 35, 2005, pp. 2165–2169.
95. Kodama, K. and T. Nishino. Observation around the cracked region due to alkali–aggregate reaction by analytical electron microscope. In P. E. Grattan-Bellew, ed., *Proceedings of the 7th International Conference on Concrete Alkali–Aggregate Reactions*, Ottawa, Canada, August 18–22, 1986, pp. 398–403.
96. Multon, S., M. Cyr, A. Sellier, N. Leklou, and L. Petit. Coupled effects of aggregate size and alkali content on ASR expansion. *Cement and Concrete Research*, v. 38, 2008, pp. 350–359.
97. Zhang, C., A. Wang, M. Tang, B. Wu, and N. Zhang. Influence of aggregate size and aggregate size grading on ASR expansion. *Cement and Concrete Research*, v. 29, 1999, pp. 1393–1396.
98. Feng, N.-Q., T.-Y. Hao, and X.-X. Feng. Study of the alkali reactivity of aggregates used in Beijing. *Magazine of Concrete Research*, v. 54, 2002, pp. 233–237.
99. Jin, W., C. Meyer, and S. Baxter. "Glascrete": Concrete with glass aggregate. *ACI Materials Journal*, v. 97, 2000, pp. 208–213.
100. Diamond, S. and N. Thaulow. A study of expansion due to alkali–silica reaction as conditioned by the grain size of the reactive aggregate. *Cement and Concrete Research*, v. 4, 1974, pp. 591–607.
101. Kawamura, M., K. Takomoto, and S. Hasaba. Application of quantitative EDXA analysis and microhardness measurements to the study of alkali–silica reaction mechanisms. In G. M. Idorn and S. Rostom, eds., *Proceedings of the 6th International Conference on Alkalis in Concrete*. Copenhagen, Denmark: Danish Concrete Association, 1983, pp. 167–174.

102. Vivian, H. E. Studies in cement aggregate reaction. *CSIRO Bulletin*, v. 256, 1950, pp. 13–20.
103. Hobbs, D. W. and W. A. Gutteridge. Particle size of aggregate and its influence upon the expansion caused by the alkali–silica reaction. *Magazine of Concrete Research*, v. 31, 1979, pp. 235–242.
104. Tomosawa, F., K. Tamura, and M. Abe. Influence of water content of concrete on alkali–aggregate reaction. In M. Kawamura, ed., *Proceedings of the 8th International Conference on Alkali–Aggregate Reactions*, 1989, pp. 881–885.
105. Ólafsson, H. The effect of relative humidity and temperature on alkali expansion of mortar bars. In P. E. Grattan-Bellew, ed., *Proceedings of the 7th International Conference on Concrete Alkali–Aggregate Reactions*, Ottawa, Canada, August 18–22, 1986, pp. 461–465.
106. Lenzner, D. Influence of the amount of mixing water on the alkali–silica reaction. *Proceedings of the 5th International Conference on Alkali–Aggregate Reaction in Concrete*, Cape Town, South Africa, March 30–April 3, 1981. Paper S252/26. Pretoria, South Africa: National Building Research Institute, 1981.
107. Krell, J. Influence of mix design on alkali–silica reaction in concrete. In P. E. Grattan-Bellew, ed., *Proceedings of the 7th International Conference on Concrete Alkali–Aggregate Reactions*, Ottawa, Canada, August 18–22, 1986, pp. 441–445.
108. Figg, J. ASR: Inside phenomena and outside effects (crack origin and pattern). In P. E. Grattan-Bellew, ed., *Proceedings of the 7th International Conference on Concrete Alkali–Aggregate Reactions*, Ottawa, Canada, August 18–22, 1986, pp. 152–156.
109. Institution of Structural Engineers. *Structural Effects of Alkali–Silica Reaction: Technical Guidance on the Appraisal of Existing Structures*. London: Institution of Structural Engineers, 1992, 48 pp.
110. Nilsson, L.-O. Pop-outs due to alkali–silica reactions: A moisture problem? *Proceedings of the 5th International Conference on Alkali–Aggregate Reaction in Concrete*, Cape Town, South Africa, March 30–April 3, 1981. Paper S252/27. Pretoria, South Africa: National Building Research Institute, 1981.
111. Bache, H. H. and J. C. Isen. Modal determination of concrete resistance to pop-out formation. *Journal of the American Concrete Institute*, v. 65, 1968, pp. 445–450.
112. Courtier, R. H. The assessment of ASR-affected structures. *Cement and Concrete Composites*, v. 12, 1990, pp. 191–201.
113. Clark, L. A. *Critical Review of the Structural Implications of the Alkali–Silica Reaction in Concrete*. TRRL Contractor Report 169. Crowthorne, United Kingdom: Transport and Road Research Laboratory, 1989, 89 pp.
114. Clark, L. A. and K. E. Ng. Some factors influencing expansion and strength of the SERC/BRE standard ASR concrete mix. *Proceedings of the SERC–RMO Conference*, 1989, p. 89.
115. Clayton, N., R. J. Currie, and R. H. Moss. The effects of alkali–silica reaction on the strength of prestressed concrete beams. *The Structural Engineer*, v. 68, 1990, pp. 287–292.
116. Poole, A. B., S. Rigden, and L. Wood. The strength of model columns made with alkali–silica reactive concrete. In P. E. Grattan-Bellew, ed., *Proceedings*

of the 7th International Conference on Concrete Alkali-Aggregate Reactions, Ottawa, Canada, August 18–22, 1986, pp. 136–140.

117. Chana, P. S. and G. A. Korobokis. *The Structural Performance of Reinforced Concrete Affected by ASR: Phase 1.* TRRL Contractor Report 267. Crowthorne, United Kingdom: Transport and Road Research Laboratory, 1991, 77 pp.

118. Abe, M., S. Kikuta, Y. Masuda, and F. Tomozawa. Experimental study on mechanical behaviour of reinforced concrete members affected by alkali–aggregate reaction. In M. Kawamura, ed., *Proceedings of the 8th International Conference on Alkali–Aggregate Reactions*, 1989, pp. 691–696.

119. Clark, L. A. and K. E. Ng. The effects of alkali–silica reaction on the punching shear strength of reinforced concrete slabs. In M. Kawamura, ed., *Proceedings of the 8th International Conference on Alkali–Aggregate Reactions*, 1989, pp. 659–664.

120. Clayton, N., R. J. Currie, and R. M. Moss. The effects of alkali–silica reaction on the strength of prestressed concrete beams. *The Structural Engineer*, v. 68, 1990, pp. 287–292.

121. Hamada, H., N. Otsuki, and T. Fukute. Properties of concrete specimens damaged by alkali–aggregate reaction: Laumontite-related reaction and chloride attack under marine environments. In M. Kawamura, ed., *Proceedings of the 8th International Conference on Alkali–Aggregate Reactions*, 1989, pp. 603–608.

122. Clayton, N. *Structural Implications of Alkali–Silica Reaction: Effect of Natural Exposure and Freeze–Thaw.* Watford, United Kingdom: Building Research Establishment, 1999, 46 pp.

123. Skalny, J. and W. A. Klemm. Alkalis in clinker: Origin, chemistry, effects. *Proceedings of the 5th International Conference on Alkali–Aggregate Reaction in Concrete*, Cape Town, South Africa, March 30–April 3, 1981. Paper S252/1. Pretoria, South Africa: National Building Research Institute, 1981.

124. Pollitt, H. W. W. and A. W. Brown. The distribution of alkalis in Portland cement clinker. *Proceedings of the 5th International Symposium on the Chemistry of Cement, Vol. 1.* Tokyo, Japan: Cement Association of Japan, 1969, pp. 322–333.

125. Svendsen, J. Alkali reduction in cement kilns. *Proceedings of the 5th International Conference on Alkali–Aggregate Reaction in Concrete.* Cape Town, South Africa, March 30–April 3, 1981. Paper S252/2. Pretoria, South Africa: National Building Research Institute, 1981.

126. Gebauer, J. Alkalis in clinker: Influence on cement and concrete properties. *Proceedings of the 5th International Conference on Alkali–Aggregate Reaction in Concrete*, Cape Town, South Africa, March 30–April 3, 1981. Paper S252/4. Pretoria, South Africa: National Building Research Institute, 1981.

127. McCarthy, G. J., K. D. Swanson, L. P. Keller, and W. Blatter. Mineralogy of Western fly ashes. *Cement and Concrete Research*, v. 14, 1984, pp. 471–478.

128. Hobbs, D. W. Deleterious expansion of concrete due to alkali–silica reaction: Influence of PFA and slag. *Magazine of Concrete Research*, v. 38, 1986, pp. 191–205.

129. Shehata, M. H. and M. D. A. Thomas. Alkali release characteristics of blended cements. *Cement and Concrete Research*, v. 36, 2006, pp. 1166–1175.

130. Hubbard, F. H., R. K. Dhir, and M. S. Ellis. Pulverised-fuel ash for concrete: Compositional characterisation of United Kingdom PFA. *Cement and Concrete Research*, v. 15, 1985, pp. 185–198.

131. de Larrard, F., J.-F. Gorse, and C. Puch. Comparative study of various silica fumes as additives in high-performance cementitious materials. *Materials and Structures*, v. 25, 1992, pp. 265–272.

132. British Standards Institution. *BS EN 1008:2002: Mixing Water for Concrete – Specification for Sampling, Testing, and Assessing the Suitability of Water, Including Water Recovered from Processes in the Concrete Industry, as Mixing Water for Concrete*. London: British Standards Institution, 2002, 22 pp.

133. World Health Organisation. *Guidelines for Drinking Water Quality. Vol. 1: Recommendations, 3rd ed.* Geneva, Switzerland: World Health Organisation, 2008, 668 pp.

134. Mizumoto, Y., K. Kosa, K. Ono, and K. Nakono. Study on cracking damage of a concrete structure due to alkali–silica reaction. In P. E. Grattan-Bellew, ed., *Proceedings of the 7th International Conference on Concrete Alkali–Aggregate Reactions*, Ottawa, Canada, August 18–22, 1986, pp. 204–208.

135. Stark, D. C. Alkali–silica reactivity: Some recommendations. *Journal of Cement, Concrete, and Aggregates*, v. 2, 1980, pp. 92–94.

136. Kawamura, M., M. Koike, and K. Nakano. Release of alkalis from reactive andesite aggregates and fly ashes into pore solutions in mortars. In M. Kawamura, ed., *Proceedings of the 8th International Conference on Alkali–Aggregate Reactions*, 1989, pp. 271–278.

137. Blight, G. E. The effects of alkali–aggregate reaction in reinforced concrete structures made with Witwatersrand, quartzite aggregate. *Proceedings of the 5th International Conference on Alkali–Aggregate Reaction in Concrete*, Cape Town, South Africa, March 30–April 3, 1981. Paper S252/15. Pretoria, South Africa: National Building Research Institute, 1981.

138. Constantiner, D. and S. Diamond. Alkali release from feldspars into pore solutions. *Cement and Concrete Research*, v. 33, 2003, pp. 549–554.

139. Goguel, R. Alkali release by volcanic aggregates in concrete. *Cement and Concrete Research*, v. 25, 1995, pp. 841–852.

140. van Aardt, J. H. P. and S. Visser. Calcium hydroxide attack on feldspars and clays: Possible relevance to cement–aggregate reactions. *Cement and Concrete Research*, v. 7, 1977, pp. 643–648.

141. Wang, H. and J. E. Gillot. The effect of superplasticisers on alkali–silica reactivity. In M. Kawamura, ed., *Proceedings of the 8th International Conference on Alkali–Aggregate Reactions*, 1989, pp. 187–192.

142. Chatterji, S. An accelerated method for the detection of alkali–aggregate reactivities of aggregates. *Cement and Concrete Research*, v. 8, 1978, pp. 647–650.

143. Uchikawa, H., S. Uchida, and S. Hanehara. Relationship between structure and penetrability of Na ion in hardened blended cement paste, mortar, and concrete. In M. Kawamura, ed., *Proceedings of the 8th International Conference on Alkali–Aggregate Reactions*, 1989, pp. 121–128.

144. Diamond, S. A review of alkali–silica reaction and expansion mechanisms: Part 2. Reactive aggregates. *Cement and Concrete Research*, v. 6, 1976, pp. 549–560.

145. Gogte, B. S. An evaluation of some common Indian rocks with special reference to alkali–aggregate reactions. *Engineering Geology*, v. 7, 1973, pp. 135–153.

146. Diamond, S. A review of alkali–silica reaction and expansion mechanisms: Part 2. Reactive aggregates. *Cement and Concrete Research*, v. 6, 1976, pp. 549–560.

147. Heaney, P. J. and J. E. Post. The widespread distribution of a novel silica polymorph in microcrystalline quartz varieties. *Science*, v. 255, 1992, pp. 441–443.

148. Deer, W. A., R. A. Howie, and J. Zussman. *An Introduction to the Rock-Forming Minerals, 2nd ed.* Harlow, United Kingdom: Longman, 1992, 712 pp.

149. Koranç, M. and A. Tuğrul. Evaluation of selected basalts from the point of alkali–silica reactivity. *Cement and Concrete Research*, v. 35, 2005, pp. 505–512.

150. Schmidt, A. and W. H. F. Saia. Alkali–aggregate reaction tests on glass used for exposed aggregate wall panel work. *Journal of the American Concrete Institute*, v. 60, 1963, pp. 1235–1236.

151. Jin, W., C. Meyer, and S. Baxter. "Glascrete": Concrete with glass aggregate. *ACI Structural Journal*, v. 97, 2000, pp. 208–213.

152. Akashi, T., S. Amasaki, N. Takagi, and M. Tomita. The estimate for deterioration due to alkali–aggregate reaction by ultrasonic methods. In P. E. Grattan-Bellew, ed., *Proceedings of the 7th International Conference on Concrete Alkali–Aggregate Reactions*, Ottawa, Canada, August 18–22, 1986, pp. 183–187.

153. Natesaiyer, K. C. and K. C. Hover. Further study of an *in situ* identification method for alkali–silica reaction products in concrete. *Cement and Concrete Research*, v. 19, 1989, pp. 770–778.

154. Guthrie, G. D. and J. W. Carey. A simple environmentally friendly and chemically specific method for the identification and evaluation of the alkali–silica reaction. *Cement and Concrete Research*, v. 27, 1997, pp. 1407–1417.

155. Regourd, M. and H. Hornain. Microstructure of reaction products. In P. E. Grattan-Bellew, ed., *Proceedings of the 7th International Conference on Concrete Alkali–Aggregate Reactions*, Ottawa, Canada, August 18–22, 1986, pp. 375–380.

156. Kojima, T., M. Tomita, K. Nakano, and A. Nakaue. Expansion behaviour of reactive aggregate concrete in thin sealed metal tube. In M. Kawamura, ed., *Proceedings of the 8th International Conference on Alkali–Aggregate Reactions*, 1989, pp. 703–708.

157. Kurihara, T. and K. Katawaki. Effects of moisture control and inhibition on alkali–silica reaction. In M. Kawamura, ed., *Proceedings of the 8th International Conference on Alkali–Aggregate Reactions*, 1989, pp. 629–634.

158. Blight, G. E. Experiments on waterproofing concrete to inhibit AAR. In M. Kawamura, ed., *Proceedings of the 8th International Conference on Alkali–Aggregate Reactions*, 1989, pp. 733–739.

159. Building Research Establishment. *Alkali–Silica Reaction in Concrete*. BRE Digest 330 Part 2. Watford, United Kingdom: Building Research Establishment, 2004, 12 pp.

160. Mo, X., C. Yu, and Z. Xu. Long-term effectiveness and mechanism of LiOH in inhibiting alkali–silica reaction. *Cement and Concrete Research*, v. 33, 2003, pp. 115–119.

161. Mitchell, L. D., J. J. Beaudoin, and P. Grattan-Bellew. The effects of lithium hydroxide solution on alkali–silica reaction gels. *Cement and Concrete Research*, v. 34, 2004, pp. 641–649.

162. Thomas, M., R. Hooper, and D. Stokes. Use of lithium-containing compounds to control expansion due to alkali–silica reaction. *Proceedings of the 11th International Conference on Alkali–Aggregate Reaction in Concrete*. Canada: Centre de Recherche Interuniversitaire sur le Beton, 2000, pp. 783–792.

163. Diamond, S. Unique response of LiNO$_3$ as an alkali–silica reaction–preventive admixture. *Cement and Concrete Research*, v. 29, 1999, pp. 1271–1275.

164. Collins, C., J. H. Ideker, G. S. Willis, and K. E. Kurtis. Examination of the effects of LiOH, LiCl, and LiNO$_3$ on alkali–silica reaction. *Cement and Concrete Research*, v. 34, 2004, pp. 1403–1415.

165. Pagano, M. A. and P. D. Cady. A chemical approach to the problem of alkali-reactive carbonate aggregates. *Cement and Concrete Research*, v. 12, 1982, pp. 1–12.

166. Gillott, J. E. and H. Wang. Improved control of alkali–silica reaction by combined use of admixtures. *Cement and Concrete Research*, v. 23, 1993, pp. 973–980.

167. Building Research Establishment. *Minimising the Risk of Alkali–Silica Reaction: Alternative Methods*. BRE Information Paper IP1/02. Watford, United Kingdom: Building Research Establishment, 2002, 8 pp.

168. Hong, S.-Y. and F. P. Glasser. Alkali binding in cement pastes: Part I. The C-S-H phase. *Cement and Concrete Research*, v. 29, 1999, pp. 1893–1903.

169. Stade, H. and D. Müller. On the coordination of Al in ill-crystallized C-S-H phases formed by hydration of tricalcium silicate and by precipitation reactions at ambient temperature. *Cement and Concrete Research*, v. 17, 1987, pp. 553–561.

170. Richardson, I. G. and G. W. Groves. The incorporation of minor and trace elements into calcium silicate hydrate (CSH) gel in hardened cement pastes. *Cement and Concrete Research*, v. 23, 1993, pp. 131–138.

171. Hobbs, D. W. Influence of pulverised-fuel ash and granulated blast-furnace slag upon expansion caused by the alkali–silica reaction. *Magazine of Concrete Research*, v. 34, 1982, pp. 83–94.

172. Duchesne, J. and M. A. Bérubé. The effectiveness of supplementary cementing materials in suppressing expansion due to ASR: Another look at the reaction mechanisms – Part 2. Pore solution chemistry. *Cement and Concrete Research*, v. 24, 1994, pp. 221–230.

173. Soles, J. A., V. M. Malhotra, and R. W. Suderman. The role of supplementary cementing materials in reducing the effects of alkali–aggregate reactivity: CANMET Investigation. In P. E. Grattan-Bellew, ed., *Proceedings of the 7th International Conference on Concrete Alkali–Aggregate Reactions*, Ottawa, Canada, August 18–22, 1986, pp. 79–84.

174. Ukita, K., S.-I. Shigematsu, M. Ishii, K. Yamamoto, K. Azuma, and M. Moteki. Effect of classified fly ash on alkali–aggregate reaction (AAR). In M. Kawamura, ed., *Proceedings of the 8th International Conference on Alkali–Aggregate Reactions*, 1989, pp. 259–264.

175. Monteiro, P. J. M., K. Wang, G. Sposito, M. C. dos Santos, and W. P. de Andrade. Influence of mineral admixtures on the alkali–aggregate reaction. *Cement and Concrete Research*, v. 27, 1997, pp. 1899–1909.

176. Gruber, K. A., T. Ramlochan, A. Boddy, R. D. Hooton, and M. D. A. Thomas. Increasing concrete durability with high-reactivity metakaolin. *Cement and Concrete Composites*, v. 23, 2001, pp. 479–484.

177. Ramlochan, T., M. Thomas, and K. A. Gruber. The effect of metakaolin on alkali ± silica reaction in concrete. *Cement and Concrete Research*, v. 30, 2000, pp. 339–344.

178. Carde, C. and R. François. Modelling the loss of strength and porosity increase due to the leaching of cement pastes. *Cement and Concrete Composites*, v. 21, 1999, pp. 181–188.

179. Carde, C. and R. François. Effect of the leaching of calcium hydroxide from cement paste on mechanical and physical properties. *Cement and Concrete Research*, v. 27, 1997, pp. 539–550.

180. Faucon, P., F. Adenot, J. F. Jacquinot, J. C. Petit, R. Cabrillac, and M. Jorda. Long-term behaviour of cement pastes used for nuclear waste disposal: Review of physicochemical mechanisms of water degradation. *Cement and Concrete Research*, v. 28, 1998, pp. 847–857.

181. Dyer, T. D. Modification of strength of wasteforms during leaching. *Proceedings of the ICE: Waste and Resource Management*, v. 163, 2010, pp. 111–122.

182. O'Connell, M., C. McNally, and M. G. Richardson. Biochemical attack on concrete in wastewater applications: A state-of-the-art review. *Cement and Concrete Composites*, v. 32, 2010, pp. 479–485.

183. Parker, C. D. The isolation of a species of bacterium associated with the corrosion of concrete exposed to atmospheres containing hydrogen sulphide. *Australian Journal of Experimental Biology and Medical Science*, v. 23, 1945, pp. 81–90.

184. Rodhe, H., F. Dentener, and M. Schulz. The global distribution of acidifying wet deposition. *Environmental Science and Technology*, v. 36, 2002, pp. 4382–4388.

185. Allahverdi, A. and F. Škvára. Acidic corrosion of hydrated cement-based materials: Part 1. Mechanisms of the phenomenon. *Ceramics-Silikáty*, v. 44, 2000, pp. 114–120.

186. de Belie, N., M. Debruyckere, S. van Nieuwenburg, and B. de Blaere. Concrete attack by feed acids: Accelerated tests to compare different concrete compositions and technologies. *ACI Materials Journal*, v. 94, 1997, pp. 546–554.

187. Türkel, S., B. Felekoğlu, and S. Dulluç. Influence of various acids on the physicomechanical properties of pozzolanic cement mortars. *Sādhanā*, v. 32, 2007, pp. 683–691.

188. Grube, H. and W. Rechenberg. Durability of concrete structures in acidic water. *Cement and Concrete Research*, v. 19, 1989, pp. 783–792.

189. de Belie, N., H. J. Verselder, B. de Blaere, D. van Nieuwenburg, and R. Verschoore. Influence of the cement type on the resistance of concrete to feed acids. *Cement and Concrete Research*, v. 26, 1996, pp. 1717–1725.

190. Torii, K. and M. Kawamura. Effects of fly ash and silica fume on the resistance of mortar to sulphuric acid and sulphate attack. *Cement and Concrete Research*, v. 24, 1994, pp. 361–370.

191. Kim, H.-S., S.-H. Lee, and H.-Y. Moon. Strength properties and durability aspects of high-strength concrete using Korean metakaolin. *Construction and Building Materials*, v. 21, 2007, pp. 1229–1237.

192. Fattuhi, N. I. and B. P. Hughes. Ordinary Portland cement mixes with selected admixtures subjected to sulphuric acid attack. *ACI Materials Journal*, v. 85, 1988, pp. 512–518.

193. Pavlík, V. and S. Unčik. The rate of corrosion of hardened cement pastes and mortars with additive of silica fume in acids. *Cement and Concrete Research*, v. 27, 1997, pp. 1731–1745.

194. Beddoe, R. E. and H. W. Dorner. Modelling acid attack on concrete: Part 1. The essential mechanism. *Cement and Concrete Research*, v. 35, 2005, pp. 2333–2339.

195. Tamimi, A. K. High-performance concrete mix for an optimum protection in acidic conditions. *Materials and Structures*, v. 30, 1997, pp. 188–191.

196. Hughes, B. P. and J. E. Guest. Limestone and siliceous aggregate concretes subjected to sulphuric acid attack. *Magazine of Concrete Research*, v. 30, 1978, pp. 11–18.

197. Hewayde, E., M. Nehdi, E. Allouche, and G. Nakhla. Effect of mixture design parameters and wetting–drying cycles on resistance of concrete to sulphuric acid attack. *Journal of Materials in Civil Engineering*, v. 19, 2007, pp. 155–163.

198. Biczok, I. *Concrete Corrosion and Concrete Protection*. Translated from the 2nd German edition. London: Collet's, 1964, 543 pp.

199. Skalny, J., J. Marchand, and I. Odler. *Sulphate Attack on Concrete*. London: Spon, 2002, 238 pp.

200. Živica, V. Deterioration of cement-based materials due to the action of organic compounds. *Construction and Building Materials*, v. 20, 2006, pp. 634–641.

201. Živica, V. and A. Bajza. Acidic attack of cement-based materials: A review – Part 1. Principle of acidic attack. *Construction and Building Materials*, v. 15, pp. 331–340.

202. Pavlik, V. Corrosion of hardened cement paste by acetic and nitric acids: Part 2. Formation and chemical composition of the corrosion products layer. *Cement and Concrete Research*, v. 24, 1994, pp. 1495–1508.

203. Bertron, A., G. Escadeillas, and J. Duchesne. Cement pastes alteration by liquid manure organic acids: Chemical and mineralogical characterization. *Cement and Concrete Research*, v. 34, 2004, pp. 1823–1835.

204. Safwan, A. K. and M. N. Abou-Zeid. Characteristics of silica fume concrete. *Journal of Materials in Civil Engineering*, v. 6, 1994, pp. 357–375.

205. Bhattacharya, V. K., K. R. Kirtania, M. M. Maiti, and S. Maiti. Durability tests on polymer cement mortar. *Cement and Concrete Research*, v. 13, 1983, pp. 287–290.

206. Vincke, E., E. van Wanseele, J. Monteny, A. Beeldens, N. de Belie, L. Taerwe, D. van Gemert, and W. Verstraete. Influence of polymer addition on biogenic sulphuric acid attack of concrete. *International Biodeterioration and Biodegradation*, v. 49, 2002, pp. 283–292.

207. Gorninski, J. P., D. C. Dal Molin, and C. S. Kazmierczak. Strength degradation of polymer concrete in acidic environments. *Cement and Concrete Composites*, v. 29, 2007, pp. 637–645.

208. Li, G., G. Xiong, and Y. Yin. The physical and chemical effects of long-term sulphuric acid exposure on hybrid modified cement mortar. *Cement and Concrete Composites*, v. 31, 2009, pp. 325–330.

209. de Belie, N., M. Debruyckere, D. van Nieuwenburg, and B. de Blaere. Attack of concrete floors in pig houses by feed acids: Influence of fly ash addition and cement-bound surface layers. *Journal of Agricultural Engineering Research*, v. 68, 1997, pp. 101–108.

210. Vipulanandan, C. and J. Liu. Glass-fibre mat-reinforced epoxy coating for concrete in sulphuric acid environment. *Cement and Concrete Research*, v. 32, 2002, pp. 205–210.

211. Aguiar, J. B., A. Camões, and P. M. Moreira. Coatings for concrete protection against aggressive environments. *Journal of Advanced Concrete Technology*, v. 6, 2008, pp. 243–250.

Chapter 4

Corrosion of steel reinforcement in concrete

4.1 INTRODUCTION

The idea of incorporating steel reinforcement in concrete structural elements originated in the mid-19th century, with a patent for the technology granted to its inventor, Joseph Monier, in 1867 [1]. The invention eventually radically extended the way in which concrete could be used in structures. The high tensile strength and ductility of steel transformed a material that could otherwise only be loaded to any great extent in compression into one that can be used structurally in flexure and, in some applications, direct tension.

One of the drawbacks of plain steel as a structural material, however, is its susceptibility to corrosion. Corrosion involves the loss of material from a metal surface as a result of a chemical reaction. It presents a problem for steel reinforcement, since the loss of material leads to a loss in cross-sectional area and a consequent loss of load-bearing capacity.

The union of concrete and steel is a mutually beneficial one, since placing a depth of concrete ('cover') between the steel surface and the environment will extend the life of the reinforcement by acting as a barrier to the substances necessary or conducive to corrosion. Additionally, the chemical environment within the pores of concrete does not encourage corrosion, thus providing further protection.

However, the protective role that concrete plays is, for several reasons, finite. This chapter examines the process of corrosion, the main mechanisms that act to limit the protection afforded to steel and the means of achieving long-term durability of reinforced concrete.

4.2 CORROSION OF STEEL IN CONCRETE

All materials can potentially undergo corrosion processes, and the corrosion of concrete exposed to acids has already been discussed in Chapter 3. However, the term is most commonly used to describe the corrosion of metals. This section will examine the process of corrosion of metals, progressively focussing on the corrosion of steel in concrete.

4.2.1 Corrosion of metals

Corrosion of metals involves an oxidation reaction, most simply expressed by the equation

$$2M + O_2 \rightarrow 2MO$$

The proportions of oxygen and metal in the resulting compound will vary depending on the oxidation state of the metal. Corrosion in this form is only of minor concern for steel in civil engineering applications at ambient temperatures, since the reaction is typically slow. Corrosion becomes a problem for steel in conventional structures where water is present. In such circumstances, galvanic corrosion can occur, which is more damaging.

4.2.2 Chemistry of galvanic corrosion

Galvanic, or 'wet', corrosion describes an electrochemical form of corrosion in which the close proximity of two different metals in contact with themselves and water containing an electrolyte leads to one of the metals corroding. Whether one or the other of the metals corrodes is dependent on the strength with which each metal's atoms are bound to each other. An indication of this is indirectly obtained in terms of the metal's 'standard electrode potential', which is the potential difference between a metal electrode and a hydrogen electrode across an electrolyte solution junction under standard conditions. A more positive standard electrode potential denotes a material that is more prone to corrosion and is thus more active or 'anodic'. Where the iron in steel is the more anodic of the two metals, it undergoes oxidation, which takes the form of ionisation at its surface:

$$Fe \rightarrow Fe^{2+} + 2e^-$$

where the metal ion dissolves. The iron can undergo further oxidation in the presence of water:

$$4Fe^{2+} + O_2 \rightarrow 4Fe^{3+} + 2O^{2-}$$

At the other metal surface, under neutral pH conditions, a reduction reaction occurs:

$$O_2 + 2H_2O + 4e^- \rightarrow 4OH^-$$

Iron hydroxides are then formed:

$$2Fe^{2+} + 4OH^- \leftrightarrows 2Fe(OH)_2$$

$$2Fe^{3+} + 6OH^- \rightarrow 2FeO(OH)H_2O$$

The hydroxides may subsequently undergo various dehydration reactions to give a mixture of hydroxides and FeO, $FeO(OH)$ and Fe_2O_3, which collectively make up the rust that is a familiar feature of the surface of plain steel that has been exposed to the elements.

From this group of reactions, it is evident that the presence of both water and oxygen is essential for galvanic corrosion to occur. Another important requirement is that the water in contact with the metal is capable of conducting electricity, which means that the presence of an electrolyte is necessary.

The overall effect of galvanic corrosion is a loss of metal from the reinforcement, leading to a decline in the load-bearing capacity of a reinforced structural element.

By compiling a list of metals in the order of their standard electrode potentials in a given electrolyte solution, a 'galvanic series' is produced, which allows the more anodic metal of a metal pairing to be identified. Consultation of galvanic series demonstrates clearly why joining plain steel sections with stainless steel bolts is not a good idea and why a sacrificial layer of zinc on galvanised steel is.

However, galvanic corrosion is possible without the presence of different metals. The circumstances under which this occurs for steel reinforcement in concrete relate to the composition of the metal, the differences in the concentration of dissolved species in the water, and the presence of regions of stress concentration in the metal.

Where an alloy forms crystals of different phases (for instance, the ferrite and cementite phases in steel), the two different phases will possess different electrode potentials, and a vast number of microscopic electrochemical cells of the sort described above can be set up. In the case of steel, cementite is cathodic, and ferrite is anodic.

Corrosion as a result of differences in concentration can be driven either by differences in electrolyte or oxygen concentration. Differences in oxygen concentration drive a common corrosion process in steel reinforcement known as 'pitting'. Where oxygen concentrations vary at different points on a steel article's surface, the part exposed to lower oxygen concentrations becomes more anodic. During pitting, small localised areas of oxygen deficiency (such as areas beneath rust patches or in cracks at the steel surface), which are nonetheless accessible by water, become anodic, leading to corrosion in this area and the formation of the beginnings of a 'pit' in the steel surface. Subsequently, the variation in oxygen concentration at the bottom of the pit compared with elsewhere on the steel surface causes the pit to grow.

Steel in contact with an electrolyte solution of varying concentration will also form an electrochemical cell, with the part exposed to lower concentrations undergoing corrosion.

Steel under stress is more anodic than steel in an unstressed state. This is clearly of significance to any steel reinforcement, whose purpose is to carry tensile stresses. However, it is of particular importance in the case of prestressed reinforcement cables, where the stresses are more substantial.

The rate of corrosion is dependent on the ratio of the surface areas of the anodic and cathodic parts of a corroding system, with a small anodic surface relative to a larger cathodic surface leading to a higher rate.

4.2.3 Passivation

The presence of concrete cover acts as a barrier to the movement of oxygen and substances capable of promoting corrosion towards the reinforcement, thus prolonging the life of the steel. However, the alkaline chemical environment in concrete also provides protection to the steel. This protection is known as 'passivation' and occurs when, under conditions of high pH, a highly impermeable oxide layer of less than 1 µm in thickness forms at the steel surface. The layer acts to limit the accessibility of the steel surface to water, oxygen and corrosive species.

The stability of the passive layer is dependent on the pH of the pore solutions of the concrete, and a decrease in pH below around 11.5 will lead to the decomposition of the layer [2]. Additionally, the passive layer can be destroyed in the presence of sufficient quantities of certain dissolved ions, with chloride ions being of greatest concern. Both of these effects will be covered in greater detail in the subsequent discussions of chloride ingress and carbonation.

4.2.4 Steel corrosion in reinforced concrete

We have already seen that the reactions involved in galvanic corrosion require both water and oxygen. Thus, the rate of reinforcement corrosion is largely dependent on the relative humidity within the concrete pores and the extent to which oxygen can access the steel surface. The importance of water is illustrated in Figure 4.1, where an increase in internal relative humidity leads to an increase in the rate of corrosion, expressed in terms of the corrosion current density (I_{corr}). I_{corr} is the current in the steel reinforcement during corrosion per unit of surface area, which gives an indication of the rate of corrosion.

The extent to which oxygen is able to reach the reinforcement is dependent on how easily oxygen can enter the concrete and how rapidly it can subsequently diffuse towards the steel. The role of both factors is illustrated in Figure 4.2, which shows the influence of water/cement (W/C) ratio

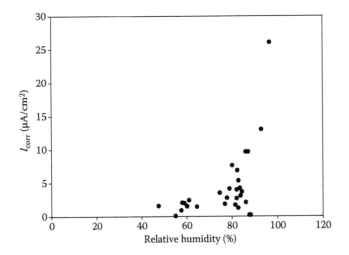

Figure 4.1 Corrosion current (I_{corr}) of steel in mortar containing 2% chloride by mass of cement versus internal relative humidity. I_{corr} is a measure of the rate of corrosion and is derived from measurements of polarisation resistance (R_p). I_{corr} is calculated using the equation $I_{corr} = B/R_p$. A value of 26 mV has been assumed for B. (From Enevoldsen, J. N. et al., *Cement and Concrete Research*, 24, 1994, 1373–1382.)

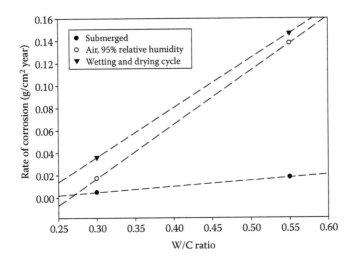

Figure 4.2 Influence of W/C ratio and exposure conditions on the rate of corrosion of steel in concrete. Dashed lines are used as guides. (From Hussain, R. R. and T. Ishida, *Construction and Building Materials*, 24, 2010, 1014–1019.)

and exposure conditions on corrosion rate. Where concrete is wholly submerged, oxygen cannot enter, leading to a low rate of corrosion. However, where oxygen can enter the concrete surface (either when continuously left exposed to air or when exposed to wetting and drying cycles), the corrosion rate is higher. Additionally, where the concrete has a higher W/C ratio, and hence the total volume of pores and the pore diameters are greater, oxygen can diffuse more rapidly, again leading to a higher rate of corrosion. Section 4.3 contains a more detailed discussion of how porosity characteristics influence diffusion rates.

Increasing temperature accelerates the rate of corrosion. Figure 4.3 shows the increase in I_{corr} with temperature.

We have seen that the electrochemical processes involved in galvanic corrosion require the transport of $Fe^{2/3+}$ and OH^- ions through solution. For this to progress at a rapid rate, the microstructure around the steel must permit the movement of these ions, and the levels of moisture present must be sufficiently high. Both of these factors determine the electrical resistivity of the concrete, and so this characteristic can be used as a measure of ion mobility. Figure 4.4 illustrates the role of moisture by plotting corrosion current versus electrical resistivity for concrete specimens with different moisture contents.

The corrosion of steel reinforcement has two detrimental influences on the performance of structural concrete. The first is that the reinforcement itself undergoes a loss in cross-sectional area, which compromises its ability (and the ability of the reinforced concrete) to carry tensile stresses. The

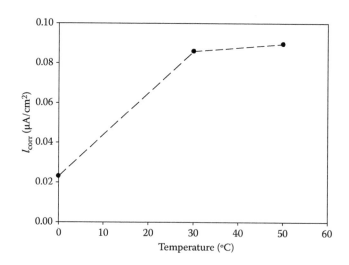

Figure 4.3 Influence of temperature on the corrosion current (I_{corr}) of steel reinforcement in concrete stored at a relative humidity of >90%. (From Lopez, W. et al., *Cement and Concrete Research*, 23, 1993, 1130–1140.)

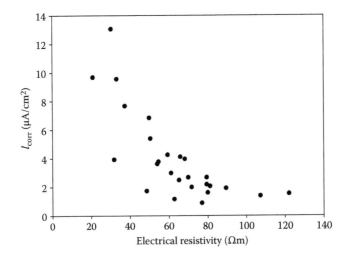

Figure 4.4 Corrosion current of steel in mortar containing 2% chloride by mass of cement versus electrical resistivity. (From Enevoldsen, J. N. et al., *Cement and Concrete Research*, 24, 1994, 1373–1382.)

second is that the formation of rust at the steel surface eventually leads to the formation of cracks in the concrete cover.

The corrosion of reinforcement will only begin to produce a loss in the load-bearing capacity of a structural element beyond a certain level of mass loss from the steel. One of the reasons for this is that the initial formation of rust at the steel surface has the effect of enhancing the bond between the steel and the concrete. This is illustrated in Figure 4.5, which shows a small gain in the flexural strength of a reinforced beam at the early stages of corrosion, followed by a decline. This increase is attributed to an increase in the frictional stress between the reinforcement and the concrete as a result of rust formation. The reason for the eventual decline in strength results has been suggested to initially result from a loss of bond strength resulting from the removal of the ribs on the reinforcement bar surface, followed by a loss of cross-sectional area [6].

The corrosion products of steel are considerably less dense than the metal, which means that rust formation leads to an expansion in volume of up to four times. As discussed in Chapter 3, in relation to sulphate attack, an increase in product volume for a reaction occurring in concrete is no guarantee of the development of expansive forces. However, in the case of rust formation, the expansion observed is probably the result of crystallisation pressures, as for ettringite. Ultimately, the development of expansive stresses leads to the formation of cracks originating at the reinforcement and extending to the concrete surface. Generally, wider reinforcing bars located closer to the concrete surface will produce cracks earlier than

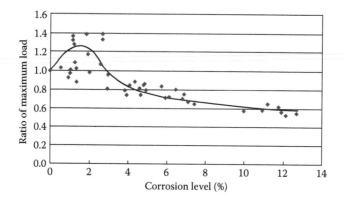

Figure 4.5 Flexural strength (expressed as a ratio of initial strength) of reinforced concrete slab specimens corroded using current-induced corrosion. The corrosion level is expressed as the percentage mass loss from the corroded reinforcement. (From Chung, L. et al., *Engineering Structures*, 26, 2004, 1013–1026.)

narrower bars located at greater depths. Cracks may also result simply from the loss of load-bearing capacity, and the resulting increased structural deflection.

The development of cracks resulting from reinforcement corrosion has the effect of easing the passage of oxygen and substances that promote corrosion, since the pores of the concrete can be bypassed in favour of a more direct route to the steel.

The overall effect of cracking is that the long-term deterioration in the load-bearing capacity of structural members typically follows the type of behaviour shown in Figure 4.6. After an initially slow rate of loss in load capacity, the deterioration accelerates in an unsustainable manner. The periods of time marked on the figure indicate key events in a structural element's path of deterioration. t_1 is the period between construction and the initiation of reinforcement corrosion. Beyond this point, corrosion continues until the performance of the element falls below the serviceability limit after a period t_2. In the case of reinforcement corrosion, the point at which the serviceability limit is reached is normally defined in terms of the development of surface cracks. *Eurocode 2* [7] defines this in terms of a maximum surface crack width (w_{max}), which, if exceeded, indicates that the serviceability limit state has been reached. Recommended values in *Eurocode 2* depend on the aggressive nature of the environment in which a structure operates, as defined by exposure classes. For lower levels of aggression, a w_{max} value of 0.4 mm is recommended, which drops to

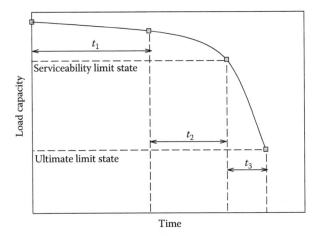

Figure 4.6 Notional deterioration of load-bearing capacity of a reinforced concrete structural element. (From Torres-Acosta, A. A. et al., *Engineering Structures*, 29, 2007, 1145–1152.)

0.3 mm for more aggressive environments. The UK National Annex to this document takes a more conservative approach and sets the limit as 0.3 mm for all exposure classes [8].

The service life of the structure is therefore equal to $t_1 + t_2$. The development of surface cracks clearly marks the start of an accelerated rate of corrosion that initiates at the beginning of period t_3, which is referred to as the 'residual life stage'. During this period, the opportunity for repair exists. Beyond t_3, the load capacity of the element falls below the ultimate limit state, at which point collapse is imminent.

On the concrete surface, corrosion of reinforcement manifests itself as cracking or even spalling of the concrete, possibly with the appearance of rust staining. It should be stressed that cracking and spalling can result from many other processes, making further investigation of deterioration necessary to reveal whether corrosion is the true cause.

Cracks running parallel to the direction of the main reinforcement may indicate the formation of expansive corrosion products. However, as discussed in Chapter 3, such crack configurations are also seen in reinforced concrete undergoing alkali–aggregate reactions.

The absence of rust staining does not necessarily indicate an absence of corrosion. Furthermore, the presence of staining is also not conclusive evidence of corrosion, as aggregate containing iron sulphide minerals may produce a similar effect [10]. Thus, it may often be necessary to remove concrete in a cracked zone to establish whether the reinforcement beneath is corroded.

4.3 CHLORIDE INGRESS INTO CONCRETE

One of the greatest threats to steel reinforcement in concrete is the chloride ion. Chlorides may enter concrete from the external environment via various mass transport processes. They can also be introduced as contaminants in constituent materials or as calcium chloride used as an accelerating admixture. The use of this compound is no longer permissible in reinforced and prestressed concrete, as a result of its corrosive nature. This section will examine the nature of the external sources of chloride, how they reach the steel reinforcement and what happens when this occurs. Additionally, some of the phenomena that hinder the progress of chlorides will be discussed, along with how these can be exploited to provide chloride resistance.

4.3.1 Chlorides in the environment

One of the main reasons that chloride ingress into concrete is of such concern to engineers is the large number of opportunities for chlorides to come into contact with reinforced concrete. In the built environment, soluble chlorides are most commonly encountered from two sources: seawater and de-icing salts on highways.

Chlorides in seawater occur largely as sodium, magnesium and calcium chloride. The concentration of chlorides varies depending on the salinity of the seawater in question, 35 g/L, chloride is present at a concentration of approximately 19,000 mg/L. De-icing salts are most commonly sodium chloride but can also be magnesium and calcium chloride.

Exposure to hydrochloric acid will provide a source of chloride ions, along with corrosion of the concrete itself (see Chapter 3).

4.3.2 Ingress mechanisms

Chloride ingress can occur in concrete as a result of a concentration gradient (diffusion), a pressure gradient causing the flow of chloride-bearing solutions through pores, and capillary action.

4.3.2.1 Diffusion

In the absence of cracks, chloride diffusion through concrete is very much dependent on the nature of the porosity. The role that porosity plays in determining the diffusion coefficient is discussed in greater detail in Chapter 5, but in essence, a low diffusion coefficient (and hence a low rate of diffusion) is achieved when the total volume fraction of porosity is low, its constrictivity is low and its tortuosity is high. The constrictivity is a measure of the extent to which changes in the width of pores along their length hinder the diffusion of chemical species. Tortuosity is a measure of the extent to which a chemical species must deviate from a direct route when diffusing from point A to point B through the pore network of concrete.

The effect of the volume fraction of porosity on chloride diffusion in concrete is best illustrated in terms of the W/C ratio. This is shown in Figure 4.7, which plots the chloride diffusion coefficient against W/C ratio for hardened Portland cement pastes. Similarly, as the degree of cement hydration increases, the total volume of porosity falls, reducing the diffusion coefficient.

It should be noted that, as chloride ingress progresses, the volume of porosity declines in the outer layer of the concrete. This is presumably the result of the formation of Friedel's salt within the pores.

As shall be seen in Chapter 5, as the maximum pore size approaches the minimum pore size, constrictivity increases, leading to higher coefficients of diffusion. Thus, pore size distribution has a significant influence on chloride diffusion coefficients. In particular, an increasing proportion of 'macropores'

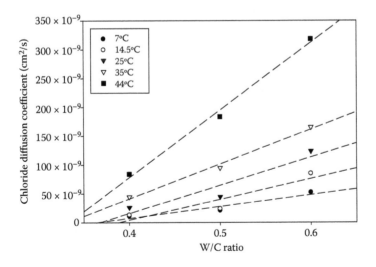

Figure 4.7 Influence of W/C ratio on chloride diffusion coefficients of hardened Portland cement pastes. (From Page, C. L. et al., *Cement and Concrete Research*, 11, 1981, 395–406.)

(>0.03 μm in diameter) in the cement matrix leads to an increase in the chloride diffusion coefficient [12].

Tortuosity is largely controlled by the particle size distributions of the materials used in the cement fraction and the volume and morphology of the cement hydration products formed. As we shall see in Section 4.3.5, the use of fine particulate materials such as silica fume (SF) has the effect of increasing tortuosity.

The main factor that will influence the rate of chloride diffusion is the presence of cracks. Diffusion through cracks can be viewed in precisely the same way as diffusion through pores, although the width of cracks can be several orders of magnitude higher than the largest pore widths in hardened cement paste. Thus, cracks present a relatively unimpeded path for chlorides through the concrete cover, and so cracked concrete displays diffusion coefficients significantly higher than the undamaged material. Measurement of concentration profiles in reinforced beams in flexural loading exposed to chloride solutions has found concentrations notably higher in the zones in tension [13]. This can be attributed to the higher density and greater width of cracks in the tensile zone. Generally, it is difficult to separate the influence of crack width and density on diffusion, since many processes that act to form cracks will produce both an increase in crack width and crack density simultaneously. Figure 4.8 shows the combined influence of crack

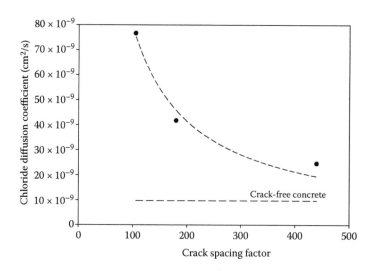

Figure 4.8 Chloride diffusion coefficient versus crack spacing factor for concrete damaged by freeze–thaw attack. (From Gérard, B. and J. Marchand, *Cement and Concrete Research*, 30, 2000, 37–43; Jacobsen, S. et al., *Cement and Concrete Research*, 26, 1996, 869–881.)

width and density on the chloride diffusion coefficient by expressing both parameters as a single term – crack spacing factor (f) – which is defined as

$$f = \frac{l}{w}$$

where l is the average distance between cracks along a straight line on the concrete surface, and w is the average crack width.

The relationship shown in Figure 4.8 can be described by the equation

$$\frac{D}{D_0} = \frac{D_1}{D_0 f} + 1$$

where D is the chloride diffusion coefficient (m²/s) of a concrete surface containing cracks with spacing factor f, D_0 is the chloride diffusion coefficient of the concrete if it did not contain cracks (m²/s) and D_1 is the chloride diffusion coefficient in free solution (m²/s) [14].

D/D_0 is the 'equivalent diffusivity' – the proportion by which the diffusion coefficient of the cracked concrete exceeds that of the uncracked concrete.

Crack widths have a greater influence over diffusion rates than crack densities [14]. The effect of crack density (in terms of crack spacing) on the rate of corrosion is shown in Figure 4.9. An interesting feature of this plot is the drop in the extent of corrosion at the lowest crack spacing. This has been attributed to the cracks being sufficiently narrow (~0.12 mm) to allow 'self-healing' to occur. 'Self-healing', or 'autogenous healing', refers to the filling of cracks as a result of the precipitation of crystals from solution [15]. The main compounds precipitated are calcium carbonate and calcium hydroxide, and the rate at which self-healing occurs depends on crack width, water pressure and temperature.

A faster rate of crystal growth is observed in narrower cracks [16]. After the initial formation of $CaCO_3$ and $Ca(OH)_2$ crystals, their subsequent growth is controlled by the rate of diffusion of calcium ions. However, where there is a flow of water through concrete as a result of a pressure gradient, the rate is accelerated, since the delivery of calcium ions becomes dependent on the rate of flow rather than diffusion. The type of cement, aggregate and the calcium content of the external water appear to play only a minor role in the self-healing process although, where seawater is present, ettringite and brucite formation also contribute to self-healing as a result of the presence of sulphate and magnesium ions. An increase in temperature leads to an increase in the rate of self-healing, at least up to 80°C [17].

Environmental factors that play a role in influencing chloride diffusion include chloride ion concentration, the cations associated with the chlorides and temperature.

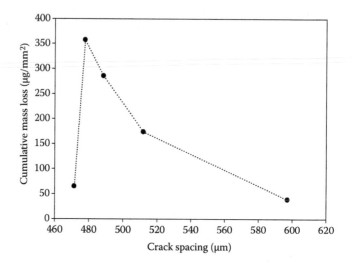

Figure 4.9 Cumulative mass loss versus crack spacing at 24 months for steel reinforcement in concrete beams containing artificial cracks. The beams were stored in high-humidity conditions and periodically sprayed with a chloride solution. (From Arya, C. and F. K. Ofori-Darko, *Cement and Concrete Research*, 26, 1996, 345–353.)

As for all diffusion processes, increased temperature leads to an increased rate of transport. The temperature dependence of diffusion is described by the Arrhenius equation:

$$D = D_0 e^{-\frac{E_A}{RT}}$$

where D_0 is a diffusion constant (m²/s), E_A is the activation energy for diffusion (J/mol), R is the gas constant (J/K mol) and T is the temperature (K).

This effect can be seen in Figure 4.7. The equation is best used in a modified form to yield an adjustment factor (F_T):

$$F_T = e^{\frac{E_A}{R}\left(\frac{1}{T_{ref}} - \frac{1}{T}\right)}$$

where T_{ref} is a specific temperature at which a diffusion coefficient has been measured (D_{ref}) [20]. Thus, a diffusion coefficient for a given temperature can be calculated using $F_T D_{ref}$.

The activation energy can be determined from diffusion data and is dependent on the W/C ratio (Figure 4.10).

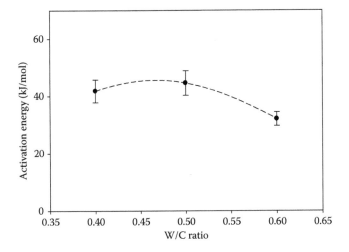

Figure 4.10 Activation energy as a function of W/C ratio. (From Page, C. L. et al., *Cement and Concrete Research*, 11, 1981, 395–406.)

Diffusion is driven by a concentration difference. However, the effect of external chloride concentration on the rate of chloride ingress is somewhat complex – there is a decrease in the chloride diffusion coefficient as the external concentration increases (Figure 4.11). This decrease results from a greater interaction between ions at higher concentrations, which hinders their movement [15,21]. Although this decline in diffusion coefficient at higher concentrations is an important phenomenon, it needs to be viewed in context. Figure 4.12 illustrates this in the form of two concentration profiles: one for a situation where there is a high external concentration and a low diffusion coefficient and the other for the opposite scenario. Although the movement of chloride ions into the material is slower for a lower diffusion coefficient, the total quantity of chloride that has entered the concrete is still much greater for the higher external concentration.

The chloride salts present influence ingress rates, with calcium chloride producing higher diffusion coefficients than sodium chloride. The reason for this has been attributed to the presence of an electrical double layer at pore surfaces within concrete [23].

An electrical double layer is formed as a result of electrostatic charge developing at the surface of hydration products as groups at their surface become ionised. The resulting negative surface charge attracts dissolved cations to create a layer of fluid at the pore surface richer in these ions. Consequently, the pore fluid further away from the pore solution contains an enriched concentration of anions. The situation for chloride ingress via diffusion is shown in a very simplified form in Figure 4.13. Where sodium

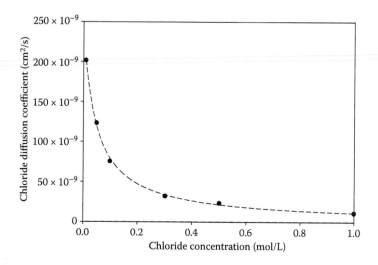

Figure 4.11 Influence of external chloride ion concentration on the chloride diffusion coefficient through concrete with a W/C ratio of 0.4. (From Tang, L., *Cement and Concrete Research*, **29**, 1999, 1469–1474.)

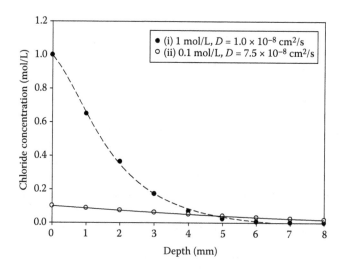

Figure 4.12 Calculated chloride concentration profiles for two scenarios: (i) with an external chloride ion concentration of 1 mol/L and a diffusion coefficient (D) through the concrete of 1.0×10^{-8} cm²/s and (ii) with a concentration of 0.1 mol/L and a diffusion coefficient of 7.5×10^{-8} cm²/s.

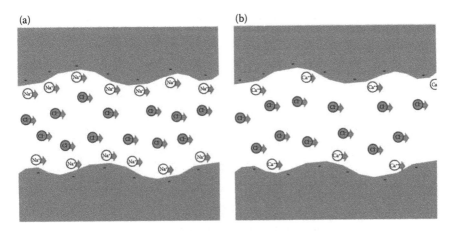

Figure 4.13 Electrical double layer configurations in concrete pores where sodium (a) and calcium (b) ions are associated with diffusing chloride ions.

ions are associated with chloride ions, a large concentration of the cations is required to produce a positive charge density to match the negative charge density of the surface layer. Where calcium ions are present, a much lower concentration is required, as a result of the ions' higher charge. This reduced concentration produces a reduced interaction between calcium ions, allowing them to diffuse at a higher rate, as discussed above. The concentration of anions in the pore fluid must be balanced with the concentration of cations such that the net charge is zero – a requirement known as the electroneutrality principle. This has the effect of also increasing the rate of chloride ion diffusion, as these ions follow the cations diffusing in the double layer.

4.3.2.2 Flow

The rate of flow of chloride-bearing solutions into concrete under a given pressure difference is dependent on the permeability of the material, which is strongly influenced by the pore structure of concrete. The relationships between pore characteristics and permeability are discussed in detail in Chapter 5. However, in short, essentially the same microstructural characteristics that influence the chloride diffusion coefficient also control the rate of flow, with a large pore volume fraction, large pore diameter and low tortuosity yielding a high permeability.

Cracking will also increase the rate of flow through concrete, as shown in Figure 4.14. As discussed for the process of chloride ingress by diffusion, self-healing of cracks will occur, particularly where the pressure gradient is high. The effect of self-healing on the rate of flow is illustrated in Figure 4.15.

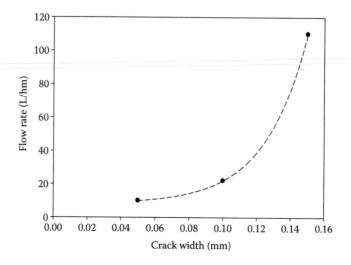

Figure 4.14 The influence of crack width on the flow of water through damaged concrete specimens. (From Reinhardt, H.-W. and M. Jooss, *Cement and Concrete Research*, 33, 2003, 981–985.)

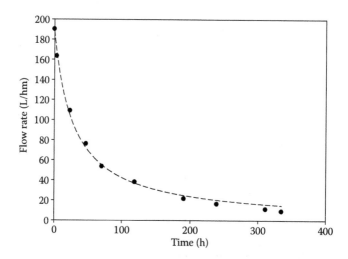

Figure 4.15 Change in flow rate with time as a result of self-healing of cracks in concrete held at 80°C. Initial average crack width is 0.15 mm; pressure gradient is 1 MPa/m. (From Reinhardt, H.-W. and M. Jooss, *Cement and Concrete Research*, 33, 2003, 981–985.)

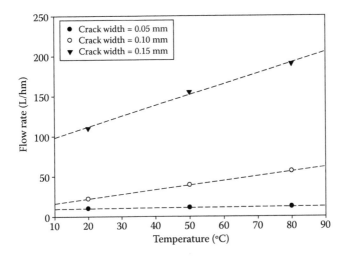

Figure 4.16 Influence of average crack width and temperature on the flow rate of water through concrete specimens under a pressure gradient of 1 MPa/m. (From Reinhardt, H.-W. and M. Jooss, *Cement and Concrete Research*, 33, 2003, 981–985.)

External environmental conditions that will influence the rate of ingress of chlorides are the pressure difference between the interior and the exterior of the concrete, the external concentration of chlorides and temperature. The viscosity of water drops with increasing temperature, reducing the resistance to its flow through porous media. However, the effect is only significant where cracks are present (Figure 4.16).

In marine environments, where magnesium chloride is present, the formation of a layer of brucite ($Mg(OH)_2$) and various calcium carbonate compounds has been observed [25]. Formation of this layer leads to a reduction in the permeability of the concrete, as illustrated in Figure 4.17.

4.3.2.3 Capillary action

When unsaturated pores at the surface of concrete come into contact with water, the process of capillary action will draw the liquid into the interior. Clearly, if the water contains dissolved chlorides, this process will also act as a further ingress mechanism.

The rate of uptake of water by concrete as a result of capillary action is dependent on the gradients of volume fraction saturation (θ) and the

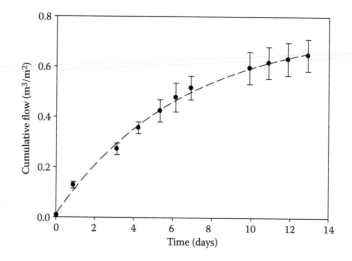

Figure 4.17 Cumulative flow versus time for seawater through concrete under a pressure differential of 5.0 MPa. A density for seawater of 1030 kg/m³ has been assumed in the conversion of these data. (From van der Wegen, G. et al., *Materials and Structures*, 26, 1993, 549–556.)

hydraulic diffusivity (D) within the concrete [27], as described by the equation

$$\frac{\partial \theta}{\partial t} = \nabla D \nabla \theta$$

where θ is the volume fraction saturation – the ratio of liquid volume to bulk concrete volume; t is the time (s); and D is the hydraulic diffusivity (m²/s).

The hydraulic diffusivity is a measure of the ability of the concrete to transmit water via capillary action. In many porous materials, the hydraulic diffusivity is described by the equation

$$D = D_0 e^{\left(B \frac{(\theta - \theta_0)}{(\theta_1 - \theta_0)}\right)}$$

where D_0 is the initial hydraulic diffusivity (m²/s), B is a constant dependent on the material, θ_0 is the initial volume fraction saturation and θ_1 is the volume fraction saturation at full saturation.

Thus, as the water content of the pores increases, so does the hydraulic diffusivity. Although the situation for concrete is somewhat more complex – the microstructure of concrete changes when it comes into contact with water – the general relationship described in the equation holds.

Capillary action plays its most significant role in situations where cyclic wetting and drying occur. Such situations include those in the tidal, splash and atmospheric zones of coastal and offshore structures (Figure 4.18) and in highway environments. The process of capillary action is relatively rapid and so plays an important role in the early ingress mechanism. Moreover, the repetition of this process where cyclic wetting and drying occurs will lead chlorides being deposited in concrete pores during drying, followed by a fresh supply of chlorides during the next period of wetting, potentially leading to the accumulation of chloride beneath the surface.

Generally, ingress resulting from wetting and drying cycles produces a slightly modified chloride concentration profile, compared with that observed for diffusion, with a peak occurring some distance beneath the surface (Figure 4.19). The reason for this profile shape is possibly the result of chloride accumulation, as discussed above. However, it is also probably the result of carbonation occurring during drying, which, as will be discussed in the following section, leads to the release of chlorides previously bound by the cement.

Aside from the chloride concentration in the external water, the period of drying appears to play the most important role in controlling the rate of ingress, resulting from more thorough drying of the porosity leading to a higher rate of uptake of water the next time the surface is wet [30]. With reference to the rate of uptake equation discussed previously, a greater degree of drying will lead to steeper hydraulic diffusivity and volume fraction saturation gradients, leading to a greater rate of absorption.

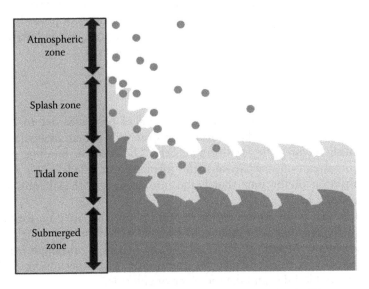

Figure 4.18 Different zones of exposure to chlorides for coastal and offshore structures.

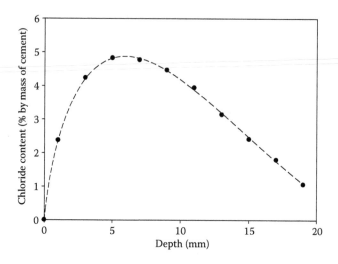

Figure 4.19 Chloride concentration profile resulting from cyclic wetting and drying. (From Polder, R. B. and W. H. A. Peelen, *Cement and Concrete Composites*, 24, 2002, 427–435.)

The rate of drying itself is partly influenced by the quality of the concrete – specifically the volume and the nature of its porosity. Concrete with a low volume of porosity and a finer pore structure will dry at a slower rate. Thus, there is a greater opportunity for rapid chloride ingress under wetting and drying conditions in concrete containing higher volumes of porosity.

It should be stressed that, in most circumstances, a combination of ingress mechanisms will be operating simultaneously.

4.3.3 Chloride binding

As chloride ions move into concrete, chemical processes act to remove some of these ions from the solution, thus rendering them unavailable for contributing towards the corrosion process. The 'chloride binding' process mainly involves two mechanisms: the formation of Friedel's salt and the immobilisation of the ions that come into contact with calcium silicate hydrate (CSH) gel.

Friedel's salt is an AFm cement hydrate phase that may be formed in hydrated cement that is brought into contact with dissolved chloride ions [25]. Its general formula is $3CaO \cdot Al_2O_3 \cdot CaCl_2 \cdot 10H_2O$, although iron and other cations can be substituted for aluminium, and other anions such as hydroxide and iodide can take the place of chloride.

Friedel's salt is normally formed as a result of a reaction with the AFm cement hydration product, monosulphate:

$$3CaO \cdot Al_2O_3 \cdot CaSO_4 \cdot 12H_2O + 2Cl^- \rightarrow 3CaO \cdot Al_2O_3 \cdot CaCl_2 \cdot 10H_2O$$

$$+ SO_4^{2-} + 2H_2O$$

Other AFm phases are also capable of forming Friedel's salt in this way. Despite also being a calcium aluminate hydrate, the AFt phase ettringite is not capable of undergoing a similar reaction, because it is more stable than Friedel's salt.

The mechanism or mechanisms that lead to chloride binding by CSH gel is still the subject of some debate. However, it is generally agreed that chloride ions can interact with the gel in a number of different ways with differing binding 'strengths'. One proposed mechanism involves chlorides interacting with CSH gel through chemisorption (adsorption involving a chemical reaction) at the gel surface, chemisorption within the spaces between the disordered layers that make up the crystal structure of the gel and incorporation into the CSH crystal lattice, with increasing strength of binding [31].

Usually, chemisorbed chlorides are present in considerably larger quantities than those incorporated into the crystal lattice. The quantity is also dependent on the composition of the gel, with a higher ratio of CaO to SiO_2 yielding a higher binding capacity. This is thought to be related to the crystal structure of CSH gel, where a lower SiO_2 content will lead to the incorporation of a larger number of hydroxide groups at the gel layer surface, which could, in turn, be occupied by chloride ions or chloride complexes.

The binding capacity of the AFm phases is higher than that of CSH gel [32]. However, a mature Portland cement is likely to contain a considerably higher proportion of CSH than AFm, which means that the majority of chloride is bound by the gel. Nonetheless, as the C_3A content and hence the potential to form AFm phases increase, the binding capacity rises significantly, particularly above levels of approximately 8% by mass (Figure 4.20). Moreover, where the proportions of hydration products deviate from those produced by a typical Portland cement (for instance, in concrete where other cementitious materials have been used), the dominant role of CSH is not necessarily maintained.

The pH of the pore fluid within concrete influences its chloride binding capacity, with a lower pH providing a greater capacity. It has been suggested that this is because OH^- and Cl^- ions compete for similar locations within and on the surface of hydration products [34]. This effect is observed within the pH ranges observed in relatively young concrete

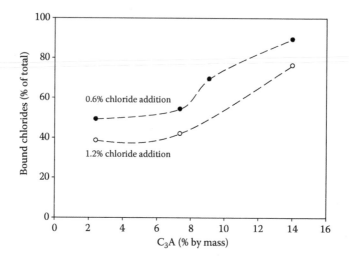

Figure 4.20 Influence of C₃A content on bound chloride in cement pastes containing chloride additions. (From Rasheeduzzafar Hussain, S. E. and S. S. Al-Saadoun, *Cement and Concrete Research*, 21, 1991, 777–794.)

(12.5–14), which has not been exposed to processes that could reduce the OH^- concentration. Where a drop in pore solution pH below 12.5 occurs, Friedel's salt will dissolve. The most common cause of this is as a result of carbonation (see Section 4.4), and the effect is evident in concrete exposed to both chloride-bearing solutions and carbonating atmospheres, where Friedel's salt is absent at the surface but appears at greater depths [35]. The stability of Friedel's salt is also compromised by the ingress of sulphates. In this case, Friedel's salt reacts to form ettringite via the following reaction:

$$3CaO \cdot Al_2O_3 \cdot CaCl_2 \cdot 10H_2O + 3SO_4^{2-} + 2Ca^{2+} + 22H_2O$$

$$\rightarrow 3CaO \cdot Al_2O_3(CaSO_4)_3 \cdot 32H_2O + 2Cl^-$$

An observation of the rate of chloride diffusion in concrete exposed to sulphates has found that the destabilising effect of sulphates only has a pronounced effect where the cement fraction of the concrete contains cementitious by-products (such as ground granulated blast-furnace slag [GGBS] or fly ash [FA]) along with Portland cement [36]. This reflects the higher aluminate content of such materials, which would produce a larger quantity of Friedel's salt in a chloride-bearing environment.

The cations associated with chloride ions also play an important role in the extent to which binding occurs. For calcium chloride ($CaCl_2$), the following reaction occurs:

$$3CaO \cdot Al_2O_3 \cdot CaSO_4 \cdot 12H_2O + CaCl_2 \rightarrow 3CaO \cdot Al_2O_3 \cdot CaCl_2 \cdot 10H_2O + CaSO_4 + 2H_2O$$

However, for sodium chloride, the reaction is

$$3CaO \cdot Al_2O_3 \cdot CaSO_4 \cdot 12H_2O + 2NaCl \rightarrow 3CaO \cdot Al_2O_3 \cdot CaCl_2 \cdot 10H_2O + Na_2SO_4 + 2H_2O$$

As discussed in the discussion of alkali–aggregate reactions in Chapter 3, sodium sulphate (Na_2SO_4) produced during the reaction will gradually convert to sodium hydroxide, which will act to increase the pH of the pore solutions. This has the effect of reducing the solubility of calcium hydroxide, limiting the amount of calcium available to form Friedel's salt and thus the binding capacity of the cement.

Where magnesium chloride ($MgCl_2$) is present, cation exchange will occur in the following manner:

$$MgCl_2 + Ca(OH)_2 \rightarrow Mg(OH)_2 + CaCl_2$$

The ultimate result is that the amount of chloride bound is the same for $CaCl_2$ [34].

As previously discussed, the formation of brucite in concrete in marine environments has been cited as the reason for a reduction in the permeability of concrete with time. On the other hand, exposure to magnesium chloride de-icing salt has been blamed for the deterioration of concrete pavements through the formation of magnesium silicate hydrate gel, as occurs during magnesium sulphate attack (Chapter 3) [37]. It is likely that both effects occur simultaneously in submerged concrete and that the brucite layer reduces the depth to which magnesium silicate hydrate formation occurs. Where concrete is permanently submerged, it is likely that the layer of brucite and other minerals will be allowed to form to an extent that a high level of protection is achieved. In applications where mechanical wearing occurs, such as on pavements or in tidal zones, the layer will be disrupted periodically, with no enhanced protection and an accumulated rate of surface deterioration.

Where $CaCl_2$ is present (either because calcium chloride is present externally or as a result of the above reaction), calcium oxychloride may be formed:

$$CaCl_2 + 3Ca(OH)_2 + 12H_2O \rightarrow 3CaO \cdot CaCl_2 \cdot 15H_2O$$

Although this reaction will contribute towards chloride binding, it has been proposed that this compound may also play a role in expansion and cracking [38].

Chloride-induced corrosion of steel can even occur in the absence of high oxygen concentrations, leading to the formation of 'green rust'. This term is used to describe fougerite, a layered double hydroxide mineral with the formula

$$\left[Fe^{2+}_{1-x} Fe^{3+}_{x} Mg_y (OH)_{2+2y} \right]^{+x} [x/nA^{-n} \cdot mH_2O]^{-x}$$

where A^{-n} can be a range of different anions, including Cl^- [39]. The formation of green rust starts when the passive layer is destroyed and iron dissolves as a result of the formation of complexes with chloride. Under pH conditions typical of concrete and where oxygen concentrations are low, the complex will precipitate as green rust. The rust acts as a further chloride-binding mechanism, although subsequent exposure to oxygen leads to its decomposition and the release of chloride back into solution [40].

4.3.4 Role of chloride in corrosion

When chloride ions reach the surface of steel reinforcement, they act to break down the passive layer at the surface and allow corrosion to progress. This 'depassivation' process almost certainly involves the formation of chloride complexes with iron from the passive layer [41], such as in the following manner:

$$Fe(OH)_2 + 6Cl^- \rightarrow FeCl_6^{-3} + 2OH^- + e^-$$

The $FeCl_6^{3-}$ complex is soluble in the concrete pore solution, and so material is removed from the passive layer, acting to compromise its protective influence. Depassivation tends to occur at localised points on the steel surface.

For depassivation to occur, and thus for corrosion to initiate, a sufficient concentration of chloride ions is necessary. This 'threshold' concentration, the form of chloride measured and the manner in which the concentration is expressed are a matter of some debate.

The least ambiguous way of measuring the concentration of chloride in concrete is to measure the total mass of chloride present. *BS 1881-124* [42] includes a technique for the determination of total chlorides through the partial digestion of a powdered sample of concrete in nitric acid, which is boiled for a short period of time and then filtered to remove insoluble aggregate particles. The concentration of chloride in the resulting solution is determined using titration. Techniques such as X-ray fluorescence spectrometry may also be used to determine total chlorides.

Although the total chloride content of concrete is easily determined, the relevance of this value is questionable, since a proportion of these chlorides are bound and hence unavailable to take part in the depassivation process. Thus, the measurement of 'free' chlorides is considered to be of greater significance. However, the measurement of free chlorides presents a number of problems.

One approach to determining the free chloride content is through water extraction–type techniques, which bring a powdered sample into contact with a volume of water and determine the chloride concentration of the resulting solution. This approach tends to overestimate the quantity of free chlorides in a sample [43]. The main reason for this is that, in the intact concrete, the cement hydration products containing chlorides will be in equilibrium with the pore solution. Any technique that brings a powdered specimen into contact with a volume of water greater than that present in the concrete pores will change this equilibrium, causing more of the hydration products to dissolve.

Regardless of whether the chlorides measured are free or total, the convention is to express the concentration as a percentage by mass of cement.

The alternative to water extraction are pore solution expression methods. These techniques use intact volumes of saturated concrete, which are compressed such that the water is expressed and subsequently analysed. It is generally agreed that the results of expression are likely to be more representative of reality. However, pore solution expression is not without its drawbacks – it requires specialised apparatus, and expression yields a very small volume of pore solution, which means that preparation and analysis must be conducted with care.

The use of a chloride threshold concentration is itself of questionable validity. A review of the literature has compiled values of total and free chloride concentrations deemed by studies from the literature to be the threshold concentration for corrosion [44]. The distributions possessed modal values for the chloride threshold at 0.8% and 0.6% chloride by mass of cement for total and free chlorides, respectively. However, the distributions cover a fairly broad range of values, suggesting that additional factors contribute towards corrosion.

Another parameter with a major influence towards corrosion is the concentration of OH^- ions, since this controls the formation of the passive layer on steel. Since chloride and hydroxide ions effectively play opposing roles, with the former destroying the passive layer and the latter forming it, it has been proposed that the $[Cl^-]/[OH^-]$ ratio provides a better means of identifying the critical point beyond which corrosion will occur [45]. Indeed, there is a school of thought that considers the presence of halides, such as chloride, as an essential prerequisite of steel corrosion with sufficient drop in pH having the effect of reducing the $[Cl^-]/[OH^-]$ ratio to levels where corrosion will occur within the range of chloride concentrations that would be encountered in drinking water.

Reviews of the range of threshold [Cl⁻]/[OH⁻] ratios in the literature also indicate a significant spread of values – values between 0.12 and 3.00 have been published [46]. Ignoring some of the more outlying results from such studies, it is evident that the [Cl⁻]/[OH⁻] threshold sits somewhere around 1.00. However, the conclusion that must be drawn from this is that a far wider range of parameters play a role in determining when corrosion starts. For this reason, and because establishing the [Cl⁻]/[OH⁻] ratio represents a much greater challenge than establishing total chloride values, it is currently the latter parameter that is used in imposing limits on chlorides introduced into concrete via materials and water. Limits are defined in *EN 206* [47], with 'chloride content' classes defined for different applications. These classes refer to different total chloride contents, and the UK interpretation of these classes is shown in Table 4.1.

It should be stressed that, although corrosion can be initiated by the presence of chloride ions, their concentration has little influence on the actual rate of corrosion. However, the presence of chloride ions appears to promote pitting.

Table 4.1 Chloride limits set for concrete in BS 8500-1

Concrete use	Maximum Cl⁻ content (% by mass of cement)
No steel reinforcement or embedded metal, with the exception of corrosion-resisting lifting devices	1.00
Containing steel reinforcement or other embedded metal, non–heat cured	0.40
Containing steel reinforcement or other embedded metal, non–heat cured, exposed to significant amounts of external chloride	0.30
Containing steel reinforcement or other embedded metal, non–heat cured, made with sulphate-resisting Portland cement conforming to BS 4027	0.20
Containing steel reinforcement or other embedded metal, heat-cured	0.10
Containing pretensioned prestressing steel reinforcement	0.10
Internal posttensioned office construction	0.40
Containing posttensioned prestressing steel or unbonded prestressing steel	No guidance provided – depends on the type of structure, construction method and exposure
Strategic structures in severe chloride environments, for example, bridges	Refer to the project specification

Source: British Standards Institution. *BS 8500-1:2006: Concrete – Complementary British Standard to BS EN 206 – BS. Method of Specifying and Guidance of the Specifier.* London: British Standards Institution, 2006.

4.3.5 Protection from chloride-induced corrosion

4.3.5.1 Mix proportions and depth of cover

There are a number of approaches in formulating concrete mixes, which can be used to reduce the risk of chloride-induced corrosion in reinforced concrete. From the discussion of chloride ingress mechanisms, it can be deduced that the following strategies will limit the rate at which chlorides penetrate concrete:

- Reducing the volume of capillary porosity
- Reducing the pore diameter
- Increasing the tortuosity, surface area and/or constrictivity of the porosity

Each strategy is best achieved through engineering of the cement fraction. The most straightforward means of reducing the capillary porosity of concrete is simply to reduce the W/C ratio. Usually, this will also reduce pore diameters and increase tortuosity, albeit to a limited extent. The last two strategies are best achieved through the combination of cement fraction particle sizes that produce a 'refined' porosity. This is most commonly achieved through the use of combinations of Portland cement and cement components that are finer than PC and that undergo either pozzolanic or latent hydraulic reactions (FA, GGBS, SF, etc.). The combined effect of improved packing of the materials in the cement fraction and the production of cement hydration products yields both a reduced pore diameter and constrictivity and usually an increased tortuosity and surface area.

The use of other cement components can potentially also increase the chloride binding capacity of the concrete. This can occur in two ways. First, where the cement component contains a higher level of Al_2O_3 than PC, higher quantities of Friedel's salt will be formed on exposure to chlorides. Materials that fall into this category include GGBS, FA, and metakaolin (MK). Second, using cementitious materials with a higher SiO_2 content will normally lead to the formation of more CSH gel during hydration, which may increase the proportion of immobilised chlorides, although it should be pointed out that the CaO/SiO_2 ratio will be reduced, thus somewhat reducing the capacity per unit mass of the gel to bind chlorides.

Figures 4.21 and 4.22 show how air permeability, chloride binding capacity, and chloride diffusion coefficient change with increasing levels of FA and GGBS. Although permeability is used in describing the behaviour of porous media under a pressure differential, it is used here as a more general means of describing the ability of a material to resist ingress in general, since, as we have seen (and as will be shown in more detail in Chapter 5), the parameters that influence flow and diffusion are closely related.

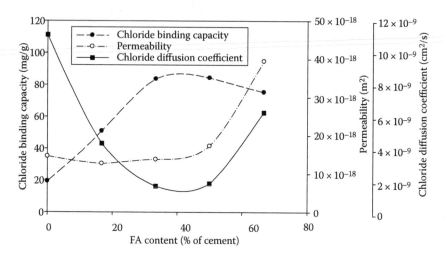

Figure 4.21 Influence of FA content on chloride binding capacity, air permeability and chloride diffusion coefficient of concrete with a fixed W/C ratio of 0.55. (From Dhir, R. K. et al., *Cement and Concrete Research*, 27, 1997, 1633–1639.)

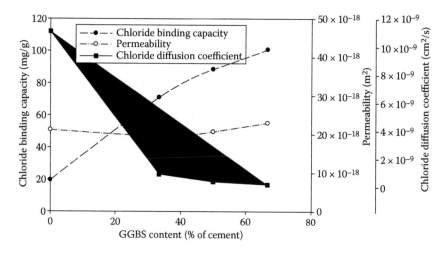

Figure 4.22 Influence of GGBS content on chloride binding capacity, air permeability and chloride diffusion coefficient of concrete with a fixed W/C ratio of 0.55. (From Dhir, R. K. et al., *Cement and Concrete Research*, 26, 1996, 1767–1773.)

In the case of FA, an increase in the quantities of this material leads to increased levels of chloride binding and a reduced diffusion coefficient up to approximately 40% of the total mass of cement. Beyond this point, chloride binding decreases and permeability increases, with a consequent increase in the diffusion coefficient. In the case of GGBS, increases in chloride binding and decreases in permeability persist to much higher levels, as do the resulting reductions in the diffusion coefficient. It is evident from both figures that chloride binding plays a significant role in reducing the rate of chloride ingress.

The situation for SF is somewhat different. Figure 4.23 plots the proportion of bound chloride in cement pastes containing increasing levels of SF. The capacity to bind ions drops as the SF content increases. The reason for this presumably partly relates to the high SiO_2 content of SF, whose presence will reduce the potential for Friedel's salt formation and the Ca/Si ratio of any CSH gel formed. However, another possible factor is that, at least according to one model for the structure of CSH gel [49], Al^{3+} ions are required to balance the incorporation of Cl^- ions within the structure. The reduction in Al_2O_3 in the cement fraction, caused by the introduction of SF, may therefore further compromise the chloride binding capacity of CSH.

Despite the drop in chloride binding capacity, SF protects steel reinforcement from chloride-induced corrosion. The conclusion must therefore be drawn that the improved performance results entirely from the refinement of porosity, reducing the rates of chloride ingress and also possibility increasing electrical resistivity.

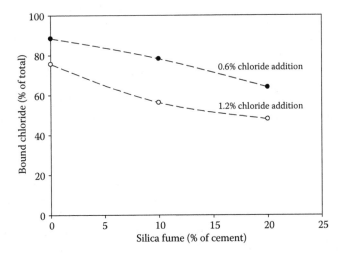

Figure 4.23 Influence of SF content on bound chloride in cement pastes containing chloride additions. (From Hussain, R. S. E. and A. S. Al-Gahtani, *Cement and Concrete Research*, 21, 1991, 1035–1048.)

The important role played by cement fraction in binding chlorides means that increased cement content will provide greater protection against chloride-induced corrosion.

In considering ways in which reinforcement can be protected from chloride-induced corrosion, it is worth bearing in mind that, in a chloride-bearing environment, it is inevitable that chlorides will eventually reach the steel. Thus, one of the most straightforward means of prolonging the period of time before corrosion is initiated is simply to place an adequate depth of concrete between the reinforcement and the external environment.

Selection of appropriate mix proportions and cover depths is detailed by one of the complementary British Standards to *BS EN 206 – BS 8500-1* [48]. The standard defines a series of exposure classes for chlorides. These are as follows:

- Corrosion induced by chlorides other than from seawater
  - XD1: moderate humidity
  - XD2: wet, rarely dry
  - XD3: cyclic wet and dry
- Corrosion induced by chlorides from seawater
  - XS1: exposed to airborne salt but not in direct contact with seawater
  - XS2: permanently submerged
  - XS3: tidal, splash and spray zones

The number in each exposure class denotes the level of aggression presented by each type of exposure, with a higher value indicating more aggressive conditions. The XS exposure classes are considered to be more aggressive than the XD classes for a given level of aggression. Environments in which cyclic wetting and drying occur are considered the most aggressive, primarily because they provide the greatest opportunity for exposure of reinforcement to both chlorides and oxygen.

BS 8500-1 provides combinations of maximum W/C ratios, minimum compressive cube (or cylinder) strengths, appropriate cement types and minimum cover depths for providing adequate protection for specified intended working lives of structures (either ≥ 50 or ≥ 100 years). More aggressive environments require lower W/C ratios, higher strengths, higher cement contents and higher cover depths. However, some flexibility is permitted through the selection of an appropriate cement. For instance, where an intended working life of ≥ 100 years is required in an XS3 environment and a cover depth of 50 mm is used, PC with 6% to 35% FA, 6% to 20% GGBS or limestone or 6% to 10% SF can be used, with a maximum W/C ratio of 0.40, a minimum characteristic cube strength of 45 N/mm^2 and a minimum cement content of 380 kg/m^3. Alternatively, a maximum W/C ratio of 0.45, a minimum characteristic cube strength of 35 N/mm^2 and a minimum cement content of 360 kg/m^3 can be used if the cement is changed to either a combination of PC with 36% to 55% FA or 66% to 80% GGBS.

4.3.5.2 Corrosion inhibitors and other admixtures

Corrosion-inhibiting admixtures are agents that increase the chloride threshold level required to cause depassivation and also slow down the rate of corrosion once depassivation has occurred. Calcium nitrite ($Ca(NO_2)_2$) was the first compound found to act as a corrosion inhibitor in concrete, although a wider range of agents have now been demonstrated to display inhibitive properties. These include other nitrite salts such as sodium nitrite; other inorganic compounds such as tin (II) sulphate; sodium molybdate and sodium fluorophosphate; and organic compounds such as malonic acid, disodium glycerophosphate, 5-hexyl-benzotriazole, dicyclohexylammonium nitrite and amine compounds [53–58].

Corrosion inhibitors are generally believed to operate either as anodic or cathodic inhibitors [59]. Anodic inhibitors include calcium and sodium nitrite. These act by oxidising $Fe(OH)_2$ in the passive layer, which (as discussed in Section 4.3.4) is vulnerable to dissolution in the presence of chlorides, to $Fe(OH)_3$, which is considerably more stable. This reaction removes nitrite ions from the solution, ultimately exhausting the quantity available. Thus, the ratio of $[Cl^-]$ to $\left[NO_2^-\right]$ is significant, with a critical $[Cl^-]/\left[NO_2^-\right]$ value proposed as being approximately 0.4, where $[Cl^-]$ is expressed in terms of kilograms per cubic metre of concrete and $\left[NO_2^-\right]$ is expressed in litres of calcium nitrite solution per cubic metre of concrete [60].

The organic inhibitors are effective via the cathodic mechanism by forming a protective layer at the steel surface, limiting the extent to which chlorides and oxygen can reach the reinforcement.

In the case of tin and molybdate compounds, it is thought that they act by precipitating a very fine layer of the metal at the steel surface, which leads to the formation of a more stable passive layer.

The use of corrosion-inhibiting admixtures is currently not covered by any British Standards, although the Concrete Society has published guidance on appropriate dosages for a given predicted chloride concentration at the reinforcement surface [60]. However, this is limited to calcium nitrite, since its longer track record of success permits greater confidence in providing guidance.

Many corrosion-inhibiting compounds can also be applied to the surface of concrete to provide protection after setting and hardening have occurred, potentially a long time after this has happened.

Damp-proofing admixtures may also provide some level of protection against the ingress of chlorides. Indeed, formulations containing a combination of inhibitors and damp-proofing agents have been shown to be effective [59]. Damp-proofing admixtures are discussed in further detail in Chapter 5.

4.3.5.3 Alternative reinforcement materials

Steel reinforcing bars for concrete are required to have an appropriate tensile and yield strength. In most cases, they are also required to be weldable and be capable of being bent (and possibly re-bent) during the assembly of the reinforcement without being damaged. In certain applications, resistance to fatigue may also be required [61]. The issue of weldability is significant, since the inclusion of higher levels of carbon and metals other than iron (principally manganese, chromium, molybdenum, vanadium, nickel and copper) will render steel unweldable. As a result, the British Standard for weldable reinforcing steel (*BS 4449*) places a limit on the carbon content of 0.24% by mass [62]. Thus, for this reason, and for reasons of economy, plain steels with lower carbon contents have tended to be favoured over other types.

However, since the issue of reinforcement corrosion results from the relative susceptibility of carbon steel, there is a clear logical argument for using reinforcement made from steels that are less prone to corrosion or for using completely different materials instead.

Corrosion-resistant steels take a number of forms, with the most obvious one being stainless steel. Stainless steel is manufactured in a variety of forms, defined by their chemical composition and the heating and cooling regimes that they have undergone. The three most common classifications are martensitic, ferritic and austenitic [63]. Martensitic stainless steels are alloys of iron and chromium, which have been quenched from above the critical range. They are typically less resistant to corrosion than other types of stainless steel and are usually deemed unsuitable for use in construction. Ferritic steels are also iron–chromium alloys but are not quenched. Again, these steels have a relatively low resistance to corrosion and can present problems when welded, unless the nitrogen content is sufficiently high. For this reason, the British Standard for stainless steel reinforcement bars (*BS 6744*) excludes martensitic and ferritic steels by requiring minimum nickel contents [64].

The most suitable stainless steels for use as reinforcement bars are austenitic, which are chromium–nickel–iron alloys. These steels are both highly resistant to corrosion and weldable. However, welding must be carried out with care, since the effect of excessive heat can weaken the steel by reversing the work hardening that is used to strengthen stainless steels. Moreover, the oxide layer formed during welding – welding scale – increases the weld's susceptibility to corrosion [65]. Various requirements need to be satisfied when welding stainless steel reinforcement – the surface should be clean, the filler rod should be of a composition as close to the steel as possible, the heat input to the weld should be carefully controlled (possibly using low-heat input techniques, such as resistance welding) and the welding scale should be removed using techniques such as shot blasting or the application of acidic pickle paste [66]. As a result, the UK guidance on the use of stainless steel reinforcement suggests that welding should only be done where these activities can be conducted in a highly controlled manner, such as in a precast factory. Joining of rods without welding is possible using stainless steel couplers.

Where stainless steel reinforcement is used, *Eurocode 2* includes provision for a reduction in the cover of concrete, $\Delta c_{dur,st}$, which is left to individual countries to define, although a recommendation of 0 mm is included. The UK National Annex [8] advises that this recommendation is adhered to unless reference to specialist literature justifies a reduction, citing the Concrete Society guidance as an example [66]. This guidance suggests that, where good workmanship can be assured, in particular, in precast concrete production, a reduction in concrete cover and possibly bar diameter (with an appropriate increase in the strength of the steel, where required) is possible. The document recommends a minimum cover of 40 mm in highly corrosive environments. The British Standard for cast stone permits the cover to be reduced from 40 to 10 mm, where austenitic stainless steel is used in place of carbon steel [67].

Stainless steel is not used for prestressing applications.

Conventional stainless steels are high-alloy steels, requiring >4% additions of other metals. However, one additional group of possible candidates for steel reinforcement are the low-alloy duplex and dual-phase steels. These are steels that have undergone sequences of heat treatment and quenching in such a manner as to produce a microstructure consisting of particles of martensite (dual-phase steel) or austenite (duplex steel) in a ferrite matrix. The resulting material is resistant to corrosion as a result of the absence of carbides (such as cementite, Fe_3C) which, as discussed previously, are cathodic in the presence of ferrite. Results of simple experiments in which dual-phase martensite steel was embedded in concrete and exposed to a corrosive environment alongside specimens containing conventional mild steel reinforcement indicate enhanced corrosion resistance [68].

Weathering steels are low-alloy steels that display enhanced resistance to atmospheric corrosion through the formation of a protective surface layer of rust. Enhanced resistance to chloride-induced corrosion has been reported, based on the absence of cracking in specimens reinforced using such a material [69]. However, weathering steel is not intended for use in applications where chloride concentrations are high. Thus, the suitability of weathering steels as reinforcement is uncertain and not covered by any British Standards.

Another means of avoiding the corrosion of reinforcing steel is to eliminate it entirely through the use of fibre-reinforced polymer (FRP) reinforcement. These are composite rods, grids and ropes that consist of continuous high-strength fibres impregnated with a polymer resin. Fibres are most commonly glass, carbon or aramid. More recently, basalt fibre, manufactured by extruding molten basalt rock, has also become more widely available. The polymer resins used are usually epoxies or vinyl esters.

The tensile strength of FRP reinforcement depends on the volume ratio of fibre to resin and the strength of the fibres, but strengths are typically similar or higher than steel reinforcing bars, which are normally produced with ultimate tensile strengths of 500 to 900 N/mm^2 [62]. Glass fibre composite reinforcing bars typically have ultimate tensile strengths between 500 and 600 N/mm^2, whereas carbon fibre bars are usually considerably stronger (2000–2500 N/mm^2). However, the stiffness of these composites is usually lower than that of steel –40 kN/mm^2 for glass fibre to 140 kN/mm^2 for carbon fibres, compared with approximately 200 kN/mm^2 for steel.

These differences in mechanical properties require changes in the approach to the design of concrete structures using FRP reinforcement, relative to those made using steel. Additionally, unlike steel, FRP reinforcement is not ductile and weaker in compression than in tension. Guidance on the design using FRP reinforcement has been published by the Institution of Structural Engineers in the United Kingdom [70].

From a durability perspective, the absence of the risk of chloride-induced corrosion means that the concrete quality and cover depth are no longer as strongly determined by the nature of the environment that the structure is exposed to, but by structural requirements, the diameter of the bar used and the maximum aggregate size. The guidance also points out that additional cover is required where protection of FRP reinforcement from fire is required, since the composites are considerably more vulnerable to the effects of heat than steel [71]. However, the document also stresses that, where fire is a significant design consideration, FRP reinforcement is not recommended.

Another difference with FRP reinforcement is that the bond between the reinforcement in its ordinary form and the cement matrix is weak in comparison to steel. This issue is overcome by many manufacturers by coating the surface with particles of sand, which improve the bond.

Joining of FRP bars is achieved through the use of couplers.

FRP composites have also been successfully used as prestressing tendons. Other than changes in the cover requirements, the main issue for concrete structures using these materials is that of creep – under tension, FRP tendons will creep to failure, with the level of loading determining the time before failure occurs. This issue, coupled with the fact that glass undergoes stress corrosion, means that glass fibre–reinforced composites are not suitable for this type of application. Aramid and carbon fibre composites are suitable, although the maximum prestress level must be limited to 50% to 60% of the ultimate capacity for carbon fibre tendons and to 40% to 50% for aramid fibres for design lives exceeding 100 years [72].

4.3.5.4 Reinforcement coatings

Steel reinforcing bars can be produced with protective coatings specifically designed to protect them from corrosion. These coatings take two forms: a layer of impermeable material that acts as a physical barrier between steel and the outer environment, and sacrificial coatings. The first type of protection is normally achieved through the application of a layer of epoxy resin. One concern with epoxy-coated reinforcement is that damage to the protective layer will allow corrosion to still occur. Figure 4.24 shows how increasing levels of epoxy layer damage lead to increasing rates of corrosion.

Damage is most commonly incurred during bending of the reinforcement during construction, although this can be minimised by ensuring that bend radii are greater than three times the bar diameter and using mandrels with a nylon facing [60]. A greater guarantee of minimal damage is achieved where cutting and bending is carried out by the manufacturer. Damage can also occur as a result of scratches during fixing, and so plastic-coated tie wire and spacers are recommended. Damage to the epoxy coating can be repaired by brush application of epoxy resin to damaged areas. This requires thorough inspection procedures to ensure that all damaged areas are located.

In precast concrete production, and possibly on site, it is also possible to prefabricate reinforcement and subsequently coat it with epoxy.

It has been proposed that the combination of connected coated and uncoated reinforcements in the same structural element presents a

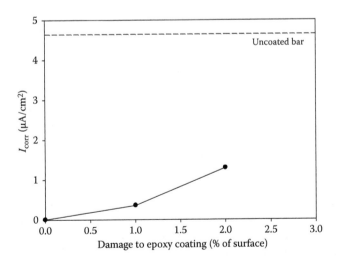

Figure 4.24 Corrosion current (I_{corr}) of epoxy-coated rebars with increasing levels of surface damage. (From Erdoğu, S. et al., *Cement and Concrete Research*, 31, 2001, 861–867.)

significant risk to durability. It has already been mentioned that a low ratio of anodic to cathodic surface leads to an accelerated rate of corrosion. The logic behind the concern is that, if chlorides reach the coated bars first and the coating is damaged, the uncoated bars will act as the cathode, and the low ratio of surface areas will lead to rapid corrosion at the damaged parts of the coated steel.

Galvanised reinforcement is produced by coating the surface of the steel with a thin layer (usually in the range of 10–250 µm in thickness) of zinc. This is done expressly to create a galvanic corrosion cell at the steel surface, but one in which the zinc is the more anodic of the two metals, which means that the steel remains uncorroded. Under conditions of relatively low chloride concentrations and within the pH range typical of concrete pore solutions, the zinc is passivated and so also acts as a physical barrier to chlorides reaching the steel. The level of passivation is significant – the chloride concentration needed to cause depassivation of zinc is higher, and the metal remains passified to slightly lower pH values. Moreover, the corrosion products of zinc are nonexpansive and appear to block pores, which means that cracking is reduced, and further protection of the steel is created [74].

Nonetheless, protection of steel by the sacrificial layer is generally finite. This is because the corrosion protection of the zinc is lost at high chloride levels (>1% by mass of cement) and at lower pH levels and, once the sacrificial layer is lost, corrosion of the steel can proceed.

4.3.5.5 Fibres

The presence of steel, polymer or glass fibre in concrete normally has the effect of enhancing surface quality, thus providing greater resistance to chloride ingress. Steel fibres, and other fibres with a high stiffness, have the additional benefit of controlling cracking, generally leading to smaller crack widths than would be formed in concrete without fibres. The use of fibres in concrete is further discussed in Chapter 5.

4.3.5.6 Surface coatings

Surface coatings provide an additional layer of protection between the external environment and the reinforcement. Although a wide range of surface coatings are available for concrete, where protection against chloride ingress is the priority – in parts of a structure where cyclic wetting and drying are likely to be experienced – the most common type are hydrophobic impregnants. These formulations are normally based on silane compounds and render the surface and near-surface pore interiors hydrophobic, thus restricting the extent to which chlorides can penetrate the surface. Silanes have the additional benefit of having little influence on the surface appearance while permitting the concrete to 'breathe' – allowing water vapour to escape from the concrete interior via the surface.

The use of silanes is compulsory on concrete bridges in the United Kingdom. Details of the requirements for protecting bridges with silanes are given in the *Highways Agency Design Manual for Roads and Bridges* [75]. Only products based on isobutyl(trimethoxy)silane are currently approved. The manual advises application to parts of bridges most likely to come into contact with chlorides, including piers, columns, crossheads, abutments, deck beams, soffits, wingwalls and retaining walls within 8 m of the edge of the carriageway; piers, columns, crossheads, abutments, bearing shelves, ballast walls and deck ends with a deck joint above; parapets and parapet plinths; and deck surfaces not protected by deck waterproofing.

The manual recommends that new structures have silanes applied prior to their first exposure to de-icing salts, since the rate of ingress on initial exposure to chloride-bearing solutions is usually relatively high due to capillary action.

Additionally, barrier coatings and pore-blocking sealers will reduce chloride ingress. These coatings are discussed in more detail in Chapter 6.

4.4 CARBONATION

Carbonation is a reaction between cement hydration products in concrete and atmospheric carbon dioxide (CO_2), which leaves steel reinforcement vulnerable to corrosion. The rising level of CO_2 in the Earth's atmosphere is currently a major concern from an environmental perspective, with the origin of the majority of this gas now deriving from human activities that involve the combustion of fossil fuels. From the perspective of concrete carbonation, the levels are still relatively low – the atmosphere currently contains approximately 390 parts per million by volume (ppmv) of CO_2. However, local concentrations of CO_2 in air sampled from urban areas have been shown to be somewhat higher (up to ~700 ppmv) [76], with locations in close proximity to various industrial and agricultural CO_2 sources being significantly higher. For instance, the fumigation of concrete grain silos using CO_2 can use levels in excess of 70% by volume [77].

4.4.1 Carbonation reaction

The chemical reactions occurring during carbonation can be broken down into two stages: the dissolution of CO_2 in water and the reaction of the product of dissolution with hydration products within the cement phase of concrete.

When CO_2 comes into contact with water, carbonic acid (H_2CO_3) is formed:

$$CO_2 + H_2O \leftrightharpoons H_2CO_3$$

When the water in question is the fluid within a concrete pore, the carbonic acid will react with Portlandite to form calcium carbonate:

$$Ca(OH)_2 + H_2CO_3 \rightarrow CaCO_3 + 2H_2O$$

CSH gel also undergoes carbonation. This initially has the effect of reducing the Ca/Si ratio of the gel as more of the calcium is converted to $CaCO_3$. However, beyond a certain point, the CSH gel is destroyed, leaving silica gel in its place [79].

$CaCO_3$ can take a number of structural forms, with the most common being calcite, vaterite and aragonite. The variety of $CaCO_3$ formed during carbonation is dependent on the conditions under which the reaction has occurred. At low CO_2 concentrations typical of current atmospheric levels, the formation of vaterite and aragonite appears to be favoured [78]. Vaterite and aragonite are metastable and will ultimately undergo a transformation to calcite, a process that is accelerated by higher CO_2 concentrations [79]. At higher CO_2 concentrations, calcite alongside smaller quantities of vaterite are observed, which presumably reflects this accelerated transformation.

The process of carbonation leads to a reduction in pH, which, as discussed previously, leads to the destruction of the passive layer around steel once pH drops below around 11.5. However, the drop in pH develops as a front that progresses into the concrete cover with time. Figure 4.25 shows how quantities of Portlandite and calcium carbonate vary with depth after a period of carbonation, along with the effect on pore solution pH.

Figure 4.26 shows the typical manner in which the carbonation front progresses with time. The progress of the carbonation front is the most common means of expressing the extent to which carbonation has occurred. This is primarily because this has the most relevance when considering carbonation in terms of cover depth from reinforcement. However, it is also because the most straightforward means of monitoring carbonation is by spraying a freshly fractured concrete surface with a solution of thymolphthalein or phenolphthalein. These are pH indicators whose colour changes at around the pH where depassivation occurs. In the case of thymolphthalein, the change in colour (from colourless to blue) occurs within the pH range of 9.3 to 10.5, whereas phenolphthalein changes from colourless to pink within the range of 8.2 to 10.0. Thus, where carbonation has occurred, the front is visible as a sharp transition from the natural colour of the concrete (carbonated) to coloured (noncarbonated).

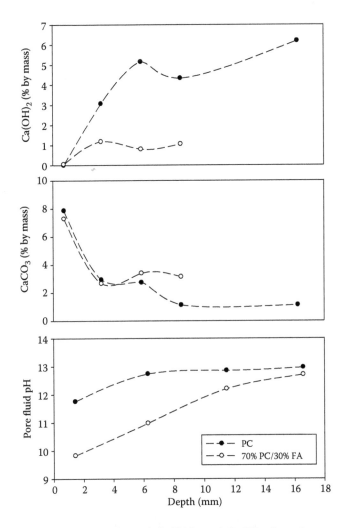

Figure 4.25 Concentration profiles of $Ca(OH)_2$ and $CaCO_3$ through cement pastes, containing Portland cement (PC) or a combination of PC and fly ash (FA) in the cement fraction, exposed to a CO_2 atmosphere and the influence on reduced $Ca(OH)_2$ levels on pH. (From McPolin, D. O. et al., *Journal of Materials in Civil Engineering*, 21, 2009, 217–225.)

The rate of advancement of the carbonation front is clearly a useful thing to be able to express. However, due to the nonlinear nature of the relationship between depth and time, expressing the rate in distance per unit time is not possible. However, because the rate of carbonation is dependent on the rate of diffusion of CO_2 into concrete, as with all diffusion-driven processes, a plot of depth versus the square root of time yields a straight

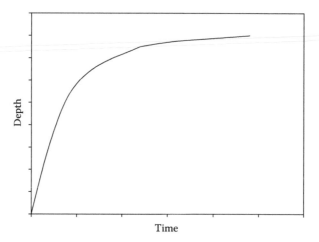

Figure 4.26 Typical plot of depth of carbonation front with time.

line. The gradient of this straight line gives a measure of the rate, whose units are millimetres per year$^{0.5}$ or similar. This has been given a number of names in the past but will be referred to here as the carbonation coefficient.

From a reinforcement corrosion perspective, the drop in pH is significant for two reasons. First, we have seen that the stability of the passive layer at the steel surface is dependent on a high pH, and once the carbonation front reaches this region, depassivation will start to occur. Second, the solubility of Friedel's salt is dependent on pH and increases with decreasing pH [35]. Thus, as carbonation progresses in concrete where chlorides are simultaneously permeating, an increase in chloride ion concentrations in the pore fluid will be observed behind the carbonation front [81]. Thus, carbonation in the presence of chlorides will lead to an increase in the $[Cl^-]/[OH^-]$ ratio (see Section 4.3.4).

4.4.2 Factors influencing rates of carbonation

The three environmental factors that have the largest impact on the rate of carbonation are CO_2 concentration, relative humidity and temperature.

To understand the influence of environmental factors on the rate of carbonation, it is first of value to consider the processes that are occurring. These can be broken down into the movement of the CO_2 molecules into concrete through the pore network and the reaction of these molecules with $Ca(OH)_2$ and other hydration products. Mass transport of CO_2 into the concrete occurs by diffusion and is thus governed largely by the nature of porosity in the concrete. A CO_2 molecule entering the concrete will initially find itself in the zone behind the carbonation front where there is an absence of hydration products capable of undergoing carbonation, and any moisture present in the concrete pores will be saturated with respect to CO_3^{2-}. Thus, there is no mechanism present to remove the CO_2 from the air in the pores, and the molecule is free to continue its process of diffusion.

As the diffusion process continues, and assuming that this particular molecule makes its way progressively further into the concrete, it will reach a zone in which the water present is not saturated with respect to CO_3^{2-}. This is the carbonation front, and the water is undersaturated because $Ca(OH)_2$ and CSH are present and CO_3^{2-} is being removed from the solution by the carbonation reaction. If we imagine at this point that our CO_2 molecule comes into contact with the surface of a droplet or film of water in a pore, it will dissolve to form H_2CO_3. This will dissociate to $2H^+$ and CO_3^{2-}, and the carbonate ion will diffuse through the water until it comes into contact with a calcium ion, which has dissolved from a hydration product, to form $CaCO_3$. The precipitation of solid $CaCO_3$ leads to the water becoming undersaturated with respect to calcium, allowing more hydration product to dissolve.

The rate at which $CaCO_3$ is produced is normally assumed to be a first-order reaction described by the equation

$$\frac{d[CaCO_3]}{dt} = k[Ca^{2+}]\left[CO_3^{2-}\right]$$

where the square brackets denote concentration (mol/L), and k is a reaction rate coefficient (L/mol s).

The overall effect of the processes described above is shown in Figure 4.27, where an idealised pore running from the exterior concrete surface into the interior contains both gas and liquid phases, with the liquid phase (water) present as a layer at the pore surface, which initially is composed of hydration products (in this simplified depiction, $Ca(OH)_2$ only). Alongside this pore, the profiles of concentrations of relevant minerals, CO_2 in the gas phase and chemical species dissolved in the pore water are plotted.

Thus, the rate at which the carbonation front moves through concrete will be influenced by the rate of diffusion of CO_2 through the pore network, the rate of diffusion of CO_3^{2-} through water and the rate of reaction. The diffusion coefficient of CO_2 through concrete typically lies in the range of 1×10^{-9} to 1×10^{-7} m²/s, whereas it is approximately 9.5×10^{-9} m²/s at 25°C for CO_3^{2-} in water.

However, given that the shortest distance between the surface of a droplet or film of water inside the concrete pores and the hydration products that line the pore walls will be extremely small relative to the distance between the exterior concrete surface and the carbonation front, even after a relatively short period of carbonation, it is the diffusion of CO_2 through the pore system that determines the rate of carbonation under a given set of ambient conditions.

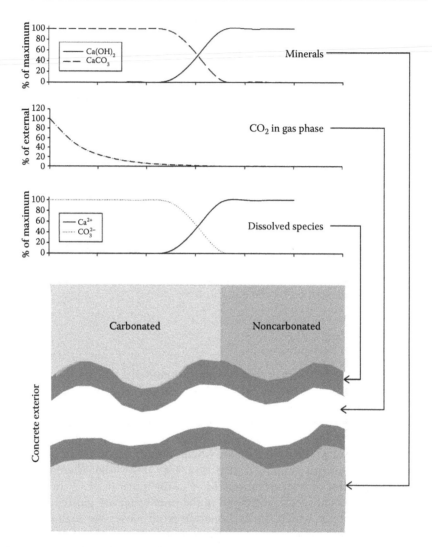

Figure 4.27 Concentration profiles of different substances through the length of a pore running from the surface of carbonating concrete.

The concentration of CO_2 in air in contact with concrete influences carbonation in two ways. First, since the movement of CO_2 into concrete is the result of diffusion, a higher concentration gradient between the exterior and the interior of concrete will lead to a greater rate of diffusion (see Chapter 5). Second, a higher concentration will yield a higher rate of reaction. As previously discussed, the concentration of CO_2 also determines the proportions of vaterite and calcite formed. However, there is little evidence to suggest that this has any significant influence on the rate at which the carbonation front progresses.

The relative humidity of the air in contact with cement will determine the amount of moisture in the concrete pores. This is significant for two reasons. First, the carbonation reaction needs water, since without water, carbonic acid cannot be formed. Thus, at low relative humidities, the lack of moisture in pores limits the extent to which the carbonation reaction can occur. Second, as relative humidity increases, so too does the quantity of condensed water in the concrete pores. This water acts to limit the air-filled volume of the pores and consequently the diffusion coefficient of the CO_2 molecules (Figure 4.28). As a result, high relative humidity also limits the carbonation rate.

The overall effect is that the optimum rate of carbonation is observed at approximately 55% (Figure 4.29).

As previously seen for chloride ions, diffusion coefficients are dependent on temperature, in a manner described by the Arrhenius equation. This is true of CO_2, although temperature also plays a role in influencing carbonation through a different means. The rate constant of a chemical reaction is given by another form of the Arrhenius equation (see Section 4.3.2):

$$k = Ae^{-E_a/RT}$$

where A is a constant. Thus, as temperature increases, it can be expected that the rate of reaction will also increase. However, the solubility of both CO_3^{2-} and $Ca(OH)_2$ (and other hydration products) decreases with increasing

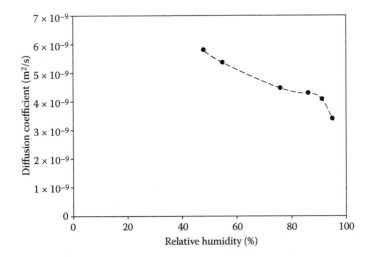

Figure 4.28 Diffusion coefficient of CO_2 through concrete with a W/C ratio of 0.4 over a range of relative humidities. (From Houst, Y. F. and F. H. Wittmann, *Cement and Concrete Research*, 24, 1994, 1165–1176.)

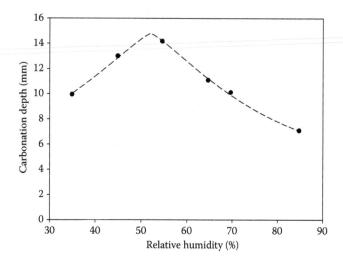

Figure 4.29 Depth of carbonation of PC concrete after 5 days of exposure to a 50% CO_2 atmosphere as a function of relative humidity. (From Papadakis, V. G. et al., *ACI Materials Journal*, 88, 1991, 363–373.)

temperature. This has the effect of reducing the rate of reaction, since this is dependent on both CO_3^{2-} and Ca^{2+} concentrations. Thus, there is a drop in the rate of carbonation at temperatures above 60°C [83].

Ultimately, however, it should be stressed that, as a result of the nature of diffusion and the nature of the influence of temperature on the diffusion coefficient, the overall impact on carbonation rates is relatively small, as illustrated in Figure 4.30.

Given the importance of diffusion in the carbonation process, the nature of porosity plays an important role in defining the rate of carbonation, with the total porosity, tortuosity and constrictivity having similar influences as those discussed for chloride diffusion. The role of the total volume fraction of porosity is best illustrated by examining the influence of W/C ratio on the diffusion coefficient of CO_2 (Figure 4.31) and the rate of advancement of a carbonation front (Figure 4.32).

As has been previously discussed, in the case of reducing the rate of chloride ingress, manipulation of tortuosity and constrictivity can be achieved through the inclusion of materials such as slag, FA and SF, whose presence acts to refine the concrete pore structure. The same is true of carbonation, but in this case, there is a significant payoff in the form of reduced $Ca(OH)_2$ content.

The quantity of $Ca(OH)_2$ present in concrete plays an important role in controlling the rate at which the carbonation front progresses. This is because higher levels of $Ca(OH)_2$ require a larger quantity of CO_2 to react with, and thus a longer period of time is required for sufficient gas molecules to be delivered by diffusion.

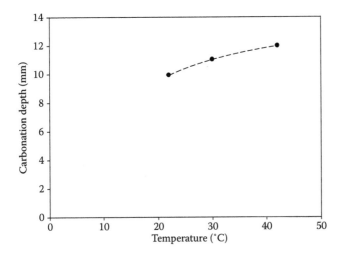

Figure 4.30 Depth of carbonation of PC concrete after 5 days of exposure to a 50% CO_2 atmosphere with a relative humidity of 65% as a function of temperature. (From Papadakis, V. G. et al., *ACI Materials Journal*, 88, 1991, 363–373.)

The reduced levels of $Ca(OH)_2$ resulting from the presence of non–Portland cement components means that the amount of CO_2 required to advance the carbonation front by a given distance is reduced. The higher tortuosity, lower constrictivity and possibly lower total porosity obtained through the use of such materials act against this, but the overall effect is, in some cases, to produce

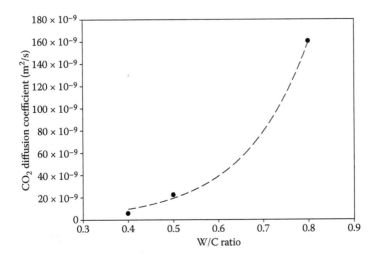

Figure 4.31 CO_2 diffusion coefficient versus W/C ratio for PC pastes as a function of W/C ratio. (From Houst, Y. F. and F. H. Wittmann, *Cement and Concrete Research*, 24, 1994, 1165–1176.)

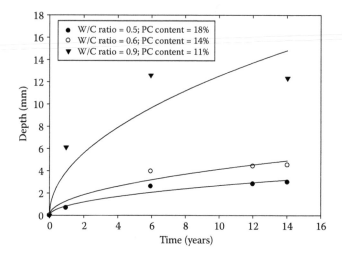

Figure 4.32 Depth of carbonation versus time for concrete exposed to a polluted atmosphere as a function of W/C ratio and cement content. (From Currie, R. J. *Carbonation Depths in Structural Quality Concrete.* Watford, United Kingdom: Building Research Establishment, 1986.)

poorer resistance to carbonation. This is shown in Figure 4.33, where the rate of carbonation is plotted against the percentage of PC in the cement fraction of concrete mixes containing GGBS, FA, SF and MK. Although the plot does not give any direct indication of the contribution of each non-PC material towards reducing the effective diffusion coefficient, it does include two different materials used at identical levels – SF and MK. Thus, from the plot, we can see that SF – the finer one of the two materials – yields a lower rate of carbonation, indicating its superiority in terms of reducing the rate of CO_2 diffusion.

Although this conflict between $Ca(OH)_2$ and pore refinement is partly dependent on the specific nature of the material used, typically where the non-PC constituent is used at levels below approximately 15% to 20% by mass of the total cement fraction, the balance between reduced diffusion coefficients and reduced $Ca(OH)_2$ is maintained, giving performance comparable or better than a PC-only mix [86].

In more general terms, for the same reason, the PC content of a concrete mix is also influential with respect to rates of carbonation, with a higher PC content giving a slower rate of advancement of the carbonation front. During mix design, to maintain a given workability, a reduction in W/C ratio to improve concrete durability will normally have the effect of increasing PC content.

As for chloride diffusion, the presence of cracks can potentially increase the rate of CO_2 diffusion considerably. Figure 4.34 shows the influence of crack width and depth on the rate of carbonation. Crack width has a much greater influence than crack depth.

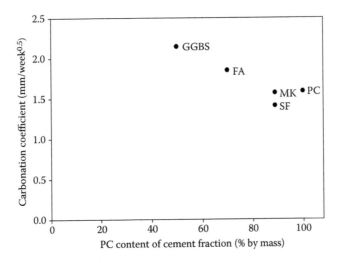

Figure 4.33 Rate of carbonation as a function of PC content in mortar mixes containing combinations of PC and GGBS, FA, SF and MK, all with a W/C ratio of 0.42. (From McPolin, D. O. et al., *Journal of Materials in Civil Engineering*, 21, 2009, 217–225.)

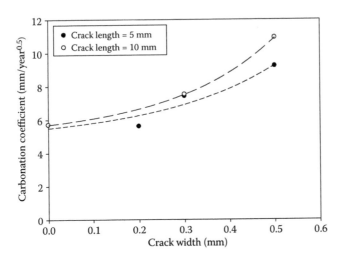

Figure 4.34 Influence of crack width and crack depth on the rate of carbonation of PC mortar containing artificial cracks exposed to a 10% CO_2 atmosphere. (From De Schutter, G., *Magazine of Concrete Research*, 51, 1999, 427–435.)

4.4.3 Changes in physical properties

Carbonation has the effect of altering the nature of the concrete within which it is occurring. In particular, the reaction of hydration products with CO_2 leads to a reduction in total porosity and average pore size, as a result of the precipitation of $CaCO_3$ crystals within pores. This effect is shown in Figures 4.35 and 4.36, which show the reduction in total porosity and average pore size, respectively. It would also appear that preferential precipitation of $CaCO_3$ occurs in larger pores [88]. The carbonation reaction yields water, which will also act to limit the mobility of CO_2 in the pores of concrete albeit until the moisture in the pores is equilibrated with respect to the external relative humidity.

The formation of $CaCO_3$ during carbonation will normally also yield a slight increase in strength. However, of greater significance from a durability perspective is carbonation shrinkage.

The mechanism behind carbonation shrinkage has not been fully resolved. It is certainly not the result of a difference in volume of the reaction product relative to the reactants, since the $CaCO_3$ produced during carbonation occupies a larger volume than the $Ca(OH)_2$ that formed it.

One of the most commonly cited explanations is that $Ca(OH)_2$ crystals in concrete are under compression as a result of drying shrinkage [89]. As $Ca(OH)_2$ dissolves during carbonation, an empty space is left behind, which allows the material surrounding the crystals to contract into this space. An alternative mechanism that has been proposed is

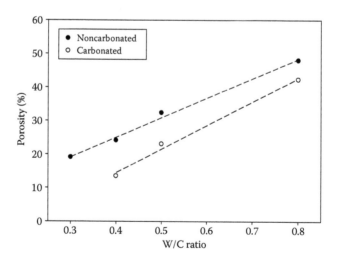

Figure 4.35 Reduction in porosity of Portland cement pastes with a range of W/C ratios. (From Houst, Y. F. and F. H. Wittmann, *Cement and Concrete Research*, 24, 1994, 1165–1176.)

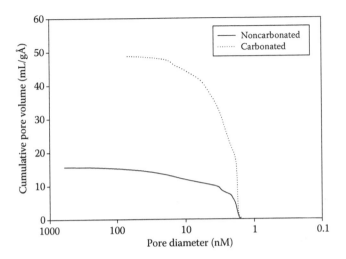

Figure 4.36 Change of pore size distribution, measured using BET analysis, in a Portland cement mortar as a result of carbonation. (From Dewaele, P. J. et al., *Cement and Concrete Research*, 21, 1991, 441–454.)

that the carbonation of CSH gel leads to its dehydration and structural changes, which lead to shrinkage [90].

Figure 4.37 plots a shrinkage curve for concrete undergoing carbonation. The magnitude of the dimensional change resulting from carbonation is potentially significant. The tensile strain capacity – the strain level capable

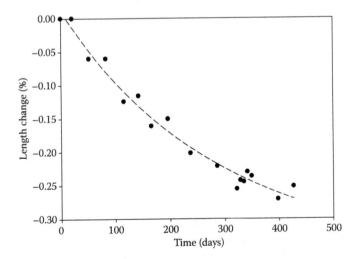

Figure 4.37 Carbonation shrinkage in PC concrete prisms exposed to a 15% CO_2 atmosphere. (From Jerga, J., *Construction and Building Materials*, 18, 2004, 645–652.)

of causing cracking – of concrete is approximately 0.015% or less, and carbonation is capable of creating such strains. Because carbonation progresses from the outside surface, cracking occurs as a result of the stresses caused by the difference in the volume of the carbonated and noncarbonated parts of the concrete. Cracking occurs in the form of crazing at the surface.

The overall effect of carbonation on the permeation characteristics of concrete is complex. The reduction in porosity means that there is a greater resistance to further ingress of CO_2 and other substances. However, where cracking occurs, the diffusion coefficient of CO_2 diffusing through the carbonated layer will increase. Generally speaking, a reduced W/C ratio will not only limit the rate of carbonation but is also likely to reduce the extent of cracking.

4.4.4 Avoiding carbonation

The diffusion coefficients of chloride ions and CO_2 through concrete are typically of a comparable order of magnitude. However, carbonation is generally considered a lesser threat. The main reason for this is that the concentrations of atmospheric CO_2 typically encountered are significantly less than can potentially be encountered in chloride-bearing environments.

Taking into account the role of $Ca(OH)_2$ and CSH in limiting the rate of advancement of the carbonation front, the conventional description of Fick's second law of diffusion can be modified to give an equation that yields the theoretical depth of carbonation (d) at time t:

$$d = \sqrt{2Dt\left(\frac{C_1}{C_0}\right)}$$

where D is the diffusion coefficient of the carbonated layer (m²/s), C_1 is the external concentration of CO_2 (m³ CO_2/m³ air) and C_0 is the amount of CO_2 required for the complete carbonation of the volume of concrete being considered (m³ CO_2/m³ concrete) [92].

As discussed previously, D is dependent on the pore characteristics of the carbonated layer, the relative humidity and the temperature. C_0 is determined primarily by the $Ca(OH)_2$ and CSH contents of the concrete and is thus dependent on the cement content and its composition and degree of hydration. In reality, D can fluctuate significantly as a result of changes in atmospheric conditions and the possible development of cracks over time.

By way of example, and assuming for the sake of simplicity that only $Ca(OH)_2$ is involved in carbonation, a concrete mix containing 350 kg/m³ of PC, which has hydrated to give it a $Ca(OH)_2$ content of 22% by mass, will need to react with approximately 25 m³ of CO_2 at 20°C and atmospheric pressure. If a value for D of 5.0×10^{-9} m²/s is used, this gives a carbonation depth of approximately 10 mm after 20 years. Thus, the rate of advancement of the carbonation front is typically slow. The carbonation coefficient in this example would be 2.2 mm/year$^{0.5}$. A review of surveys of structures in which carbonation depth was measured indicates a wide range of carbonation coefficients between <0.1 and 15 mm/year$^{0.5}$, although the higher values were typically observed in structures where very weak or poor-quality concrete was present [85].

Although it has been shown that many parameters influence the rate of advancement of the carbonation front, some influence it more than others. By establishing the parameters that are the most influential, the best approaches to protecting reinforcement from carbonation-induced corrosion can be identified. Examining the equation above indicates that reducing the diffusion coefficient of CO_2 through the concrete and increasing the level of hydration products available for carbonation are achieved by reducing the W/C ratio and increasing the PC content, respectively. However, in terms of the sensitivity of the value of d to these parameters, the W/C ratio will have a much greater influence. This is illustrated in Figure 4.38, which plots the depth of carbonation at 20 years for a range of cement contents and W/C ratios, calculated using the above equation and the same approach as the previous calculation. The relationship between W/C ratio and the

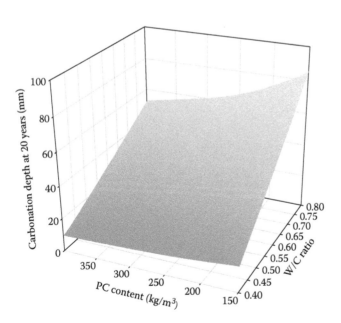

Figure 4.38 Theoretical influence of PC content and W/C ratio on depth of carbonation after a period of 20 years.

diffusion coefficient shown in Figure 4.31 was used. The influence of concrete cover depth can also be seen by rearranging the equation to obtain a description of the time necessary to reach a depth d:

$$t = \frac{d^2}{2Dt\left(\dfrac{C_1}{C_0}\right)}$$

Using this equation, the influence of cover depth can be seen, as shown in Figure 4.39, which plots the time taken to reach a given depth using fixed values of W/C ratio and PC content. Thus, W/C ratio and cover depth are the parameters under the control of the engineer, which are most effective in protecting against carbonation-induced corrosion. This is reflected in the manner in which carbonation protection is dealt with in the standards.

BS EN 206 [47] defines four exposure classes for situations in which concrete is exposed to air and corrosion induced by corrosion is a possibility. These are as follows:

- XC1: dry or permanently wet
- XC2: wet, rarely dry
- XC3: moderate humidity
- XC4: cyclic wet and dry

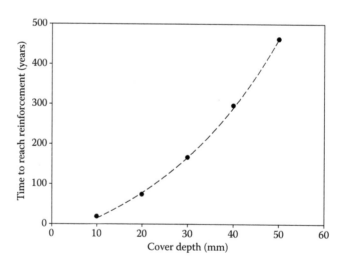

Figure 4.39 Theoretical influence of cover depth on the time taken to reach the reinforcement for concrete containing 350 kg/m³ of PC with a W/C ratio of 0.4.

XC3 and XC4 are the more aggressive exposure conditions, as a result of the importance of the role played by relative humidity. No additional guidance is provided for concrete encountering these types of exposure. However, the UK annex to this standard, *BS 8500-1* [48], provides appropriate cement types, maximum W/C ratios and minimum cover depths, cement contents and strengths. These requirements are less demanding than for chloride exposure classes. For example, for the most aggressive carbonating conditions with a working life requirement for a structure of 100 years, the lowest depth of cover is 30 mm, for which the following requirements are made:

- Any cement type other than a combination of Portland cement and 36% to 55% FA
- A minimum cement content of 340 kg/m
- A maximum W/C ratio of 0.45
- A minimum cube strength of 50 N/mm²

In comparison, the lowest depth of cover for the most aggressive chloride exposure class (XS3) is 45 mm. Two possible sets of requirements are defined, but the less demanding, at least from an economic perspective, are as follows:

- Portland cement with 21% to 35% FA or 36% to 65% GGBS
- A minimum cement content of 380 kg/m
- A maximum W/C ratio of 0.40
- A minimum cube strength of 40 N/mm²

It should be noted that carbonation and chloride ingress can occur simultaneously. This is most likely to be the case under XD1, XD3, XS1, and XS3 exposure classes discussed previously for chloride-induced corrosion, since these are likely to permit moisture levels in the concrete to be within the range where carbonation is most pronounced. However, the higher quality of concrete required for resistance to chloride ingress and the greater cover depths mean that carbonation is unlikely to present a major problem in such conditions.

BS 8500-1 requires higher cement contents for increasingly aggressive environments. Given the lesser role played by this parameter, the argument has been made for removing the requirement for a minimum cement content with regards to a number of aspects of durability including carbonation resistance [93].

Other options for protection against carbonation include many of those detailed for chloride-induced corrosion – corrosion-inhibiting admixtures, corrosion-resistant reinforcement, fibres and surface coatings. However, although hydrophobic impregnants are favoured for chloride-bearing environments, such coatings are of little value in carbonating conditions. What are required in the case of carbonation are coatings that are capable of blocking the pores at the concrete surface to prevent the movement of CO_2 molecules inwards. Many coatings and paints are marketed specifically as being anticarbonation and are typically based around elastomers that are capable of remaining intact despite cracking of the concrete underneath.

The most commonly encountered anticarbonation coatings contain acrylate elastomers, which have the additional benefit of allowing water vapour to pass through [94]. By allowing the concrete to 'breathe' in this way, problems with freeze–thaw attack can potentially be reduced.

REFERENCES

1. Hudson, K. *Building Materials*. London: Longman, 1972, pp. 535–554.
2. Curtis, K. E. and K. Mehta. A critical review of deterioration of concrete due to corrosion of reinforcing steel: Durability of concrete. *Proceedings of the 4th CANMET/ACI International Conference*, 1997.
3. Enevoldsen, J. N., C. M. Hansson, and B. B. Hope. The influence of internal relative humidity on the rate of corrosion of steel embedded in concrete and mortar. *Cement and Concrete Research*, v. 24, 1994, pp. 1373–1382.
4. Hussain, R. R. and T. Ishida. Influence of connectivity of concrete pores and associated diffusion of oxygen on corrosion of steel under high humidity. *Construction and Building Materials*, v. 24, 2010, pp. 1014–1019.
5. Lopez, W., J. A. Gonzalez, and C. Andrade. Influence of temperature on the service life of beams. *Cement and Concrete Research*, v. 23, 1993, pp. 1130–1140.
6. Chung, L., S.-H. Cho, J.-H. J. Kim, and S.-T. Yi. Correction factor suggestion for ACI development length provisions based on flexural testing of RC slabs with various levels of corroded reinforcing bars. *Engineering Structures*, v. 26, 2004, pp. 1013–1026.
7. British Standards Institution. *BE EN 1992-1-1:2004: Eurocode 2 – Design of concrete structures: Part 1-1. General rules and rules for buildings*. London: British Standards Institution, 2004, 230 pp.
8. British Standards Institution. *U.K. National Annex to Eurocode 2: Design of concrete structures – Part 1-1. General rules and rules for buildings*. London: British Standards Institution, 2005, 26 pp.
9. Torres-Acosta, A. A., S. Navarro-Gutierrez, and J. Terán-Guillén. Residual flexure capacity of corroded reinforced concrete beams. *Engineering Structures*, v. 29, 2007, pp. 1145–1152.
10. Everett, L. H. and K. W. J. Treadaway. Deterioration due to corrosion in reinforced concrete. Building Research Establishment Information Paper IP12/80. Garston, United Kingdom: Building Research Establishment, 1980, 4 pp.
11. Page, C. L., N. L. Short, and A. El Tarras. Diffusion of chloride ions in hardened cement pastes. *Cement and Concrete Research*, v. 11, 1981, pp. 395–406.
12. Moukwa, M. Penetration of chloride ions from seawater into mortars under different exposure conditions. *Cement and Concrete Research*, v. 19, 1989, pp. 894–904.

13. Francois, R. and J. C. Maso. Effect of damage in reinforced concrete on carbonation or chloride penetration. *Cement and Concrete Research*, v. 18, 1988, pp. 961–970.
14. Gérard, B. and J. Marchand. Influence of cracking on the diffusion properties of cement-based materials: Part 1. Influence of continuous cracks on the steady-state regime. *Cement and Concrete Research*, v. 30, 2000, pp. 37–43.
15. Chatterji, S. Transportation of ions through cement-based materials: Part 1. Fundamental equations and basic measurement techniques. *Cement and Concrete Research*, v. 24, 1994, pp. 907–912.
16. Edvardsen, C. Water permeability and autogenous healing of cracks in concrete. *ACI Materials Journal*, v. 96, 1999, pp. 448–454.
17. Reinhardt, H. -W. and M. Jooss. Permeability and self-healing of cracked concrete as a function of temperature and crack width. *Cement and Concrete Research*, v. 33, 2003, pp. 981–985.
18. Jacobsen, S., J. Marchand, and L. Boisvert, L. Effect of cracking and healing on chloride transport in OPC concrete. *Cement and Concrete Research*, v. 26, 1996, pp. 869–881.
19. Arya, C. and F. K. Ofori-Darko. Influence of crack frequency on reinforcement corrosion in concrete. *Cement and Concrete Research*, v. 26, 1996, pp. 345–353.
20. Slater, D. and B. N. Sharp. The design of coastal structures. In R. T. L. Allen, ed. *Concrete in Coastal Structures*. London: Thomas Telford, 1998, pp. 189–246.
21. Zhang, T. and O. E. Gjørv. Effect of ionic interaction in migration testing of chloride diffusivity in concrete. *Cement and Concrete Research*, v. 25, 1995, pp. 1535–1542.
22. Tang, L. Concentration dependence of diffusion and migration of chloride ions: Part 2. Experimental evaluations. *Cement and Concrete Research*, v. 29, 1999, pp. 1469–1474.
23. Chatterji, S. Transportation of ions through cement-based materials: Part 3. Experimental evidence for the basic equations and some important deductions. *Cement and Concrete Research*, v. 24, 1994, pp. 1229–1236.
24. Reinhardt, H.-W. and M. Jooss. Permeability and self-healing of cracked concrete as a function of temperature and crack width. *Cement and Concrete Research*, v. 33, 2003, pp. 981–985.
25. Buenfield, N. R. and J. B. Newman. The permeability of concrete in a marine environment. *Magazine of Concrete Research*, v. 36, 1984, pp. 67–80.
26. van der Wegen, G., J. Bijen, and R. van Selst. Behaviour of concrete affected by seawater under high pressure. *Materials and Structures*, v. 26, 1993, pp. 549–556.
27. Hall, C. Barrier performance of concrete: A review of fluid transport theory. *Materials and Structures*, v. 27, 1994, pp. 291–306.
28. Polder, R. B. and W. H. A. Peelen. Characterisation of chloride transport and reinforcement corrosion in concrete under cyclic wetting and drying by electrical resistivity. *Cement and Concrete Composites*, v. 24, 2002, pp. 427–435.
29. Hong, K. and R. D. Hooton. Effects of cyclic chloride exposure on penetration of concrete cover. *Cement and Concrete Research*, v. 29, 1999, pp. 1379–1386.

30. Birnin-Yauri, U. A. and F. P. Glasser. Friedel's salt, $Ca_2Al(OH)_6(Cl,OH).2H_2O$: Its solid solutions and their role in chloride binding. *Cement and Concrete Research*, v. 28, 1998, pp. 1713–1723.
31. Beaudoin, J. J., V. S. Ramachandran, and R. F. Feldman. Interaction of chloride and C-S-H. *Cement and Concrete Research*, v. 20, 1990, pp. 875–883.
32. Hirao, H., K. Yamada, H. Takahashi, and H. Zibara. Chloride binding of cement estimated by binding isotherms of hydrates. *Journal of Advanced Concrete Technology*, v. 3, 2005, pp. 77–84.
33. Rasheeduzzafar Hussain, S. E. and S. S. Al-Saadoun. Effect of cement composition on chloride binding and corrosion of reinforcing steel in concrete. *Cement and Concrete Research*, v. 21, 1991, pp. 777–794.
34. Tritthart, J. Chloride binding in cement: Part 2. The influence of the hydroxide concentration in the pore solution of hardened cement paste on chloride binding. *Cement and Concrete Research*, v. 19, 1989, pp. 683–691.
35. Suryavanshi, A. K. and R. N. Swamy. Stability of Friedel's salt in carbonated concrete structural elements. *Cement and Concrete Research*, v. 26, 1996, pp. 729–741.
36. Tumidajski, P. J. and G. W. Chan. Effect of sulphate and carbon dioxide on chloride diffusivity. *Cement and Concrete Research*, v. 26, 1996, pp. 551–556.
37. Lee, H., R. D. Cody, A. M. Cody, and P. G. Spry. Effects of various de-icing chemicals on pavement concrete deterioration. *Proceedings of the Mid-Continent Transportation Symposium*, 2000, pp. 151–155.
38. Sutter, L., K. Peterson, S. Touton, T. van Dam, and D. Johnston. Petrographic evidence of calcium oxychloride formation in mortars exposed to magnesium chloride solution. *Cement and Concrete Research*, v. 36, 2006, pp. 1533–1541.
39. Trolard, F., G. Bourrié, M. Abdelmoula, P. Refait, and F. Feder. Fougerite: A new mineral of the pyroaurite–iowaite group: Description and crystal structure. *Clays and Clay Minerals*, v. 55, 2007, pp. 323–334.
40. Sagoe-Crentsil, K. K. and F. P. Glasser. "Green rust," iron solubility, and the role of chloride in the corrosion of steel at high pH. *Cement and Concrete Research*, v. 23, 1993, pp. 785–791.
41. Hoar, T. P. and W. R. Jacob. Breakdown of passivity of stainless steel by halide ions. *Nature*, v. 216, 1967, pp. 1299–1301.
42. British Standards Institution. *BS 1881-124:1988: Testing Concrete – Part 124. Methods for Analysis of Hardened Concrete*. London: British Standards Institution, 1988, 24 pp.
43. Haque, M. N. and O. A. Kayyali. Determination of the free chloride in concrete is not that simple. *Transactions of the Institution of Engineers, Australia, Civil Engineering*, v. 37, 1995, pp. 141–148.
44. Alonso, M. C. and M. Sanchez. Analysis of the variability of chloride threshold values in the literature. *Materials and Corrosion*, v. 60, 2009, pp. 631–637.
45. Page, C. L., N. R. Short, and W. R. Holden. The influence of different cements on chloride-induced corrosion of reinforced steel. *Cement and Concrete Research*, v. 16, 1986, pp. 79–86.
46. Alonso, C., C. Andrade, M. Castellote, and P. Castro. Chloride threshold values to depassivate reinforcing bars embedded in standardized OPC mortar. *Cement and Concrete Research*, v. 30, 2000, pp. 1047–1055.

47. British Standards Institution. *BS EN 206: Concrete. Specification, Performance, Production, and Conformity*. London: British Standards Institution, 2013, 98 pp.
48. British Standards Institution. *BS 8500-1:2006: Concrete – Complementary British Standard to BS EN 206-1: Part 1. Method of Specifying and Guidance of the Specifier*. London: British Standards Institution, 2006, 66 pp.
49. Richardson, I. G. and G. W. Groves. The incorporation of minor and trace elements into calcium silicate hydrate (CSH) gel in hardened cement pastes. *Cement and Concrete Research*, v. 23, 1993, pp. 131–138.
50. Dhir, R. K., M. A. K. El-Mohr, and T. D. Dyer. Developing chloride-resisting concrete using PFA. *Cement and Concrete Research*, v. 27, 1997, pp. 1633–1639.
51. Dhir, R. K., M. A. K. El-Mohr, and T. D. Dyer. Chloride-binding in GGBS concrete. *Cement and Concrete Research*, v. 26, 1996, pp. 1767–1773.
52. Hussain, R. S. E. and A. S. Al-Gahtani. Pore solution composition and reinforcement corrosion characteristics of microsilica-blended cement concrete. *Cement and Concrete Research*, v. 21, 1991, pp. 1035–1048.
53. Tommaselli, M. A. G., N. A. Mariano, and S. E. Kuri. Effectiveness of corrosion inhibitors in saturated calcium hydroxide solutions acidified by acid rain components. *Construction and Building Materials*, v. 23, 2009, pp. 328–333.
54. Sagoe-Crentsil, K. K., F. P. Glasser, and V. T. Yilmas. Corrosion inhibitors for mild steel: Stannous tin (SnII) in ordinary Portland cement, v. 24, 1994, pp. 313–318.
55. Andrade, C., C. Alonso, M. Acha, and B. Malric. Preliminary testing of Na_2PO_3F as a curative corrosion inhibitor for steel reinforcements in concrete. *Cement and Concrete Research*, v. 22, 1992, pp. 869–881.
56. Sagoe-Crentsil, K. K., V. T. Yilmaz, and F. P. Glasser. Corrosion inhibition of steel in concrete by carboxylic acids. *Cement and Concrete Research*, v. 23, 1993, pp. 1380–1388.
57. Monticelli, C., A. Frignani, and G. Trabanelli. A study on corrosion inhibitors for concrete application. *Cement and Concrete Research*, v. 30, 2000, pp. 635–642.
58. Nmai, C. K. Multifunctional organic corrosion inhibitor. *Cement & Concrete Composites*, v. 26, 2004, pp. 199–207.
59. Hansson, C. M., L. Mammoliti, and B. B. Hope. Corrosion inhibitors in concrete: Part 1. The principles. *Cement and Concrete Research*, v. 28, 1998, pp. 1775–1781.
60. The Concrete Society. *Enhancing Reinforced Concrete Durability*. Technical Report Number 61. Camberley, United Kingdom: The Concrete Society, 2004, 208 pp.
61. British Standards Institution. *BS EN 10080:2005: Steel for the Reinforcement of Concrete – Weldable Reinforcing Steel: General*. London: British Standards Institution, 2005, 74 pp.
62. British Standards Institution. *BS 4449:2005+A2:2009: Steel for the Reinforcement of Concrete – Weldable Reinforcing Steel: Bar, Coil, and Decoiled Product – Specification*. London: British Standards Institution, 2005, 34 pp.
63. Building Research Establishment. *Stainless Steel as a Building Material*. BRE Digest 349. Garston, United Kingdom: Building Research Establishment, 1990, 28 pp.

64. British Standards Institution. *BS 6744:2001+A2:2009: Stainless Steel Bars for the Reinforcement of and Use in Concrete – Requirements and Test Methods.* London: British Standards Institution, 2005, 28 pp.
65. Nurnberger, U. Corrosion behaviour of welded stainless reinforced steel in concrete: Corrosion of reinforcement in concrete construction. In Page, C. L., P. Bamforth, and J. W. Figg, eds., *Proceedings of the 4th International Symposium,* Cambridge, United Kingdom, July 1–4, 1996, pp. 623–629.
66. The Concrete Society. *Guidance on the Use of Stainless Steel Reinforcement.* Technical Report 51. Slough, United Kingdom: The Concrete Society, 1998, 56 pp.
67. British Standards Institution. *BS 1217:2008: Cast Stone – Specification.* London: British Standards Institution, 2008, 14 pp.
68. Trejo, D., P. Monteiro, and G. Thomas. Mechanical properties and corrosion susceptibility of dual-phase steel in concrete. *Cement and Concrete Research,* v. 24, 1994, pp. 1245–1254.
69. Winslow, D. N. High-strength low-alloy, weathering, steel as reinforcement in the presence of chloride ions. *Cement and Concrete Research,* v. 16, 1986, pp. 491–494.
70. The Institution of Structural Engineers. *Interim Guidance on the Design of Reinforced Concrete Structures Using Fibre Composite Reinforcement.* London: The Institution of Structural Engineers, 1999, 116 pp.
71. Wang, Y. C., P. M. H. Wong, and V. Kodur. Mechanical properties of fibre-reinforced polymer reinforcing bars at elevated temperatures. *SFPE/ASCE Specialty Conference: Designing Structures for Fire.* Baltimore, Maryland, September 30–October 1, 2003, pp. 1–10.
72. Burke, C. R. and C. W. Dolan. Flexural design of prestressed concrete beams using FRP tendons. *PCI Journal,* v. 46, 2001, pp. 76–87.
73. Erdoğu, S., T. W. Bremner, and I. L. Kondratova. Accelerated testing of plain and epoxy-coated reinforcement in seawater and chloride solutions. *Cement and Concrete Research,* v. 31, 2001, pp. 861–867.
74. Yeomans, S. R. *Corrosion of the Zinc Alloy Coating in Galvanized Reinforced Concrete.* Corrosion/98 Paper 653. Houston, Texas: NACE, 1998, 10 pp.
75. The Highways Agency. *Design Manual for Roads and Bridges: BD 43/03 – The Impregnation of Reinforced and Prestressed Concrete Highway Structures Using Hydrophobic Pore-Lining Impregnants.* Norwich, United Kingdom: HMSO, 2003, 12 pp.
76. Grimmond, C. S. B., T. S. King, F. D. Cropley, D. J. Nowak, and C. Souch. Local-scale fluxes of carbon dioxide in urban environments: Methodological challenges and results from Chicago. *Environmental Pollution,* v. 116, 2002, pp. S243–S254.
77. Banks, H. J. Recent advances in the use of modified atmospheres for stored product pest control. *Proceedings of the 2nd International Working Conference on Stored Product Entomology,* 1978, pp. 198–217.
78. Anstice, D. J., C. L. Page, and M. M. Page. The pore solution phase of carbonated cement pastes. *Cement and Concrete Research,* v. 35, 2005, pp. 377–383.
79. Šauman, Z. Carbonization of porous concrete and its main binding components. *Cement and Concrete Research,* v. 1, 1971, pp. 645–662.

80. McPolin, D. O., P. A. M. Basheer, and A. E. Long. Carbonation and pH in mortars manufactured with supplementary cementitious materials. *Journal of Materials in Civil Engineering*, v. 21, 2009, pp. 217–225.

81. Kayyali, O. A. and M. N. Haque. Effect of carbonation on the chloride concentration in pore solution of mortars with and without fly ash. *Cement and Concrete Research*, v. 18, 1988, pp. 636–648.

82. Houst, Y. F. and F. H. Wittmann. Influence of porosity and water content on the diffusivity of CO_2 and O_2 through hydrated cement paste. *Cement and Concrete Research*, v. 24, 1994, pp. 1165–1176.

83. Liu, L., J. Ha, T. Hashida, and S. Teramura. Development of a CO_2 solidification method for recycling autoclaved lightweight concrete waste. *Journal of Material Science Letters*, v. 20, 2001, pp. 1791–1794.

84. Papadakis, V. G., C. G. Vayenas, and M. N. Fardis. Fundamental modelling and experimental investigation of concrete carbonation. *ACI Materials Journal*, v. 88, 1991, pp. 363–373.

85. Currie, R. J. *Carbonation Depths in Structural Quality Concrete*. Watford, United Kingdom: Building Research Establishment, 1986, 19 pp.

86. Malami, Ch., V. Kaloidas, G. Batis, and N. Kouloumbi. Carbonation and porosity of mortar specimens with pozzolanic and hydraulic cement admixtures. *Cement and Concrete Research*, v. 24, 1994, pp. 1444–1454.

87. De Schutter, G. Quantification of the influence of cracks in concrete structures on carbonation and chloride penetration. *Magazine of Concrete Research*, v. 51, 1999, pp. 427–435.

88. Dewaele, P. J., E. J. Reardon, and R. Dayal. Permeability and porosity changes associated with cement grout carbonation. *Cement and Concrete Research*, v. 21, 1991, pp. 441–454.

89. Powers, T. C. A hypothesis on carbonation shrinkage. *Journal of the Portland Cement Association Research & Development Laboratories*, v. 4, 1962, pp. 40–50.

90. Swenson, E. G. and P. J. Sereda. Mechanism of the carbonation shrinkage of lime and hydrated cement. *Journal of Applied Chemistry*, v. 18, 1968, pp. 111–117.

91. Jerga, J. Physicomechanical properties of carbonated concrete. *Construction and Building Materials*, v. 18, 2004, pp. 645–652.

92. De Ceukelaire, L. and D. van Nieuwenburg. Accelerated carbonation of a blast-furnace cement concrete. *Cement and Concrete Research*, v. 23, 1993, pp. 442–452.

93. Dhir, R. K., P. A. J. Tittle, and M. J. McCarthy. Role of cement content in specifications for durability of concrete – A review. *Concrete*, v. 34, 2000, pp. 68–76.

94. Zafeiropoulou, T., E. Rakanta, and G. Batis. Performance evaluation of organic coatings against corrosion in reinforced cement mortars. *Progress in Organic Coatings*, v. 72, 2011, pp. 175–180.

Chapter 5

Specification and design of durable concrete

5.1 INTRODUCTION

In previous chapters, mechanisms of the deterioration of concrete and approaches to controlling or avoiding problems associated with deterioration have been examined. Although this is useful, it is hopefully evident that it is quite possible that a concrete structure will be threatened simultaneously by more than one type of deterioration mechanism. Moreover, the design of concrete does not solely revolve around durability – other characteristics will be required, such as compressive strength, to satisfy the structural and other functional requirements of a concrete element. The process of defining what characteristics concrete must have for a particular application is 'specification', and this chapter aims to examine the approach taken in specifying concrete and how the specification for durability fits into this process.

However, before this can be done, it is useful to look first at the manner in which substances may pass through the surface of concrete and move into its interior, since these processes are fundamental to how many aspects of specification for durability have been developed. Additionally, because specification of concrete is focussed very much on the selection and proportioning of constituent materials, an overview of these materials is also presented.

From the specification, a concrete mix can be designed. Although this book does not provide a detailed methodology for mix design, it outlines the general process and how achieving the requirements for durability fits into it.

5.2 CONCRETE AS A PERMEABLE MEDIUM

Concrete is made from particles whose distribution of sizes covers a wide range. Once mixed, compacted and hardened, the product appears to be a single monolithic article. However, much of the original structure of this collection of particles remains despite the transformation it has undergone. Specifically, the space between particles remains. This space takes the form of an interconnected network of pores, which, to some extent, permits substances to enter the concrete and move through its volume. Moreover, as we have seen in Chapter 2, stresses that the concrete experiences have the potential to cause the formation of cracks, which will also act as a route beneath the concrete surface. This section will examine the permeable nature of concrete, the mechanisms that cause substances to ingress into concrete and the factors that influence the rate of ingress.

5.2.1 Porosity

Concrete is a porous material – in a dry state, a proportion of its interior is occupied by an empty space. The majority of the porosity is, under normal circumstances, in the cement matrix. This porosity can be divided in terms of dimensions into two main categories: capillary porosity and gel porosity.

Capillary porosity is the space left between cement particles and typically has dimensions around the order of 1 μm in diameter. The capillary porosity is defined mainly by the water/cement (W/C) ratio of the concrete mix, the degree of cement hydration and the particle size distribution of the cement. The greater the volume of water relative to that of cement, the larger the space between cement particles, as illustrated in Figure 5.1.

The reaction of cement with water causes the formation of hydration products at the cement grain surfaces while the diameter of the unreacted grains decreases. These hydration products occupy a larger volume than the unreacted cement and, as a consequence, the formation of hydration products acts to reduce the capillary porosity (also shown in Figure 5.1).

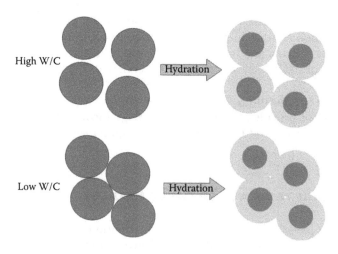

Figure 5.1 Influence of W/C ratio on capillary porosity.

Thus, a low W/C ratio and a high degree of cement hydration will lead to lower volumes of total porosity and smaller pore diameters.

Particle size also plays an important role, since larger particles will have greater volumes of capillary porosity and pore diameters, even when they are closely packed, as in Figure 5.2. This figure also illustrates the importance of particle size distribution, since the presence of finer particles in the spaces between larger particles acts to 'refine' the porosity.

The main product of cement hydration, calcium silicate hydrate (CSH) gel, contains pores with diameters mainly in the 1- to 100-nm range. Thus, as cement hydration proceeds, the proportion of capillary porosity is reduced, whereas gel porosity increases.

Understanding the nature of gel porosity – and, indeed, capillary porosity – is complicated by the problem of measurement – different techniques for measuring pore size distribution provide somewhat different results, and there has been much debate as to what is really being measured. The results of measurement with one of the most common techniques – mercury porosimetry – are shown in Figure 5.3, which shows peaks in the pore size distribution, corresponding to capillary and gel porosities.

Another factor that influences the porosity of concrete is the nature of the interface between the cement paste and the aggregate particles. Typically, this region, known as the 'interfacial transition zone' or ITZ, has a higher porosity than the bulk cement paste. This is the result of the wall effect, illustrated in Figure 5.4, where cement particles located against the surface of an aggregate particle are packed less efficiently than elsewhere, leading to a higher volume of capillary porosity and larger pore diameters in this

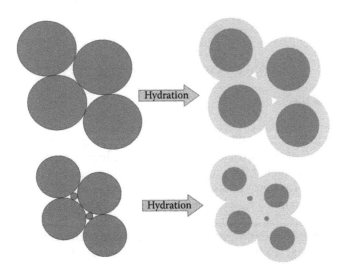

Figure 5.2 Influence of particle size on capillary porosity.

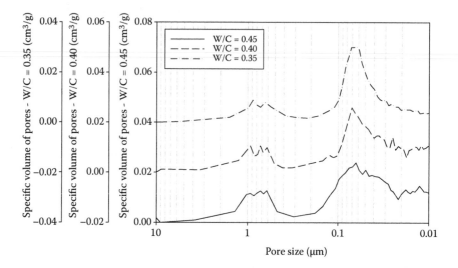

Figure 5.3 Pore size distribution of three concrete mixes of different W/C ratios.

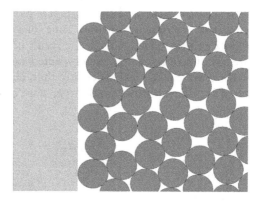

Figure 5.4 Wall effect.

region (Figure 5.5). This illustration shows a very simplified example of the effect, since all of the particles are of equal size. In reality, another feature of the wall effect is that finer particles are usually present in larger proportions in close proximity to the wall. The repercussions of the wall effect are shown in Figure 5.6, which plots pore size distributions for concrete containing increasing quantities of aggregate.

Although the ITZ certainly makes a contribution to the permeation properties of concrete, studies of its overall influence compared to that from the porosity in the cement in general and from microcracks indicate that the contribution is relatively small [1].

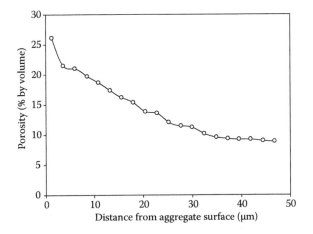

Figure 5.5 Volume of cement matrix porosity versus distance from an aggregate particle in a 1-year-old concrete specimen. (From Scrivener, K. L. and K. M. Nemati, *Cement and Concrete Research*, 26, 1996, pp. 35–40.)

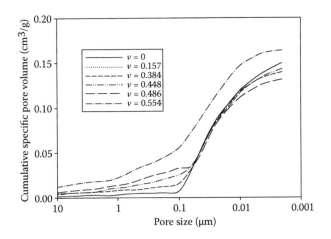

Figure 5.6 Cumulative pore size distribution plot obtained using mercury porosimetry on mortars containing a range of volume fractions (v) of aggregate. (From Winslow, D. N. and D. C. Cohen, *Cement and Concrete Research*, 24, 1994, pp. 25–37.)

5.2.2 Cracks

Concrete is a brittle material, which means that it is prone to cracking. Furthermore, it is heterogeneous, being made up of a mixture of different solid materials with pores that may contain either air or liquid. The heterogeneous nature of concrete means that there are potentially points in the material that are considerably weaker than others, and it is from these points that crack growth usually initiates.

Cracks are normally classified in terms of whether they are visible to the naked eye (macrocracks) or only visible under a microscope (micro-cracks). This system is, in some respect, less than ideal – cracks will be present over a continuous range of widths but will essentially all behave in a similar fashion, making this means of classification some-what arbitrary. Additionally, depending on how they are formed, many cracks in concrete will have a wide end and taper to a narrower crack width, or to nothing. However, the distinction is useful in describing how cracks tend to propagate in concrete – microcracks will be present in a volume of concrete from the point at which the cement has set. The application of a sufficiently high stress will cause more microcracks to form and existing microcracks to grow. Thus, microcracks are future macrocracks. Some of the causes of cracking have been discussed in Chapters 2, 3, and 4.

From a durability perspective, cracks are important for two main rea-sons. First, the presence of cracks will permit substances detrimental to the concrete or its reinforcement to ingress at a faster rate than would other-wise be the case. This is discussed in further detail in the subsequent sec-tions. Second, the appearance of macrocracks of sufficient width marks the end of the serviceability of a structure. This event is discussed in further detail in Chapter 7.

5.2.3 Absorption

When a narrow tube (capillary) is brought into contact with the surface of a liquid, intermolecular forces between the two substances will act to draw the liquid up into it. The liquid rises in the capillary until it reaches a height where a balance is reached between the intermolecular forces and the gravitational force from the mass of the liquid:

$$h = \frac{2\gamma \cos\theta_C}{\rho g r}$$

where h is the height of the liquid (m), γ is the air–liquid surface tension (N/m), θ_C is the contact angle between the liquid and the capillary wall material, ρ is the density of the liquid (g/m^3), g is the gravitational field strength (N/g) or the acceleration caused by gravity (m/s^2) and r is the capil-lary radius (m).

This process is referred to as 'capillarity' or 'capillary action'. The hardened cement matrix of concrete (and aggregate, where it is porous) is more complex than the idealised tube described above, since the pores are highly interconnected, and of variable surface composition and radius. Nonetheless, capillary action will cause concrete to absorb water in the manner described by the equation

$$V = AS\sqrt{t}$$

where V is the cumulative volume of water absorbed (m³) at time t (s), A is the area that is wetted (m³) and S is the sorptivity of the concrete (m/s$^{0.5}$).

5.2.4 Flow

Where a fluid (usually water) pressure gradient exists across a depth of concrete, flow will occur. Mass transport by flow is described by Darcy's law:

$$Q = kA\left[\frac{(p_1 - p_2)}{\mu L}\right]$$

where Q is the total discharge through the concrete (m³/s), k is the permeability (m²), A is the cross-sectional area through which flow is occurring (m²), $p_1 - p_2$ is the pressure difference (Pa), μ is the viscosity of the fluid (Pa s) and L is the distance over which the pressure difference exists (m).

The rate of flow of a fluid into concrete under a given pressure difference is dependent on the permeability of the material. The permeability, whose units are meters squared, can be defined in terms of the pore structure of the material using the equation

$$k = \frac{r_e P_e^2}{S_e}$$

where r_e is the effective pore radius of the material (m), P_e is the effective porosity and S_e is the effective internal surface area of a porous solid (m^{-1}) [4].

'Effective' porosity in this context means porosity that is able to contribute to fluid flow. Clearly, the equation represents an oversimplification of the situation in concrete, where porosity is present over a range of pore sizes.

Where a crack is present, the crack's permeability, k_{crack}, can be described using the equation

$$k_{crack} = \xi l w$$

where ξ is a parameter of roughness, l is the crack length at the surface (m) and w is the crack width (m) [5].

Thus, as a crack becomes wider, crack permeability increases. Crack permeability is somewhat different to bulk permeability, since in the equation for flow rate, A is equal to lw. Thus, flow through concrete is significantly influenced by the presence of a crack.

5.2.5 Diffusion

Diffusion occurs when a concentration gradient exists for a given substance between the interior of concrete and the external environment. The process of diffusion is described by Fick's first law:

$$-j = D\frac{dc}{dz}$$

where j is the flux per unit area (g m^{-2} s^{-1}), D is the coefficient of diffusion (m^2 s^{-1}) and dc is the concentration difference (g/L) across a distance dz (m).

In reality, the situation for concrete is not quite as simple, since many substances of concern from a durability perspective interact with the concrete constituents. Moreover, these interactions may cause changes to the porosity, which, in turn, may cause changes in the concrete microstructure (cracking, precipitation of reaction products, etc.). These changes will produce a change in the coefficient of diffusion with time.

In the absence of cracks, the diffusion coefficient of a substance into concrete is very much dependent on the nature of its porosity. The effective diffusion coefficient of a substance through a porous medium can be described using the equation

$$D_e = \varepsilon \frac{\delta}{\tau^2} D^*$$

where D_e is the effective diffusion coefficient (m^2/s), ε is the volume fraction of porosity, δ is the constrictivity, τ is the tortuosity and D^* is the self-diffusion coefficient of the substance (m^2/s) [6].

The constrictivity term is necessary in the above equation because the diffusion of an ion will be slowed where the pores in a porous solid become narrower at points along their length. This is an inevitable situation in concrete, where the pores formed between cement grains will fluctuate in diameter significantly. Constrictivity can be described using the equation

$$\delta = \frac{(A_{max} A_{min})^{1/2}}{A_{mean}}$$

where A_{max} is the maximum cross-sectional area of a pore (m²), A_{min} is the minimum cross-sectional area of a pore (m²) and A_{mean} is the mean cross-sectional area of a pore (m²) [7].

As the maximum pore size approaches the minimum pore size, the constrictivity increases, leading to higher coefficients of diffusion.

Tortuosity refers to the extent to which the pathway of interconnected pores through the cement matrix of concrete between two points deviates from a straight line. It is described by the equation

$$\tau = \frac{l_e}{l}$$

where l_e is the average pore length running from one side of a volume of material to the other (m), and l is the distance across the volume of material (m) [7].

In reality, this means of expressing tortuosity is very much oversimplified, since the equation treats pores as individual channels running through a material, whereas pores in hardened cement paste are in fact interconnected networks composed of the interstices between hydrated cement grains. Tortuosity is largely controlled by the original particle size distribution of the materials that make up the cement matrix and the nature of the cement hydration products formed. Generally, the presence of finer material will lead to a greater tortuosity.

5.3 CEMENT

The principal properties of the materials used as constituents of the cement fraction of concrete which influence durability are their chemical composition, particle size distribution, and reaction kinetics, and the contribution towards strength that these reactions produce. The emphasis placed in the Eurocodes and related standards is on the combinations of cementitious materials in which Portland cement (PC) clinker is universally present. This reflects the fact that, although other hydraulic cements (such as calcium aluminate and calcium sulfoaluminate cements) are available, the vast majority of concrete construction in Europe is conducted by using PC clinker as the fundamental cementitious component. For this reason, PC and materials that can be used in conjunction with PC clinker are exclusively covered in this section.

Cement for concrete is covered by *BS EN 197-1* [8]. The standard defines 27 common cements, which are grouped into five main types:

1. CEM I
   - PC
2. CEM II
   - Portland–slag cement: PC clinker and between 6% and 35% ground granulated blast-furnace slag (GGBS)
   - Portland–silica fume (SF) cement: PC clinker and between 6% and 10% SF
   - Portland–pozzolana cement: PC clinker and between 6% and 25% of either a natural or a natural calcined pozzolana
   - Portland–fly ash (FA) cement: PC clinker and between 6% and 35% of either a siliceous or a calcareous FA
   - Portland–burnt shale cement: PC clinker and between 6% and 35% of burnt shale
   - Portland–limestone cement: PC clinker and between 6% and 35% limestone
   - Portland–composite cement: PC clinker and a multiple combination of any of the materials mentioned above totalling between 6% and 35%
3. CEM III
   - Blast-furnace cement: PC clinker and between 36% and 95% GGBS
4. CEM IV
   - Pozzolanic cement: PC clinker and between 11% and 55% of a multiple combination of SF, natural pozzolana, natural calcined pozzolana, siliceous FA and calcareous FA
5. CEM V
   - Composite cement: PC clinker and between 18% and 50% of a multiple combination of natural pozzolana, natural calcined pozzolana and siliceous FA

The CEM II cements are further subdivided into products that are described using a naming system with the general form CEM II/W–X Y Z. W is the letter A, B, and, in the case of slag cement, C, which indicate the extent to which the non–PC component is present, with A denoting a lower content. X is a letter indicating what the non–Portland constituent is

S	GGBS
D	SF
P	natural pozzolana
Q	natural calcined pozzolana

V	siliceous FA
C	calcareous FA
L and LL	limestone
M	composite

Y is the strength class of the cement and can be 32.5, 42.5 or 52.5, where the number represents the minimum strength of the cement in newtons per square millimetre (when tested in accordance with *BS EN 196-1* [9]) at 28 days. Z denotes whether the cement is normal early strength (N) or high early strength (R).

In the case of CEM III, IV and V, the naming system simply takes the form CEM III/W Y Z.

Where multiple combinations of materials other than PC clinker are present (i.e., Portland composite, pozzolanic cements, and composite cements), a list of these materials in the form of their allocated letters (e.g., S-V-L) must be provided before Y in the name.

The cements described above are products that can be purchased from the manufacturer. However, such combinations of materials can also be produced at the concrete production stage, simply by combining them in the mixer. This approach is covered by *BS EN 8500-2* [10] and covers combinations including FA, GGBS, and limestone.

In the following sections, PC (CEM I) and a number of the other cementitious materials most commonly encountered in the United Kingdom are discussed.

5.3.1 Portland cement

PC is manufactured by burning a powdered mixture of limestone and clay or shale in a rotary kiln where temperatures normally reach up to 1450°C. The resulting clinker is ground with a small quantity of gypsum (or another form of calcium sulphate) and possibly an even smaller quantity of a minor additional constituent, such as limestone.

BS EN 197-1 requires that, for the product to be considered PC, two-thirds of the clinker (by mass) must comprise a combination of tricalcium silicate ($3CaO \cdot SiO_2$) and dicalcium silicate ($2CaO \cdot SiO_2$), which are referred to, in the shorthand of the cement chemist, as C_3S and C_2S, respectively. Additionally, the ratio of CaO to SiO_2 must be greater than 2.0. The typical chemical composition of PC is indicated on a SiO_2–Al_2O_3–CaO ternary diagram in Figure 5.7.

In addition to C_3S and C_2S, the clinker also contains tricalcium aluminate ($3CaO \cdot Al_2O_3$) and tetracalcium aluminoferrite ($4CaO \cdot Al_2O_3 \cdot Fe_2O_3$). In shorthand terms, tricalcium aluminate and tetracalcium aluminoferrite are referred to as C_3A and C_4AF, respectively.

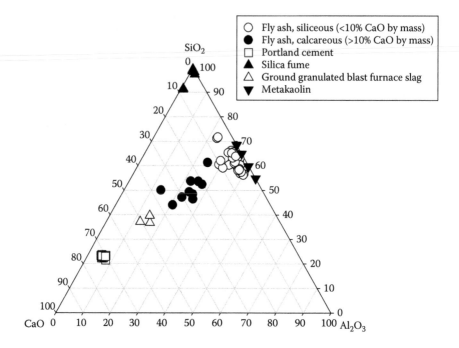

Figure 5.7 Ternary diagram showing the results of the chemical analyses of various cement components. (From Hubbard, F. H. et al., *Cement and Concrete Research*, 15, 1985, 185–198; Shehata, M. H. and M. D. A. Thomas, *Cement and Concrete Research*, 36, 2006, 1166–1175; de Larrard, F. et al., *Materials and Structures*, 25, 1992, 265–272; Buchwald, A. et al., *Journal of Materials Science*, 42, 2007, 3024–3032; Velosa, A. L. et al., *Acta Geodynamica et Geomaterialia*, 6, 2009, 121–126; Hobbs, D. W., *Magazine of Concrete Research*, 34, 1982, 83–94; Dhir, R. K. et al., *Development of a Technology Transfer Programme for the Use of PFA to EN 450 in Structural Concrete. Technical Report CTU/1400*. London: Department of Environment, 2000; plus the author's own data.)

PC is hydraulic, which means that its constituent compounds (or 'phases') undergo a reaction with water to produce products that act to bind the solid constituents of concrete together. These hydration reactions are outlined in Figure 5.8, with a graph indicating the typical quantities of the products that develop over a 28-day period shown in Figure 5.9.

The calcium silicate phases undergo a reaction with water to produce CSH gel and calcium hydroxide ($Ca(OH)_2$). CSH gel makes up the largest proportion of mature hardened PC (~50%–70% by mass in a mature paste) and makes the largest contribution to its strength and stiffness. Typically, the CSH gel forms around the original cement grain, extending outwards into the space between each cement grain. CSH gel's chemical composition is somewhat variable. The ratio of Ca to Si can vary considerably – between

$$\left.\begin{array}{l} C_3S \\ C_2S \end{array}\right\} + \text{water} \longrightarrow \text{CSH gel} + \text{calcium hydroxide}$$

$$\left.\begin{array}{l} C_3A \\ C_4AF \end{array}\right\} + \text{water} + \text{gypsum} \longrightarrow \text{AFt}$$

$$\text{AFt} + C_3A + C_4AF \longrightarrow \text{AFm}$$

Figure 5.8 Outline of the reactions of the four PC clinker phases with water (CSH = calcium silicate hydrate).

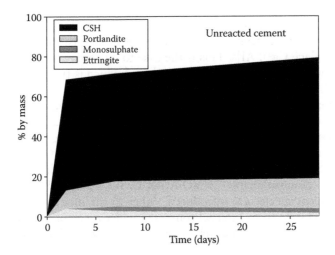

Figure 5.9 Development of cement hydration products in a PC paste over a 28-day period (CSH = calcium silicate hydrate). (From Dyer, T. D. et al., *Journal of Materials in Civil Engineering*, 23, 2011, 648–655.)

approximately 0.6 and 1.7 [11]. This parameter plays a role in defining some aspects of chemical durability (see Chapters 3 and 4). Typically, the Ca/Si ratio of the majority of CSH gel in a mature PC, which has not undergone chemical interactions with other materials, will be between 1.0 and 1.6, although there will be local variation [12].

$Ca(OH)_2$ also plays important roles in the durability of concrete – it is the source of hydroxide ions that lead to the potentially high pH in concrete pore fluids. This has relevance in a number of areas of chemical durability, particularly the corrosion of steel (Chapter 4) and alkali-silica reaction (Chapter 3). Although calcium hydroxide is only slightly soluble in water, it acts as a source of calcium during sulphate attack (Chapter 3). $Ca(OH)_2$ is normally present at approximately 20% to 25% by mass in mature Portland cement pastes that have not undergone chemical reactions with external substances.

C_3A and C_4AF, in the presence of gypsum, react with water to form the AFt phase. AFt stands for aluminoferrite-tri and has the general chemical formula $[Ca_3(Al,Fe)(OH)_6 \cdot 12H_2O]_2 \cdot X_3 \cdot xH_2O$. X_3 is the constituent that lends the 'tri' term to the name and can be CO_3^{2-}, $H_2SiO_4^{2-}$ or SO_4^{2-}, although in PC hydrating in the absence of external chemical influences, the sulphate supplied by gypsum means that ettringite ($[Ca_3(Al,Fe)(OH)_6 \cdot 12H_2O]_2 \cdot 3CaSO_4 \cdot 7H_2O$ or $3CaO \cdot (Al,Fe)_2O_3(CaSO_4)_3 \cdot 32H_2O$) is the AFt phase formed. The formation of ettringite in mature concrete is associated with sulphate attack, which can present a severe threat to concrete durability (see Chapter 3). However, its formation in the early life of concrete is desirable and does not present any problems.

Other ions that comprise the AFt phase may be replaced by others. Al^{3+} and Fe^{3+} can be replaced by chemically similar metals. Furthermore, silicon can replace Ca^{2+}, Al^{3+}. This is the case for thaumasite, which can also play an important role in sulphate attack and which is further discussed in Chapter 3.

Once the source of the X component or components is exhausted and cement hydration continues, the AFt phase begins to convert to aluminoferrite-mono (AFm) phases. This has the general formula $[Ca_2(Al,Fe)(OH)_6] \cdot X \cdot xH_2O$. Here, X can be a wider range of constituents, including $1/2SO_4^{2-}$, $1/2CO_3^{2-}$, Cl^- and $1/2H_2SiO_4^{2-}$. During the 'normal' process of cement hydration, ettringite converts to monosulphate – $3CaO \cdot Al_2O_3 \cdot CaSO_4 \cdot 12H_2O$. However, where other chemical species are introduced, either in the mix constituents or in the hardened state from external sources, a wider range of AFm phases will be encountered. Of particular relevance to concrete durability is Friedel's salt

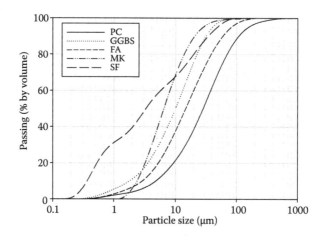

Figure 5.10 Particle size distributions of PC, GGBS, FA, MK and SF.

([Ca$_2$(Al,Fe)(OH)$_6$]·Cl·xH$_2$O or 3CaO·Al$_2$O$_3$·CaCl$_2$·10H$_2$O), which acts to immobilise chloride ions in concrete (Chapter 4). Typically, the total mass of the AFt and AFm phases present will not be more than 10% to 15% in a mature PC paste.

The particle size of PC will depend partly on the strength class, since a finer particle size distribution will produce a faster rate of reaction. However, cement particle distributions typically have a mean particle size of approximately 20 μm. An example of a PC particle size distribution is shown in Figure 5.10.

5.3.2 Ground granulated blast-furnace slag

Blast-furnace slag is a coproduct of the smelting of iron ore to produce iron metal. It results from the addition of limestone to the blast furnace to remove silicon, magnesium and aluminium impurities present in the ore. The slag floats on top of the molten iron and can be tapped off separately. When this is done, it is typically rapidly cooled by spraying it with water. This produces glass granules, the subsequent grinding of which produces GGBS, which can be used as a cement component in concrete.

BS EN 197-1 requires that two-thirds of the material be glassy. In chemical terms, it also requires that the slag has a CaO, MgO and SiO$_2$ content, which, in total, comprises at least two-thirds of the material by mass and that the ratio (CaO + MgO)/SiO$_2$ is greater than one. Typical chemical compositions of GGBS are shown in Figure 5.7.

GGBS undergoes what is known as a 'latent hydraulic reaction'. This means that it will react with water to produce both CSH gel and the AFm and AFt phases discussed in Section 5.3.1, but the reaction is said to be latent because it will not initiate without exposure to high-pH conditions – 'activation'. Thus, the presence of PC is required to provide these conditions. The reaction of GGBS is slower than that of PC – typically, it does not reach a significant rate until several days after mixing. This means that the strength development of concrete containing slag is typically slower than that of a mix containing PC as the only cement component. However, at later ages (>28 days), the strength will typically be comparable with that of PC.

In a mature cement paste composed of PC and GGBS, the quantity of CSH gel is typically higher than for PC alone, with a lower average Ca/Si ratio. Because PC is the sole source of Ca(OH)$_2$, the quantity of this compound will be lower as a result of the 'dilution' of the PC by GGBS. Additionally there will be somewhat higher quantities of AFm and possibly AFt phases.

The requirements of GGBS are covered by *BS EN 15167-1* [21]. In terms of material properties, the standard defines the maximum levels of certain chemical constituents, including chlorides and sulphates, as well as the requirements for setting time and 'activity index'. The activity index is a measure of how reactive the slag is and is measured by determining the 7- and 28-day strengths a mortar (prepared and tested in accordance with *BS EN 196-1*) containing a 50%:50% by mass mixture of GGBS and PC, and another containing 100% PC as the cement fraction, both with a W/C ratio of 0.5. The activity index is the ratio of the strength of the GGBS–PC mortar to that of the PC mortar expressed as a percentage and is required to be 45% at 7 days and 70% at 28 days.

GGBS is usually somewhat finer than PC, as illustrated in Figure 5.10.

5.3.3 Fly ash

Coal burnt in the furnaces of coal-fired power stations contains impurities of inorganic minerals, such as clays. The temperatures involved are sufficiently high to melt these clay particles, which then solidify as they are rapidly cooled with the exhaust gases of the plant. The release of the resulting fine hollow, spherical particles into the atmosphere is not acceptable from an environmental perspective, and so pollution control technologies, such as bag filters, cyclones and electrostatic precipitators, are used to capture this material – FA.

The spherical FA particles ('cenospheres') consist of a mixture of aluminosilicate glass and crystalline minerals such as mullite and quartz. Additionally, particles of crystalline minerals may exist alongside the spheres. The chemical composition of FA is indicated on the ternary diagram in Figure 5.7. The plot indicates that the material can be subdivided into two subcategories, based on calcium content. Calcareous FA is ash containing more than 10% CaO and results from burning lignite ('brown coal') or subbituminous coals. Siliceous FA contains less than 10% CaO and results from burning bituminous coal and anthracite. In the United Kingdom, siliceous FA is the norm, although calcareous ashes are commonly encountered elsewhere, including North America and several European countries.

5.3.3.1 Siliceous FA

It is generally accepted that the glassy part of FA is the constituent with cementitious properties. In the case of siliceous FA, this results from a pozzolanic reaction – a reaction with $Ca(OH)_2$ produced during the hydration of PC. The reaction is slower than that of GGBS, although, again, at later ages, strengths can potentially match those of PC for a given W/C ratio. As for GGBS, the reaction products are CSH gel and calcium aluminate hydrates. A mature cement paste containing PC and FA will contain more CSH gel (with a lower average Ca/Si ratio) and higher quantities of

AFm. The $Ca(OH)_2$ content will be lower, both as a result of the dilution of PC and because $Ca(OH)_2$ is consumed in the pozzolanic reaction.

BS EN 450-1 [22] covers the requirements of siliceous FA for use in concrete. The standard defines the limits for material characteristics. Many of the limits placed on chemical composition (Table 5.1) relate to durability: Cl content is limited to avoid issues relating to reinforcement corrosion (Chapter 4); SO_3 content is limited to avoid sulphate attack; and alkali limits are partly set to avoid problems with alkali-silica reaction (Chapter 3).

The standard also places limits on various other characteristics, namely, fineness, activity index, soundness, water requirement and loss on ignition. In the case of FA, fineness is measured in terms of the mass of ash retained on a 45-μm mesh when wet sieved. The standard defines two categories of ash: Category N (fineness ≤ 40% by mass) and Category S (≤12% by mass). A particle size distribution for a Category N ash is shown in Figure 5.10.

The activity index is determined in the same manner as for GGBS, except that a combination of 75% PC:25% FA is used. The limits defined by the standard are 75% and 85% of the reference strength at 28 and 90 days, respectively.

One feature of FA that makes its lower strength at earlier ages tolerable is its beneficial influence on workability. To achieve the same workability characteristics as concrete containing only PC as a cement component, FA requires less water, thus allowing for a reduction in the W/C ratio, with a consequent improvement in strength. For this reason, BS EN 450-1 defines a water requirement for Category S ash. This is determined by establishing the water required to achieve the same flow as a PC-only reference mortar (measured using a flow table in accordance with BS EN 1015-3) in a mortar containing 70% PC:30% FA by mass as the cement components. The standard requires that ≤95% by mass of water is required relative to the reference mortar.

The carbon content of FA limits the benefits to workability. For this reason, the standard also limits the loss on ignition (mass loss after heating at

Table 5.1 Limits on chemical composition for FA in BS EN 450-1

Constituent	Limit (% by mass)	Method
Cl	≤0.10	BS EN 196-21
SO_3	≤3.0	BS EN 196-2
Free CaO	≤2.5	BS EN 451-1
Reactive CaO	≤10.0	BS EN 197-1
Reactive SiO_2	≥25	BS EN 197-1
$SiO_2 + Al_2O_3 + Fe_2O_3$	≥70	BS EN 196-2
Total alkalis	≤5.0 Na_2O equivalent	BS EN 196-21
Magnesium oxide	≤4.0	BS EN 196-2
Soluble phosphate	≤0.01	BS EN 450-1

975°C for 1 h) to less than 9.0% by mass, and in the United Kingdom, ash must have a loss-on-ignition value of less than 7.0% to be suitable for use in concrete.

BS EN 450-1 has been revised to include the more recent practice of co-combustion – the burning of waste products – specifically, biomass, animal meal, sewage and paper sludge, petroleum coke and liquid and gas waste fuels. The main aim of this aspect of the update has been to ensure that the composition of FA derived from plants where co-combustion is practiced does not deviate too far from the compositions of purely coal-derived ash. The standard limits the contribution to the total mass of ash coming from the co-combustion fuels to 20% by mass and that liquid and gas fuels (which make no contribution to the final mass of ash) make up only 20% of the power plant's total fuel mix on a calorific basis.

5.3.3.2 Calcareous FA

The requirements of calcareous FA are covered under *BS EN 197-1*. It defines the ash as having a reactive CaO content of ≥10.0% by mass and a SiO_2 content of ≥25.0% when the reactive CaO content lies between 10.0% and 15.0%.

The cementitious reactions of calcareous FA are somewhere between that of a pozzolana, such as FA, and a hydraulic material. For this reason, the standard requires that if the CaO content of ash exceeds 15.0%, it should be capable of achieving a strength of 10.0 N/mm² by 28 days when used as the sole cement component in a mortar prepared in accordance with *BS EN 196-1*. The products of hydration will be similar with those of siliceous FA and GGBS.

5.3.4 Silica fume

SF is a by-product of the manufacture of silicon, for the electronics industry, and ferrosilicon alloys, which are used in the iron and steel industries. The manufacturing process involves the reduction of quartz (SiO_2) using a source of carbon, such as petroleum coke or charcoal, in a furnace. A proportion of the raw material is volatilised in the furnace as SiO, which is oxidised in contact with air to form fine particles of amorphous SiO_2. These particles are captured by pollution control systems.

The composition of SF is very close to pure SiO_2 (Figure 5.7). SF undergoes a pozzolanic reaction like FA. However, the absence of Al_2O_3 means that the product of the reaction is exclusively CSH gel, with an average Ca/Si ratio lower than that in a mature, hardened PC.

The particle size of SF is extremely fine, with mean particle sizes typically in the 100 to 1000 nm range (Figure 5.10).

SF for use in concrete is dealt with by *BS EN 13263-1* [23]. In chemical terms, it requires that the material contains ≥85% by mass of SiO_2. It also places limits on elemental Si (Si that is not chemically combined with other elements), free calcium oxide, sulphate, loss on ignition and chloride. Total alkalis must be declared to allow the total alkali content of concrete to be determined where alkali–aggregate reactions are a concern.

There are also requirements in terms of fineness (expressed in terms of specific surface) and activity index. SF is typically highly active and, consequently, the activity index requirements are that mortar bars made with 90% PC and 10% SF as the cement fraction should achieve at least the same strength as the 100% PC reference mix by an age of 28 days.

5.3.5 Limestone

Limestone powder can be obtained as a by-product from limestone processing, ground specifically for use as a cement component or interground with PC clinker. The particle size distribution obtained depends on the processing used and on the requirements of the application, but mean particle sizes a little below those of PC are typical.

Limestone consists largely of $CaCO_3$ but may contain other minerals. Limestone is sometimes referred to as an inert filler, which means that it does not undergo any cementitious reactions. Although this is essentially true – it does not make any contribution to the development of strength through chemical reaction – its presence leads to the formation of AFm phases containing carbonate ions – monocarbonate and hemicarbonate – instead of monosulphate.

The requirements placed on limestone in *BS EN 197-1* are that it should contain at least 75% calcium carbonate, less than 1.20% clay and less than 0.50% organic carbon. Two categories of limestone, dependent on the organic carbon content, are defined: LL ($\leq$0.20%) and L ($\leq$0.50%).

5.3.6 Pozzolanas

Pozzolanas are subdivided into two subcategories: natural pozzolanas and natural calcined pozzolanas. Natural pozzolanas are materials that undergo reactions similar to those undergone by FA and SF, but which have obtained these characteristics through natural processes. Typically, such materials originate from volcanic processes and may be volcanic glasses or materials deriving from the action of weathering or hydrothermal processes on these materials in such a manner as to lead to the formation of zeolites. Another example of a natural pozzolana is diatomaceous earth, whose particles are individual skeletons of algae – diatoms and similar. The skeletons consist of amorphous hydrated silica, which is pozzolanic. There are no viable sources of natural pozzolanas in the United Kingdom, although there are many sources in other areas, including locations around the Mediterranean, Canada and the Middle East.

Natural calcined pozzolanas are minerals, or mixtures of minerals, that have been purposely rendered pozzolanic by thermal processing. In the United Kingdom, metakaolin (MK) is the most commonly encountered of these. It is produced by calcining kaolin clay at temperatures around which kaolinite begins to thermally decompose (600°C–900°C). The resulting material is largely amorphous and has a composition essentially the same as the clay that was used to produce it (see Figure 5.7). In the presence of hydrating PC, it produces hydration products similar to those produced by FA, although it is typically used in smaller quantities than FA.

The particles of MK produced are essentially the same size as the clay particles that formed them, and is, thus, typically a relatively fine material (Figure 5.10).

BS EN 197-1 requires that pozzolanas have a reactive SiO_2 content of ≥25% by mass.

5.4 AGGREGATES

In many aspects of concrete durability, aggregate plays an important role. Not only is it involved directly in certain processes that compromise durability (for instance, abrasion and ASR), but it can also introduce chemical substances (in particular, chlorides, sulphates and alkalis) that can contribute to other problems. In addition, as discussed previously, the presence of aggregate creates an ITZ, which can assist the ingress of harmful substances.

Standards relating to aggregates are discussed in the following sections, with an emphasis placed on requirements for durability.

5.4.1 Natural aggregate

The main standard covering aggregates for use in concrete is *BS EN 12620* [24]. It covers natural, manufactured and recycled aggregates (RAs) and includes limits on, and requirements to provide information relating to, certain material characteristics. It also contains details on the actions necessary to ensure the conformity of an aggregate with the standard. The characteristics are as follows:

- Grading
- Particle shape
- Shell content
- Fines content
- Fines quality
- Resistance to fragmentation
- Resistance to wear
- Resistance to polishing and abrasion

- Particle density and water absorption
- Bulk density
- Freeze–thaw resistance
- Drying shrinkage
- Alkali–silica reactivity
- Chemical requirements

BS EN 12620 defines how characteristics should be measured and expressed, but only provides actual limits in certain cases. Thus, it is up to the specifier or concrete producer (see Section 5.6) to determine what characteristics are required. However, the British Standards Institution has also produced a Published Document, *PD 6682-1* [25], which provides suggested limits for certain characteristics for concrete produced in the United Kingdom.

Grading requirements are defined for coarse aggregate, fine aggregate, 0/8-mm natural graded aggregate, all-in aggregate and filler aggregate.

The characteristics relevant from a durability perspective are resistance to fragmentation, wear, polishing and surface abrasion; freeze–thaw resistance; drying shrinkage; alkali–silica reactivity; and several of the chemical requirements.

Resistance to fragmentation, wear, polishing and surface abrasion all provide some indication of the abrasion resistance (AR) that an aggregate material will impart to concrete. The AR requirements of concrete floors and screeds are covered by *BS 8204-2* [26]. The standard takes the approach of defining the requirements for aggregate AR in terms of resistance to fragmentation.

Resistance to fragmentation is measured using the Los Angeles (LA) test, which involves loading a revolving drum with a quantity of aggregate whose particles are within a narrow size range (10–14 mm) and 11 steel balls and revolving the drum at a constant, standard rate. The drum has a 'shelf' on its inside circumference, which acts to agitate the contents, leading to violent collisions between the aggregate particles and the steel balls. At the end of the test, the aggregates are sieved using a sieve with an aperture size of 1.6 mm that, prior to spending time in the drum, would have retained all of the aggregate particles. The amount still retained on the sieve after testing is used to calculate an LA coefficient. A high level of retention gives a low LA coefficient, which denotes a high resistance to fragmentation. The LA test is described in *BS EN 1097-2* [27].

PD 6682-1 suggests that a maximum LA coefficient of 40 is appropriate for most concrete applications, and lower values should only be required for very high-performance applications, if at all.

Resistance to polishing and surface abrasion apply to AR of aggregates used in highway surface courses. Both types of resistance are measured using the techniques described in *BS EN 1097-8* [28].

Measurement of polished stone value (PSV) and aggregate abrasion value (AAV) are discussed in Chapter 2. In the United Kingdom, the Highways Agency includes requirements for PSVs or AAVs for different road and traffic conditions in the *Design Manual for Roads and Bridges* [29].

The freeze–thaw resistance of aggregate is quantified in *BS EN 12620* either in terms of the results of the *EN 1367-1* or *EN 1367-2* test methods [30,31]. Both methods involve measuring the amount of fragmentation resulting from exposing aggregates to conditions that cause expansion within porosity within the aggregate particles. In the case of the *EN 1367-1* method, this is achieved through exposure to freezing and thawing cycles, whereas *EN 1367-2* uses crystallisation pressure caused by the precipitation of magnesium sulphate from the solution. It is the second of these tests that has the most significance in terms of the specification of concrete in the United Kingdom, as will be discussed later.

Drying shrinkage is expressed in terms of the value obtained using a test described in *EN 1367-4* [32]. The test does not involve a direct measurement of the shrinkage of the aggregate but instead measures the shrinkage of concrete prisms containing a fixed mass of aggregate whose grading lies within a set of limits. Shrinkage from a saturated condition is measured after drying in an oven at 110°C. *BS EN 12620* requires that, where the specifier requires it, drying shrinkage should be $\leq 0.075\%$.

The susceptibility of sources of aggregate to ASR is not dealt with directly by *BS EN 12620*, which refers the user to relevant local standards. In the United Kingdom, guidance on ASR is provided by *BRE Digest 330* [33]. The guidance recommends the ASR test method described in *BS 812-123* [34], which involves the preparation of cement-rich concrete prisms containing the aggregate under investigation and measuring their expansion at a temperature of 38°C at high humidity. Expansion in excess of 0.20% after 12 months of testing indicates an expansive reaction, whereas expansion below 0.05% means that the aggregate is nonexpansive. An area of uncertainty exists between these limits, which is tentatively subdivided and labelled 'possibly expansive' (0.10%–0.20%) and 'probably nonexpansive' (0.05%–0.20%). *BRE Digest 330* also recommends further investigation of test prisms using petrographic methods to confirm that ASR has occurred. A modified version of the *BS 812-123* method has been devised specifically for greywacke aggregates [35].

Where an aggregate is found to be expansive or possibly expansive, this does not mean that the material cannot be used in concrete but that measures must be taken to control the reaction (see Section 5.6.8).

The chemical requirements covered by *BS EN 12620* relating to aggregates, in general, are chlorides, acid-soluble sulphates, total sulphur, 'constituents which alter the rate of setting and hardening of concrete' and 'carbonate content of fine aggregates for concrete pavement surface courses'. Of these chlorides, sulphates, sulphur and carbonate content have direct relevance to durability. Testing for the chemical characteristics of aggregates is carried out using the methods described in *BS EN 1744-1* [36].

No limit is placed on chlorides, but limits on the total chloride content of concrete are set in the specification standards (see Chapter 4). *PD 6682-1* recommends that acid-soluble sulphate is limited to ≤0.80% by mass. Other than for air-cooled blast-furnace slag (see later), *BS EN 12620* limits total sulphur to ≤1% by mass.

The carbonate content is included because it may be necessary to limit the content of carbonate minerals present in fine aggregate used for surface courses, as these constituents are more prone to abrasion.

5.4.2 Recycled aggregate

RA is defined as 'aggregate resulting from the processing of inorganic material previously used in construction'. *BS EN 12620* contains special requirements and guidance for the use of these types of materials. The main reason for this is that there are potentially greater risks associated with some constituents of RAs, which means that a more thorough approach to characterisation is necessary.

The standard includes a system for categorising RA based on its constituent materials (e.g., concrete, masonry, bituminous materials, etc.).

The chloride content of RA is required to be determined in terms of acid-soluble chloride rather than water-soluble chloride for conventional aggregate. This is likely to give a higher value and therefore reflects a more conservative approach to the material's acceptability for use in reinforced concrete. The approach is felt to be sound on the grounds that chlorides bound in hydrated cement attached to RA particles are unlikely to be fully released on exposure to water alone.

BS EN 12620 makes the point that, with regards to ASR, RA should be viewed as being potentially reactive, unless otherwise demonstrated, and that users should be aware that unpredictable variability of the material is a possibility.

In the United Kingdom, *BS 8500-2* [10] makes a distinction between RA (which is defined as aggregate resulting from the reprocessing of inorganic material previously used in construction) and recycled concrete aggregate (RCA, which is defined as RA principally composed of crushed concrete). The standard provides limits for the proportion of other materials that can be present (masonry, lightweight material less dense than water, asphalt, glass, plastics and metal) as well as limits on the fines content and the acid-soluble sulphate (1.0% by mass for RCA).

The standard only covers the use of RCA and RA as coarse aggregate, since there are concerns that fine aggregate may contain excessive quantities of gypsum particles from plaster. However, it stresses that the use of fine RCA and RA can be included in a project specification, if particular sources of sulphate-free materials can be assured.

The standard limits the use of RCA to concrete with a characteristic cube strength of ≤ 50 N/mm^2. It also excludes RCA from use in environments where chloride-induced corrosion, chemical attack and more aggressive forms of freeze–thaw attack are likely threats.

When specifying concrete, *BS 8500-2* states that when the use of RA is permitted, it is required that, as well as conforming to the requirements detailed previously, the maximum acid-soluble sulphate content, the method used for the determination of chloride content, a classification with respect to alkali–aggregate reactivity, the method for the determination of the alkali content and any limitations on the use of materials must be specified.

BS EN 206 [37], to which *BS 8500-2* is a complementary standard, is now more specific with regards to limits for RA. The standard defines requirements for two different types of recycled aggregate – one only suitable for concrete with a compressive strength class (in terms of cube strength) of ≤ 37 N/mm^2 (Type B), and one suitable for higher strength concrete (Type A). Limits are placed on the characteristics and composition of RA, with characteristics relevant from a durability perspective being resistance to fragmentation (a maximum LA coefficient of 50 required) and a water-soluble sulphate content of $\leq 0.7\%$ by mass. Recommended limits on constituents are set to maximise the content of recycled concrete aggregate and/or unbound/hydraulically bound aggregate and natural stone, whilst limiting levels of constituents which are likely to be more problematic: fired clay, bituminous materials, glass and gypsum (see Chapter 3), and other contaminants (timber, plastic etc.). Stricter limits are set for Type A. Details of categories for different RA constituents are listed in *BS EN 12620*.

For non-aggressive conditions (Exposure Class X0 – see Section 5.7.4) the standard recommends a limit of 50% replacement of coarse aggregate. Under the least aggressive carbonating conditions (XC1 and XC2), a limit of 30% is set for Type A RA and 20% for Type B. For the more aggressive carbonating conditions (XC3 and XC4) plus the lowest levels of aggression for freeze-thaw attack, corrosion induced by chlorides (not from seawater) and chemical attack (XF1, XD1 and XA3) a limit of 30% is set for Type A RA, whilst Type B is not recommended. Above these exposure classes RA is not recommended for any type of RA.

5.4.3 Air-cooled blast-furnace slag

As previously discussed, the granulation of blast-furnace slag produces a granular, glassy material. By allowing slag to cool slowly, crystallisation of calcium aluminosilicates and calcium magnesium silicates occurs. The cooled material is then crushed and screened to produce aggregate.

BS EN 12620 contains specific requirements for air-cooled blast-furnace slag, partly as a result of historical problems resulting from aggregate expansion. Thus, this standard includes additional and modified requirements to provide confidence that volume stability will not be an issue for a given source of slag.

The issue of volume instability is covered by a specific section in the standard entitled *Constituents Which Affect the Volume Stability of Air-Cooled Blast-Furnace Slag*. These two components are β-dicalcium silicate and iron sulfides. β-Dicalcium silicate is a metastable form of dicalcium silicate ($2CaO·SiO_2$), which will over time undergo a phase transformation to the more stable γ-dicalcium silicate [38]. This transformation involves an increase in volume, which leads to the disintegration of slag particles. A method for testing for the presence of the β form is described in *BS EN 1744-1*, which involves the inspection of split slag particles under ultraviolet light.

Iron sulphide (and magnesium sulphide) also causes disintegration, this time via the hydrolysis of the sulphide. An example of such a hydrolysis reaction is

$$FeS + 2H_2O \rightarrow Fe(OH)_2 + H_2S$$

A method for determining the potential for sulphide disintegration is also described in *BS EN 1744-1*, which involves submerging particles of slag in water for a number of days and observing the extent to which disintegration occurs. Sulphide may also be present in natural aggregates. In the past, problems have been encountered with aggregates deriving from mining wastes in Devon and Cornwall in the UK, which contain quantities of the sulphide mineral pyrites.

The sulphate limits placed on air-cooled blast-furnace slag aggregates are less stringent than for other aggregates. This is because sulphur is typically present in the material in a form that is less available for reaction. As a result, *BS EN 12620* sets a higher limit of 2% by mass of total sulphur, and *PD 6682-1* follows a similar strategy for acid-soluble sulphate (≤1.00% by mass).

5.4.4 Lightweight aggregates

BS EN 12620 does not include aggregate with a density less than 2000 kg/m³, thus excluding lightweight aggregates. Instead, they are covered by *BS EN 13055-1* [39], where the intended use is in concrete. The standard divides the requirements of lightweight aggregate into two categories: physical and chemical. The physical characteristics are as follows:

- Density
- Aggregate size and grading

- Particle shape
- Fines
- Grading of fillers
- Water absorption
- Water content
- Crushing resistance
- Percentage of crushed particles
- Resistance to disintegration
- Freezing and thawing resistance

From the perspective of concrete durability, only resistance to disintegration and freezing/thawing are of direct relevance.

Disintegration of lightweight aggregate may result from the expansion of oxide constituents that are potentially present in certain materials. A method for evaluating any potential problem with disintegration is included in an annex of the standard. The method involves exposing the aggregate to high humidity at an elevated temperature and pressure. Sieve analysis is used to establish the extent to which disintegration has occurred.

A test method for freeze–thaw resistance is also included in an annex of the standard. It involves exposing the aggregates to a sequence of freeze–thaw cycles, followed by a sieve analysis to determine the level of deterioration. In the case of both disintegration and freeze–thaw resistance, limits are not set by the standard – the performance of the aggregates is simply required to be declared by the producer.

In terms of determination of the chemical characteristics of chloride, acid-soluble sulphate and total sulphur are required in the same manner as for normal and heavyweight aggregates. BS 8500-2 requires that lightweight aggregate should have an acid-soluble sulphate content of ≤1%, measured using a technique described in BS EN 1744-1. There is also a requirement for the declaration of the presence of 'organic contaminators' that are equivalent to the 'constituents which alter the rate of setting and hardening of concrete' in BS EN 12620 and determined in the same manner.

Additionally, loss on ignition must be determined and declared for lightweight aggregates that are recycled ashes, such as furnace bottom ash from coal-fired power stations. BS 8500-2 requires that the loss-on-ignition of the material at 950°C (again following a technique in BS EN 1744-1) must be less than 10% by mass.

The standard states that the reactivity of lightweight aggregate with respect to ASR should be determined in a manner appropriate for the place of use, where required. Thus, the use of the BS 812-123 test method also applies to lightweight aggregate in the United Kingdom.

In certain cases, where freeze–thaw attack is a potential problem, it is necessary to specify freeze–thaw-resistant aggregate. Where the aggregate is lightweight, *BS 8500-2* requires that the producer is able to demonstrate that freeze–thaw-resistant concrete can be produced from the material.

5.5 ADMIXTURES

A number of admixtures for use in concrete have specific capabilities to enhance the durability performance of concrete. Others impart characteristics to concrete which have more general applications, but which can be used to improve durability in a certain case. These admixtures are described below. Where admixtures are intended to enhance a specific aspect of concrete durability, further details can be found in the relevant preceding chapters.

5.5.1 Water reducers and superplasticisers

Water reducers and superplasticisers do not impart any specific quality to concrete to make it more resistant to physical or chemical deterioration. However, they require discussion in this chapter because of the very important role that they can play in producing durable concrete.

Both admixtures have the effect of reducing the viscosity of the fresh mix. This effect can be used in concrete construction in a number of ways – it can simply be used to produce a more workable mix or, by reducing the water and cement content of a concrete mix simultaneously, a material with the same workability and strength, but with a potentially lower cost. However, where durability is concerned, the substances are of greatest benefit in their water-reducing abilities.

Water provides concrete with its fluidity, but the greater the water requirement, the greater the quantity of cement needed to achieve a given W/C ratio. Water reducers break this dependency and essentially allow the required workability to be achieved over a wide range of cement contents and W/C ratios. As we shall see, specification of concrete for durability often involves setting limits in terms of maximum cement content and minimum W/C ratio. Thus, this group of admixtures can play an important role in producing concrete within these limits.

Water-reducing admixtures are typically based on either lignosulphonates (which derive from the paper manufacturing process), hydroxycarboxylic acids (produced either by chemical or biochemical synthesis) or hydroxylated polymers (manufactured from natural polysaccharides). All work by the same mechanism – they are absorbed onto the surface of cement particles in fresh concrete and limit the extent to which these particles are attracted to each other by van der Waals forces, thus reducing the viscosity of the cement paste.

Many superplasticisers operate by a similar mechanism but are considerably more effective. Typically, a conventional water reducer will reach a saturation concentration (the dosage above which no further reduction in water is possible) at approximately 1% by mass of the cement. At such a dosage, the level of water reduction will be approximately 10% compared with a mix containing no admixture. In the case of superplasticisers, the influence on viscosity is of a longer range, which means that higher dosages can potentially achieve levels of water reduction of up to almost 30%. Superplasticisers are based on sulphonated naphthalene formaldehyde, sulphonated melamine formaldehyde and polyacrylates. A more recent additional group of superplasticisers – polycarboxylate ethers (PCEs) – work by a different mechanism of 'steric stabilisation'. This involves chains of polymers being adsorbed at cement particle surfaces and preventing particles coming sufficiently close to allow van der Waals interactions.

In certain circumstances, water-reducing admixtures can have detrimental effects on durability in the form of cracking resulting from drying shrinkage, which is typically exaggerated in comparison with mixes with the same cement content without admixture. The effect is still larger where formulations contain an accelerating admixture to counteract the retarding effect of water reducers [20]. This is not usually an issue for superplasticisers.

5.5.2 Air-entraining agents

Air-entraining agents help introduce and stabilise microscopic bubbles of air within the cement matrix of concrete. The presence of air bubbles acts to significantly limit the damage caused by the cyclic freezing and thawing of water in concrete pores. The nature, mechanism and performance of air-entraining agents are discussed in detail in Chapter 2.

5.5.3 Damp proofers

Damp proofers are admixtures that render the surface of pores hydrophobic. This is achieved through a number of different mechanisms. The most commonly encountered damp proofers contain fatty acids such as oleic ($C_{17}H_{33}COOH$) and stearic ($C_{17}H_{35}COOH$) acid. These react with hydration products at pore surfaces in the manner shown in Figure 5.11. The attachment reaction leaves a layer of hydrophobic hydrocarbon chains at the surface, which cause water coming into contact with the pore surface to adopt a high contact angle (Figure 5.12). As discussed in Section 5.2.3, this significantly reduces the extent to which capillary action will draw water into the concrete pores.

Damp proofing can also be achieved by the introduction of admixtures consisting of emulsions of waxes. These emulsions are stabilised using emulsifiers such as sorbitan monostearate such that they begin to coalesce in the high-pH conditions produced by the hydrating cement, thus depositing a thin film of hydrophobic wax at pore surfaces some time after mixing [40].

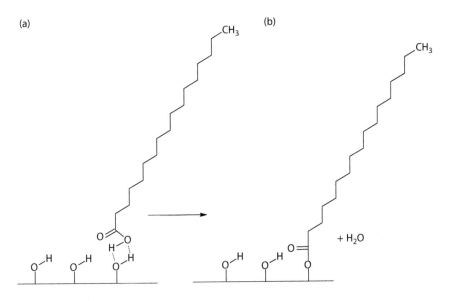

Figure 5.11 Attachment of an oleic acid to a hydration product surface. (a) Electrostatic interactions bring the oleic acid molecules into close proximity to hydroxide groups at cement hydration product surfaces; (b) reaction between hydroxides groups and oleic acid leads to molecules becoming attached to the surface via a chemical bond, with the formation of a water molecule.

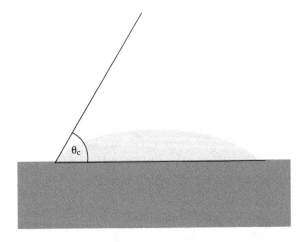

Figure 5.12 Contact angle (θ_c) between a water droplet and a solid surface.

Additionally, admixtures containing fine powders of hydrophobic particles may also be used.

The use of damp proofers can also limit the flow of water into concrete acting under a pressure differential. Before water can enter concrete, the pressure acting on the concrete surface must exceed the capillary entry pressure (p_{ce}). This pressure is required to overcome the difference in pressure between the fluid on the outside of the concrete and the fluid within the pores. Where concrete durability is concerned, the external fluid is normally water and the internal fluid is air. The capillary entry pressure is described by the equation

$$p_{ce} = \frac{-2\gamma \cos\theta_c}{r}$$

where r is the pore radius (m), γ is the surface tension of water (N/m) and θ_c is the contact angle of the water with the pore surface.

It should be noted that the processes controlling the capillary entry pressure are essentially the same as those for absorption, and the equation is therefore closely related to the equation for capillary action in Section 5.2.3. Where there is a low contact angle, the capillary entry pressure is negative and capillary action occurs. However, where damp proofers are used and the contact angle is >90°, the concrete surface resists infiltration.

Results of calculations using this equation are shown in Figure 5.13. Although these results appear impressive, they should be viewed in context – the pressure exerted by a raindrop in winds travelling at 80 km/h would

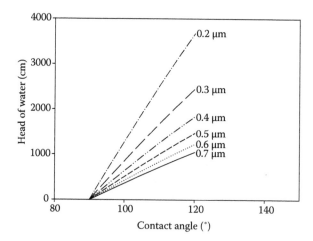

Figure 5.13 Calculated head of water required for water to pass through a concrete surface with pore openings of various radii.

be sufficient to allow the water to penetrate the surface considered in the calculation in Figure 5.13 with a pore radius of 0.7 µm, even if the contact angle were 120°. Thus, damp proofers are likely to offer limited resistance to water at high hydrostatic pressures. It should also be noted that cracks that form in the concrete in its hardened state will not be lined with hydrophobic agent to the same extent as pores and so are likely to compromise the damp-proofing effect.

5.5.4 Corrosion inhibitors

Corrosion inhibitors are admixtures that increase the threshold chloride concentration needed to initiate corrosion and that also limit the rate of corrosion after initiation. They are discussed in further detail in Chapter 4.

5.5.5 Alkali–aggregate expansion–reducing admixtures

Alkali–aggregate expansion–reducing admixtures are admixtures that limit the extent to which expansion results from alkali–aggregate reactions. They are most commonly based around lithium compounds whose presence in concrete acts to prevent the formation of the gel formed during this reaction. These admixtures are discussed in further detail in Chapter 3.

5.6 FIBRES

Fibres for use in concrete can take a number of different forms. A wide range of materials can be used including steel, polymers, carbon and glass. Fibres are available with a range of different dimensions in terms of width, length and uniformity of profile, and may also be manufactured with different morphological features including crimping and hooked ends.

The nature of properties imparted to concrete is dependent on the material and dimensions of the fibres. Steel fibres (with diameters in the range of 0.4 to 1.4 mm and lengths between 10 and 60 mm) and the so-called 'macrosynthetic fibres' – polymer fibres with similar dimensions – are capable of providing additional toughness to concrete [41]. This toughness takes the form of a stress–strain relationship in compression and tension, which more closely resembles that of a ductile material. It should be stressed that this does not mean that the concrete is rendered ductile by the presence of fibres – the concrete is cracking as it deforms, but the presence of the fibres delays crack growth and transmits stress across cracks. Thus, it is more accurate to say that the 'postcracking' load capacity of the concrete is improved.

This enhancement typically leads to smaller crack widths forming in a concrete element, which has benefits for protecting concrete against the ingress of chemical substances that can damage either the concrete or its reinforcing steel. Furthermore, in certain structural elements, these types of fibres can be used in place of some of the steel reinforcement. Suitable structural elements are those supported by a continuous substrate, such as slabs supported by the soil or piles or composite slabs on steel decking. Sprayed concrete elements are also suitable. Such structures require minimal ('nominal') quantities of reinforcement, largely to control cracking, and it is this reinforcement that can effectively be replaced by fibres. Typically, the maximum level of dosage of steel fibres is approximately 1.00% by volume, whereas for macrosynthetic fibres, it is approximately 1.35%.

Toughening effects are not observed to any great extent with glass or microsynthetic fibres (typically, tens of micrometers in diameter and up to 25 mm in diameter), as a result of their lower stiffness [42]. These are added in smaller quantities – microsynthetic fibres are typically used at a maximum dosage of 0.1% by volume, whereas for glass fibres, the maximum dosage is slightly higher – to control cracking resulting from drying shrinkage. 'Control', in this context, means an increase in the number of cracks, but with smaller crack widths compared with concrete without fibres.

The presence of all types of fibre normally has the effect of enhancing surface quality, through the reduction of bleeding at the surface and through the reduction of the extent to which plastic cracking occurs.

Additionally, as discussed in Chapter 2, enhanced freeze–thaw resistance is often observed.

The properties of steel fibres for use in concrete are covered by *BS EN 14889-1* [43]. Since steel fibres do not provide structural reinforcement, their corrosion is usually only of major concern from an aesthetic perspective. In such cases, stainless steel fibres can be used.

The properties of polymer fibres are covered by *BS EN 14889-2* [44]. *Eurocode 2* does not cover the use of fibres in concrete. *BS EN 206* [37] contains basic requirements for the use of fibres (and declaration of their use) in concrete. It also contains testing requirements to confirm fibre type and content.

5.7 SPECIFYING DURABLE CONCRETE

The specification of concrete is covered by the European Standard *EN 206: Concrete: Specification, Performance, Production, and Conformity* [37]. The standard defines the responsibilities of the specifier (in terms of specifying concrete), the concrete producer (in terms of the conformity of the product and production control) and the user (in terms of placing the concrete). The standard has national complementary standards, *BS 8500-1* and *BS 8500-2* [10,45], which cover the method of specifying concrete and the requirements of the concrete and its constituent materials, respectively.

BS EN 206 defines three approaches to specifying concrete: 'designed concrete', 'prescribed concrete' and 'standardised prescribed concrete'. *BS 8500-1* adds two other approaches to specification: 'designated concrete' and 'proprietary concrete'. Before dealing with the specification of durable concrete, it is useful to examine what these approaches are.

5.7.1 Designated concrete

Designated mixes are mixes in which the specifier defines what role the concrete is to play in a structure to the concrete producer, who then develops a mix to satisfy the performance requirements for that purpose. *BS EN 8500-1* lists 21 designated concretes. Some of these concretes are intended for specific purposes, such as in unreinforced foundations and paving, whereas others are intended for more general application.

The standard defines a minimum strength class, a required consistence in the form of a consistence class, the maximum W/C ratio, the minimum cement content and the permissible cement types for each mix. The strength class is the characteristic cube or cylinder compressive strength at 28 days. The characteristic strength is the strength below which 5% of strength test results are expected to fall. Strength classes are defined in *BS EN 206*, with additional classes defined in *BS 8500-2*. The consistence class is defined in terms of the results required from a specific measurement of concrete consistence (such as slump). Consistence classes are also defined in *BS EN 206*.

The use of designated mixes for structural elements exposed to chloride-bearing environments is not appropriate, and specification via designed concrete, prescribed concrete or standardised prescribed concrete is required.

A producer wishing to supply designated concrete must have accredited production control and product conformity certification from a third party.

5.7.2 Designed concrete

Designed concrete is described by *BS EN 206* as 'concrete for which the required properties and additional characteristics are specified to the producer who is responsible for providing a concrete conforming to the required properties and additional characteristics'.

Designed mixes are typically used for applications where the specification of a designated mix is not possible. This includes concrete for structural elements exposed to chloride-bearing environments, concrete that requires a compressive strength outside the range of strengths defined for designated mixes (a cube strength between 8 and 50 N/mm^2), lightweight and heavyweight concretes and concrete using special cements for specific purposes.

The basic properties that may be specified are as follows:

- A requirement to conform to *EN 206*
- Compressive strength
- Exposure conditions that the concrete will be subjected to during service
- Maximum aggregate size
- Chloride content

In the case of ready- and site-mixed concrete, it is also necessary to define the required consistence of the fresh mix and the density for lightweight and heavyweight concrete mixes.

Aside from these basic requirements, the specifier may have other requirements beyond the scope of the basic properties. These can include requirements for special materials, such as aggregates or low-heat cements, air content (where freeze–thaw attack is an issue), temperature in the fresh state, the rate of strength development, the extent to which heat is evolved during cement hydration, retardation of stiffening (which may be required where long periods of time are required between mixing and placing), resistance to water penetration, AR and tensile strength. Also included are technical requirements relating to how the concrete is used once it reaches the construction site (placing, surface finish, etc.).

5.7.3 Prescribed concrete, standardised prescribed concrete and proprietary concrete

Prescribed concrete is concrete whose composition in terms of the materials used and their proportions in the mix are provided to a concrete producer by the specifier. The design of the mix may have been carried out by the specifier or by another party. A variant of prescribed concrete is standardised prescribed concrete, where the composition is provided in the form of a standard valid in the place of use.

Proprietary concrete is concrete whose performance is defined in terms of measurements made using test methods stipulated by the specifier. The specifier will consult with the producer to identify a suitable concrete mix that the producer can supply, without any requirement on the part of the producer to disclose the mix proportions or materials used.

The vast majority of concrete specified in the United Kingdom is either designated or proprietary. Although durability is dealt with in *BS EN 206*, it allows for specification for durability to be covered through complementary standards for individual countries. This is the case for the United Kingdom, where *BS 8500-1* explicitly covers specification for durability in terms of designated and designed mixes. Thus, the actions required to specify durable concrete are discussed in the next sections exclusively in these terms. It should be pointed out that the majority of proprietary concrete will be designed by a producer taking into account the durability-related content of *BS 8500-1*.

5.7.4 Specifying for durability: Designated concrete

The first stage of specifying for durability for both designated and designed concrete is to establish the exposure conditions that the concrete will be exposed to. These are defined in both *BS EN 206* and *BS 8500-1* as a series of classes defined by the type of environment that concrete will be exposed to (Table 5.2). These classes are divided into six categories: 'no risk of corrosion or attack' (X0), 'corrosion induced by carbonation' (XC), 'corrosion induced by chlorides' (XD), 'corrosion induced by chlorides from seawater' (XS), 'freeze–thaw attack' (XF) and 'chemical attack' (XA). These exposure classes are self-explanatory, with the exception of 'chemical attack', which refers to either sulphate attack, acid attack or a combination of both. These categories are then divided into individual classes, with a higher number denoting greater aggression – that is, XS1 is less aggressive than XS3.

It is potentially the case that concrete may experience more than one type of exposure, in which case specification should be conducted for each exposure class. The exception to this is concrete experiencing a chloride-bearing environment where carbonation is also a possibility. As discussed in Chapter 4, it is likely that the requirements for chloride resistance sufficiently outweigh those for carbonation, and so chloride exposure should be specified for exclusively.

Designated concrete can be specified for resistance against carbonation, freeze–thaw attack and chemical attack. The procedures required to do this are outlined below. The general approach taken is to determine the appropriate minimum designation for the concrete based on durability requirements. This minimum designation is then compared against the other requirements of the concrete, such as compressive strength. Where the designation for requirements other than durability indicates a higher quality designation, this designation is selected, since this will offer adequate protection.

5.7.4.1 Carbonation

The procedure for specifying designated concrete for resistance to carbonation is outlined in Figure 5.14. The process is relatively straightforward, starting with establishing the exposure class and the manner in which the concrete element will be used. This is defined in various terms, such as whether reinforcement is present, the configuration of the element (e.g., if it is horizontal) and the manner in which water is likely to come into contact with it. In some cases, the exposure class may not only be just an XC class

Table 5.2 Exposure classes defined in BS EN 206

Exposure class	Environment
No risk of corrosion or attack	
X0	For concrete without reinforcement or embedded metal: all exposures, except where there is freeze–thaw, abrasion or chemical attack For concrete with reinforcement or embedded metal: very dry
Corrosion induced by carbonation	
XCI	Dry or permanently wet
XC2	Wet, rarely dry
XC3	Moderate humidity
XC4	Cyclic wet and dry
Corrosion induced by chlorides	
XDI	Moderate humidity
XD2	Wet, rarely dry
XD3	Cyclic wet and dry
Corrosion induced by chlorides from seawater	
XSI	Exposed to airborne salt but not in direct contact with seawater
XS2	Permanently submerged
XS3	Tidal, splash and spray zones
Freeze–thaw attack	
XFI	Moderate water saturation, without de-icing agent
XF2	Moderate water saturation, with de-icing agent
XF3	High water saturation, without de-icing agent
XF4	High water saturation, with de-icing agents or sea water
Chemical attack	
XAI	Slightly aggressive chemical environment
XA2	Moderately aggressive chemical environment
XA3	Highly aggressive chemical environment

(as defined in Table 5.2) but may also be a freeze–thaw (XF) class where both types of exposure apply.

This information allows a suitable minimum concrete designation to be identified along with the nominal depth of cover. Where possible, the standard presents a number of different combinations of concrete designation and cover depth, with higher quality designations requiring less cover. Where prestressed reinforcement is present, the specifier is required to check the relevant design code for details of whether an additional amount of cover (Δc) is required.

At this point, the specifier has all the information that they require, but details of the requirements of the concrete designation in terms of strength

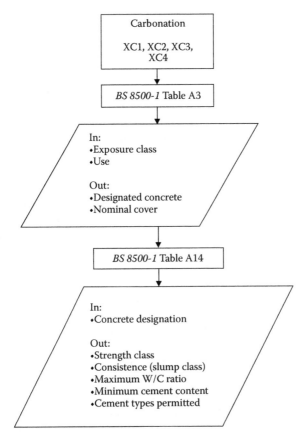

Figure 5.14 Approach to specifying designated concrete for protection against carbonation-induced corrosion in *BS 8500-1*.

class, consistence, maximum W/C ratio, minimum cement content and permitted cement types are provided to the producer in a table (Table A14) in *BS 8500-1*.

5.7.4.2 Freeze–thaw attack

The specification of designated concrete for freeze–thaw resistance is slightly more complex than that for carbonation, at least if reinforcing steel is present. Where reinforcement is not present, the approach is simple – the exposure class is used to identify the minimum concrete designation. Cover is not an issue in these circumstances.

However, where steel is present and the concrete is exposed to a carbonating environment, reference to Table A3 in the standard is required to identify a nominal cover depth, along with a possible modification of the minimum concrete designation.

The procedure for specification is outlined in Figure 5.15.

5.7.4.3 Chemical attack

Although chemical attack (XA) classes are defined in *BS EN 206*, *BS 8500-1* takes a slightly different approach by translating exposure conditions into an aggressive chemical environment for concrete (ACEC) class, which gives a more precise measure of aggression than the exposure classes. Threat of chemical attack may come from sulphates in seawater or sulphates and/or acidic conditions in soil.

Where seawater exposure is the concern, no system for selecting a specific designated mix is provided. However, the standard does contain requirements for unreinforced designed concrete (discussed in the next section), which can potentially be translated into a concrete designation.

Where contact with aggressive conditions in soil is the source of chemical attack, the specifier is asked to provide details of soil and groundwater conditions in terms of magnesium and sulphate concentrations – whether the groundwater is static or mobile, whether the site is a brownfield site, and the groundwater pH. This information is used to identify the site's ACEC class.

In the present form of the standard, if the concrete is reinforced, the specifier refers to a table (A3) that provides nominal cover depths and the minimum concrete designation for less aggressive environments. However, the specifier is then referred to a larger table (A9) that covers all ACEC classes in more detail and repeats the requirements of the previous table, making Table A3 unnecessary.

In Table A9, the ACEC class, the intended working life and the groundwater hydraulic gradient are used to identify the minimum concrete designation, along with the lowest nominal cover permitted. In certain cases, additional protective measures (APMs) are also required. These are discussed in more detail in Chapter 3.

The specification procedure is outlined in Figure 5.16.

5.7.5 Specifying for durability: Designed concrete

The specification of designed concrete for durability, in most cases, follows a procedure similar with that for designated concrete and is conducted alongside specification structural and other requirements. The main difference is that the outputs of the process are a series of criteria that must be met during the design process, rather than a designation. The criteria will always include a maximum W/C ratio, a minimum cement content and the types of cement permitted. Depending on the type of exposure involved,

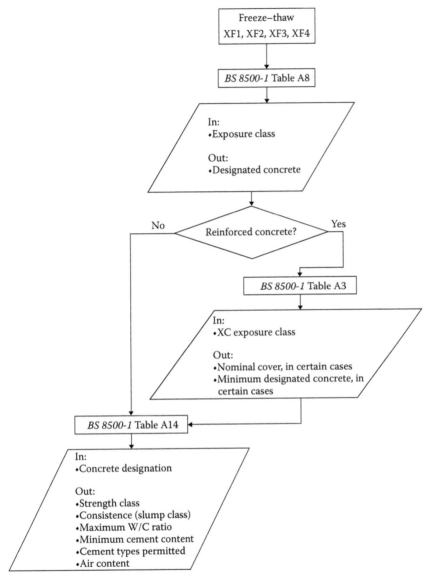

Figure 5.15 Approach to specifying designated concrete for protection against freeze–thaw attack in *BS 8500-1*.

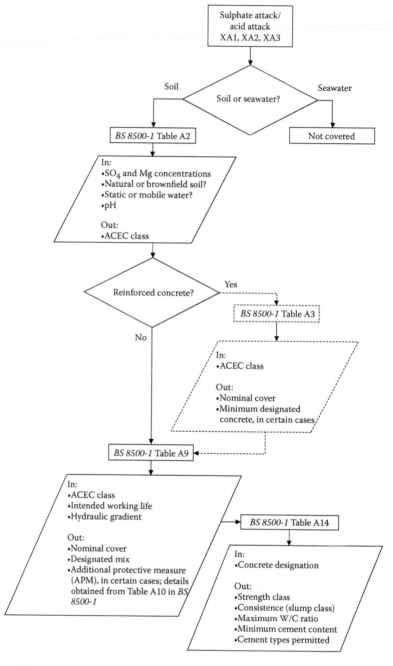

Figure 5.16 Approach to specifying designated concrete for protection against chemical attack in *BS 8500-1*.

other parameters may also need to be specified including the minimum compressive strength class and the nominal cover depth.

Specification of designed concrete for durability in *BS 8500-1* covers chloride exposure, carbonation, freeze–thaw attack and chemical attack. The procedures are discussed below.

5.7.5.1 Chlorides and carbonation

The procedure for specifying designated concrete for resistance to chloride- and carbonation-induced corrosion is shown in Figure 5.17.

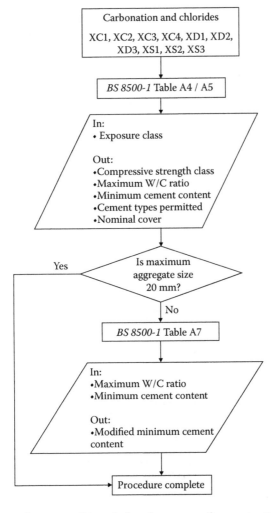

Figure 5.17 Approach to specifying designed concrete for protection against both chloride- and carbonation-induced corrosion in *BS 8500-1*.

The exposure class is used to determine appropriate maximum W/C ratios, minimum cement contents and minimum compressive strength classes, nominal cover depths and cement types. The approach taken is, wherever possible, to offer the specifier a range of different options in terms of cement types and cover depths such that practical, economic and aesthetic considerations can be used to select the most appropriate cover. Where a lower cover depth is selected, more resistant cement types, lower W/C ratios, higher cement contents, and higher strength classes are required. Two different tables may be used for this procedure (A4 or A5) depending on whether the intended service life of the structure is ≥ 50 years or ≥ 100 years.

Where maximum aggregate sizes other than 20 mm are to be used, a modification of the minimum cement content is then made.

5.7.5.2 Freeze–thaw attack

Specification of designed concrete for freeze–thaw attack involves interrogation of just one table (Figure 5.18). The exposure class is used to obtain the required minimum strength class, maximum W/C ratio, minimum cement content, and cement types permitted and, in certain cases, the requirement for freeze–thaw-resisting aggregates. The standard takes the approach of offering the specifier one of the two options for each exposure class – either the concrete is required to contain entrained air (the volume of which is

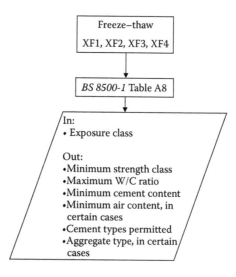

Figure 5.18 Approach to specifying designed concrete for protection against freeze–thaw attack in *BS 8500-1*.

defined by the maximum aggregate size) or a lower W/C ratio and a higher cement content and compressive strength class are required.

Freeze–thaw-resisting aggregates are required for the two most aggressive exposure classes – XF3 and XF4. Resistance requirements are in terms of magnesium sulphate soundness. Loss in mass of ≤18.0 using this test is required for the XF4 exposure class, whereas a loss in mass of ≤25% is required for XF3 [10].

5.7.5.3 Chemical attack

The approach taken to specifying designed concrete for chemical attack in the soil follows the general procedure used for designated concrete (Figure 5.19) – the ACEC class is established from the results of the chemical analysis of the site and from other details, which are used (along with the hydraulic gradient of the groundwater and the intended working life of the concrete) to identify the specification required. This is done by first establishing the nominal cover depth, a design chemical (DC) class and any APMs required. The DC class and the intended maximum aggregate size are then used to determine the maximum W/C ratio, the minimum cement content and the cement types permitted.

5.7.6 Specifying for AR

BS EN 206 and *BS 8500-1* do not cover AR. Instead, this is dealt with by a separate standard – *BS 8204-2:2003: Screeds, Bases, and In Situ Floorings – Part 2. Concrete Wearing Surfaces: Code of Practice* [25]. The specification process involves identifying the level of abrasive action on the concrete surface in terms of one of four AR classes. For the least abrasive conditions (AR2 and AR4), the standard gives the minimum strength class, the minimum cement content, the permitted aggregate and the method of surface finishing. For the most abrasive conditions (AR0.5 and AR1), the standard requires a specially designed proprietary concrete. For each AR class, the maximum permitted abrasion depth measured using the AR test method described in *BS EN 13892-4* is also provided.

BS 8204-2 requires that aggregate for applications where AR is required must have an LA coefficient of ≤40. Where concrete is directly finished – in other words, where the concrete itself provides the wearing surface – this is all that is required. However, the option to upgrade the concrete's wearing ability through the application of a surface treatment exists. In the case of the three lesser abrasive conditions (AR1, AR2 and AR3), a wearing screed is applied, with the guidance on the proportions for which are given in the standard. Wearing screeds are discussed in further detail in Chapter 6.

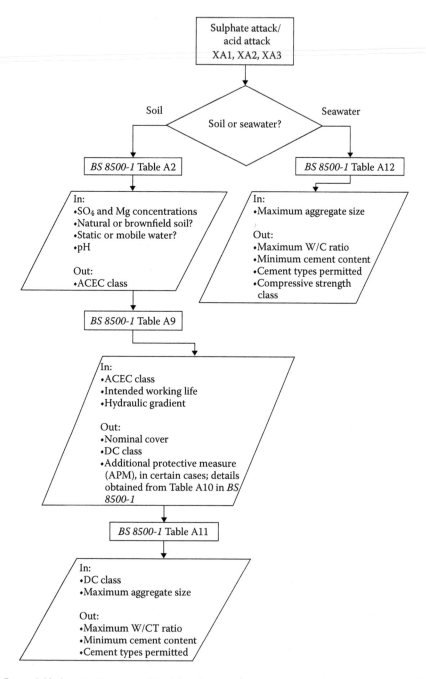

Figure 5.19 Approach to specifying designed concrete for protection against chemical attack in *BS 8500-1*.

Where the highest level of AR is required, dry shake/sprinkle finishes are required to be applied to the surface of a concrete with a maximum wear depth of 0.3 mm measured using the *BS EN 13892-4* test.

5.7.7 Specification for the control of alkali–silica reaction

For designed, designated and standardised prescribed concrete, *BS EN 206* places responsibility for ensuring that alkali–silica reaction is minimised on the concrete producer. In the case of prescribed concrete, the standard states that it is the specifier's responsibility to ensure that adequate measures have been taken. However, *BS 8500-1* argues that a producer is in a much better position to achieve this and requires that the producer is exclusively responsible. *BS 8500-1* states that a producer following the guidance in *BRE Digest 330* [33] is deemed as satisfying this responsibility. This guidance is discussed in Chapter 3.

Aspects of the measures required for minimising ASR require information relating to the alkali content of the various concrete constituents. *BS EN 8500-2* defines how the alkali content of various materials should be determined, declared and monitored over time.

5.7.8 Specification for control of drying shrinkage

BS 8500-2 [10] requires that aggregate should not display shrinkage of more than 0.075% using the *EN 1367* test outlined in Section 5.4. Where aggregate displays shrinkage in excess of this limit, provided that the concrete has been designed in a manner that takes this shrinkage into account, or that the concrete is not expected to undergo drying, this requirement is not necessary. As discussed in Chapter 2, most aggregates do not undergo significant shrinkage. Approaches to designing concrete containing shrinkage-susceptible aggregates are the subject of *BRE Digest 357* [46].

5.8 CONCRETE MIX DESIGN

A number of methodologies have been published on the design of concrete. Most follow a common sequence, based on the fact that the strength of concrete is largely determined by its W/C ratio and that the consistence is mainly determined by the free water content of the mix. The sequence starts by using the required compressive strength to establish the W/C ratio, then the required consistence is used to establish the water content, and various parameters are then used to establish first the total quantity of aggregate required followed by the relative proportions of fine and coarse aggregate.

Although it is not the intention of this chapter to provide a detailed description of the concrete design process, it is relevant to examine the manner in which specification for durability fits into the design process. The design methodology around which this discussion is based is the Building Research Establishment (BRE) method [47], although the same principles apply to other methods. The design sequence using the BRE method is outlined in Figure 5.20.

The specified characteristics of the concrete required for the basic design process are the compressive strength class and the consistence class. Because of the inherent variability of concrete, the BRE method takes the approach of defining a margin on top of a specified characteristic strength above which the running mean strength obtained from the routine testing of the concrete production output must lie. In the BRE method, the margin is either directly specified or calculated based on the standard deviation of compressive testing and the specified proportion of 'defective' results. BS EN 206 defines the margin as being about twice the expected standard deviation of test results coming from a given production source.

Where durability is an issue, the maximum W/C ratio, the minimum cement content and the cement type will also have been specified, as discussed previously. Other information relating to the materials is also needed – the aggregate size (both in terms of the maximum coarse aggregate size and the fineness of the fine aggregate) and the type of aggregate (specifically whether it is crushed or uncrushed). It is also useful to possess a value for the density of the coarse aggregate.

The crucial points where the design is shaped by durability requirements are where the W/C ratio and the cement content are established. At these points, it is necessary to compare the values obtained from the design process against the specified maximum W/C ratio and the minimum cement content. Where the W/C ratio exceeds the maximum specified value or the cement content lies below the minimum specified value, the specified values must be adopted.

The BRE method also permits the design of concrete containing air entrainment. To compensate for the reduction in strength experienced when entrained air is present, a downward adjustment of the W/C ratio is made. Air-entrained concrete is typically more workable than an equivalent mix without air, and so an adjustment of water content is also possible. Finally, an adjustment of the density of the concrete is required when the quantity of aggregate required is calculated.

The use of cements and cement combinations containing siliceous FA is also dealt with. The approach taken to design using FA is that of the k-value concept. This concept is described in BS EN 206 and is a means of selecting

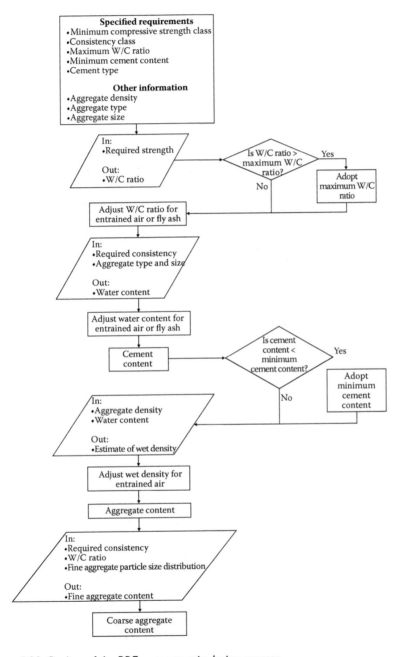

Figure 5.20 Outline of the BRE concrete mix design process.

an appropriate cement content (where 'cement' here means a combination of PC and FA), which takes into account the lower contribution that FA makes to strength. This is done by stating that the W/C ratio obtained from the design process is equal to

$$\frac{\text{water}}{(\text{cement} + k \times \text{addition})}$$

where the addition is the quantity (by mass) of FA, and k is the k-value. The k-value is a measure of the contribution that the FA is able to make toward strength development. In the case of siliceous FA conforming to BS 3892 (a now withdrawn standard), the method proposes a k-value of 0.3. Thus, where the design method identifies a W/C ratio of, for instance, 0.45, using this k-value and an FA content of 35% of the cement fraction by mass, the PC content of the mix (C) will be

$$C = \frac{(100 - f)W}{(100 - 0.7f)\left(\dfrac{W}{C}\right)}$$

where f is the percentage by mass of FA in the cement fraction, W is the mass of water in the mix and W/C is the water–cement ratio obtained from the design process.

The FA content (F) will be

$$F = \frac{fC}{100 - f}$$

As a result, the actual water–cement ratio (where cement is composed of both PC and FA) of the mix in the example would be 0.34.

An additional adjustment of the water content is also required, resulting from the water-reducing capabilities of FA.

The BRE method also provides guidance on the design of concrete containing GGBS. However, because aspects of the performance of GGBS are more source-specific, less explicit guidance is provided. Additionally, it is possible to use the method to design concrete using other materials, such as SF and MK, as long as adequate information about the material is available. This should include a k-value.

One effect of BS EN 450 being adopted in the United Kingdom for FA is that the k-value can potentially vary from source to source to a greater extent than it had previously. This means that the use of a single k-value for FA, as advised in the BRE method, is no longer sound. *BS EN 206* provides minimum limits for *k*-values for FA used in combination with CEM I cement. It also requires that where the FA–PC ratio (by mass) exceeds 0.33, then the excess FA should be discounted in the *k*-value equation.

The standard also advises that, where the concrete is subject to exposure classes XA2 and XA3 and the aggressive substance is sulphate, the *k*-value concept is not recommended. Although there is no easy means of translating XA classes into the ACEC classes used by *BS 8500-1*, it is probably sound to say that this means any ACEC class where the sulphate concentration from a 2:1 water–soil extract is greater than 500 mg/L.

BS EN 206 also contains advice relating to *k*-values for SF, which is essentially that a *k*-value of 2.0 should be used (reflecting the greater activity of SF). Where the concrete will be exposed to carbonation or freeze–thaw attack, a *k*-factor of 1.0 is used. As for FA, where the SF–PC ratio exceeds 0.11, the excess is discounted in the *k*-value equation.

One important element of the design process, which is not covered in Figure 5.20, is the need for the production of trial mixes. All of the rules and relationships used to form the design method are based on generalisations about the manner in which materials behave in concrete in both the fresh and hardened states, and it is almost certain that a mix that has been designed using these relationships will not entirely satisfy the specified requirements. Thus, a trial mix must be made to establish deviations from the predicted concrete properties to allow modifications. The method suggests making not just the designed mix but also mixes with W/C ratios slightly below and above the design W/C ratio. This hopefully avoids the lengthy process of producing a second trial mix after the first. Testing of the trial mix will always include a measure of consistence (such as the slump test) and measurement of cube strengths. However, it may also be necessary to test for durability characteristics, such as abrasion resistance.

5.9 SPECIAL CONCRETE

The specification and concrete design processes described apply to 'normal' concrete, which does not possess special properties suitable for more specialised applications. Where such materials are required, additional requirements for specification and design are necessary. Some of these special types of concrete are discussed below.

5.9.1 Self-compacting concrete

After concrete is placed, it must be compacted such that the still-fluid material fully fills the formwork or space that it is intended to occupy, and air pockets in the material are driven to the surface, leaving a dense and homogeneous material. Conventional concrete construction normally uses vibration as a means of compaction, which can present practical problems where access to the concrete is difficult. The noise and transmitted vibration from compaction also presents health and safety issues. Moreover, complex and congested configurations of reinforcement can present a significant obstacle to efficient compaction.

The reason for the need for compaction is that the fresh concrete is sufficiently viscous to resist complete consolidation once placed, and additional kinetic energy must be provided to encourage it to move. Self-compacting (or self-consolidating) concrete is a material whose viscosity has been reduced sufficiently to allow it to flow around reinforcement and to fill the required space under its own weight.

This reduction in viscosity is achieved through the use of superplasticisers in combination with a higher proportion of fine material than would be used in conventional concrete – up to approximately 600 kg/m³. This fine material can include PC, but its exclusive use would be economically prohibitive. For this reason, a wide range of other powders is used in combination with PC. These can include inert mineral powders, such as limestone, along with cementitious materials such as FA, GGBS and SF.

The significant levels of superplasticisers used can create problems with the segregation of the different constituents. For this reason, admixtures known as viscosity modifiers are used to improve the cohesivity of the fresh concrete.

Specification of self-compacting concrete is not covered by either the *BS EN 206* or the *BS 8500* standards. However, guidance exists in the form of *The European Guidelines for Self-Compacting Concrete*, which has been developed by members of the European concrete industry [48]. This also provides some general guidance on mix design. The basic specified requirements are similar with those for normal concrete. However, a range of additional aspects of performance relating to consistence and segregation are also either basic or additional requirements.

With respect to durability, the guidelines are compatible with the procedures of *EN 206* and *BS 8500* and refer specifically to *EN 206* on these matters.

5.9.2 High-strength concrete

High-strength concrete is defined by *BS EN 206* as normal-weight concrete having a minimum characteristic cube strength of 60 N/mm² or heavy-weight concrete of 55 N/mm² at 28 days. Typically, this is achieved through high powder contents (where the powder is PC and other cementitious and inert materials, often including SF) and through very low W/C ratios. Superplasticisers are usually needed to keep water contents as low as possible.

Specification of high-strength concrete is covered by *BS EN 206*, in which compressive strength classes go as high as a minimum characteristic cube strength of 115 N/mm². The standard includes additional requirements for tests on and inspection of constituent materials, equipment and the production procedures used when high-strength concrete is manufactured. However, specification for durability remains the same.

The design of high-strength concrete is briefly covered in a Concrete Society technical report [49]. In general, the guidance recommends using conventional design methods, such as the BRE method, pointing out that the W/C ratio versus strength relationships used in these methods still applies. However, because of the high cohesivity of high-strength mixes, it recommends higher target consistences (100-mm slump is suggested) to ensure that placing and compaction are possible.

5.9.3 Foamed concrete

Foamed concrete is used in applications where lightweight material possessing some load-bearing capacity is required to fill a space. Such applications include bulk filling of decommissioned underground space, such as sewers and tanks, backfill behind retaining walls, infill in certain structural features and reinstatement of highway trenches. The material is delivered from a pump, possibly after pumping over a distance, and is designed to flow into the space it is required to fill without assistance.

The material consists of fine aggregate; cement in the form of PC, possibly combined with FA, GGBS or SF; inert fine fillers; water; and a foaming agent. The foaming agent is a surfactant from which foam is manufactured by mixing with water and either forcing air through narrow openings in contact with the solution or spraying the solution through baffles. The foam is then combined with the other ingredients to produce a slurry that is then pumped into place.

Specification of foamed concrete is currently not covered by any British or European standards, but guidance on specification has been published by the Transport Research Laboratory in the United Kingdom [50]. The basic requirements include permitted materials, plastic (wet) density, cube strength and pour depth. Additional optional requirements include workability, maximum cube strength, resistance to segregation and durability. Under durability, the specification states that it is normally not necessary to specify for freeze–thaw resistance, as the high void content of foamed concrete makes it extremely resistant to this type of deterioration. The guidance suggests that, where chemical attack from substances in soil may be an issue, specification for resistance should be carried out in line with *BS 5328-1*. This standard is no longer current, but specification to *BS 8500-1*, as for conventional concrete, is equally applicable.

Resistance of foamed concrete to carbonation and chloride ingress is generally lower than for conventional concrete, but the use of steel reinforcement in combination with foamed concrete is uncommon.

The document also provides additional guidance for mix design.

5.9.4 Lightweight and heavyweight concrete

In certain cases, the production of concrete with substantially lower or higher density than normal concrete is desirable. Lightweight concrete is a poor conductor of heat and thus has applications as an insulating material and as a fire-resistant component in structures. Its low density may also be useful where structural loads are to be kept low. Heavyweight concrete can be used in applications including as a ballast in offshore engineering applications and as a shield in nuclear power plants. In both cases, the modification of density is normally achieved through the use of either lightweight or heavyweight aggregates, although lightweight concrete can, in some cases, be produced through other means (such as foamed concrete).

The specification of lightweight and heavyweight concrete is covered by the *BS EN 206* and the *BS 8500* standards. For heavyweight concrete, there is no difference to the manner in which the concrete is specified. In the case of lightweight concrete, a density class must be specified, and a different set of compressive strength classes specifically for this material is used. No special requirements for durability are required.

Guidance on lightweight concrete has been published in a document developed by the Institution of Structural Engineers and The Concrete Society [51]. This includes a design methodology based around a combination of two methods – the *FIP Manual of Lightweight Aggregate Concrete* [52] and *ACI 211-2-69: Recommended Practice for Selecting Proportions for Structural and Lightweight Concrete*, which has subsequently been revised [53].

Despite stating that it is not intended for the design of heavyweight concrete, the BRE design method is largely applicable. The only significant shortfall is that the graphical system used to estimate the wet density of a mix is configured only to handle aggregate densities of up to 2900 kg/m³. The density of aggregates used in heavyweight concrete is commonly as high as 3900 kg/m³ and in certain cases can potentially be as high as 8900 kg/m³. There is, however, potential for estimating the wet density through other means or through extension of the graphical system. Alternatively, the American Concrete Institute document *ACI 211.1-91: Standard Practice for Selecting Proportions for Normal, Heavyweight, and Mass Concrete* [54] can be used.

REFERENCES

1. Wong, H. S., M. Zobel, N. R. Buenfeld, and R. W. Zimmerman. Influence of the interfacial transition zone and microcracking on the diffusivity, permeability, and sorptivity of cement-based materials after drying. *Magazine of Concrete Research*, v. 61, 2009, pp. 571–589.

2. Scrivener, K. L. and K. M. Nemati. The percolation of pore space in the cement paste/aggregate interfacial zone of concrete. *Cement and Concrete Research*, v. 26, 1996, pp. 35–40.

3. Winslow, D. N. and D. C. Cohen. Percolation and pores structure in mortars and concrete. *Cement and Concrete Research*, v. 24, 1994, pp. 25–37.

4. Meng, B. Calculation of moisture transport coefficients on the basis of relevant pore structure parameters. *Materials and Structures*, v. 27, 1994, pp. 125–134.

5. Reinhardt, H.-W. and M. Jooss. Permeability and self-healing of cracked concrete as a function of temperature and crack width. *Cement and Concrete Research*, v. 33, 2003, pp. 981–985.

6. van Brackel, J. and P. M. Heertjes. Analysis of diffusion in macroporous media in terms of a porosity, a tortuosity, and a constrictivity factor. *International Journal of Heat and Mass Transfer*, v. 17, 1974, pp. 1093–1103.

7. Curie, J. A. Gaseous diffusion in porous media: Part 2. Dry granular materials. *British Journal of Applied Physics*, v. 11, 1960, pp. 318–324.

8. British Standards Institution. *BS EN 197-1:2011: Cement: Part 1. Composition, Specifications, and Conformity Criteria for Common Cements*. London: British Standards Institution, 2011, 50 pp.

9. British Standards Institution. *BS EN 196-1:2005: Methods of Testing Cement. Determination of Strength*. London: British Standards Institution, 2005, 36 pp.

10. British Standards Institution. *BS 8500-2:2006: Concrete: Complementary British Standard to BS EN 206-1 – Part 2. Specification for Constituent Materials and Concrete*. London: British Standards Institution, 2006, 52 pp.

11. Chen, J. J., J. J. Thomas, H. F. W. Taylor, and H. M. Jennings. Solubility and structure of calcium silicate hydrate. *Cement and Concrete Research*, v. 34, 2004, pp. 1499–1519.

12. Taylor, H. F. W. *Cement Chemistry, 2nd ed.* London: Thomas Telford, 1997, 480 pp.

13. Hubbard, F. H., R. K. Dhir, and M. S. Ellis. Pulverised-fuel ash for concrete: Compositional characterisation of United Kingdom PFA. *Cement and Concrete Research*, v. 15, 1985, pp. 185–198.

14. Shehata, M. H. and M. D. A. Thomas. Alkali release characteristics of blended cements. *Cement and Concrete Research*, v. 36, 2006, pp. 1166–1175.

15. de Larrard, F., J.-F. Gorse, and C. Puch. Comparative study of various silica fumes as additives in high-performance cementitious materials. *Materials and Structures*, v. 25, 1992, pp. 265–272.

16. Buchwald, A., H. Hilbig, Ch. Kaps. Alkali-activated metakaolin-slag blends: Performance and structure in dependence of their composition. *Journal of Materials Science*, v. 42, 2007, pp. 3024–3032.

17. Velosa, A. L., F. Rocha, and R. Veiga. Influence of chemical and mineralogical composition of metakaolin on mortar characteristics. *Acta Geodynamica et Geomaterialia*, v. 6, 2009, pp. 121–126.

18. Hobbs, D. W. Influence of pulverised-fuel ash and granulated blast-furnace slag upon expansion caused by the alkali–silica reaction. *Magazine of Concrete Research*, v. 34, 1982, pp. 83–94.

19. Dhir, R. K., M. J. McCarthy, and K. A. Paine. *Development of a Technology Transfer Programme for the Use of PFA to EN 450 in Structural Concrete.* Technical Report CTU/1400. London: Department of Environment, 2000, 73 pp.

20. Dyer, T. D., J. E. Halliday, and R. K. Dhir. Hydration chemistry of sewage sludge ash (SSA) used as a cement component. *Journal of Materials in Civil Engineering*, v. 23, 2011, pp. 648–655.

21. British Standards Institution. *BS EN 15167-1:2006: Ground Granulated Blast-Furnace Slag for Use in Concrete, Mortar, and Grout – Part 1. Definitions, Specifications, and Conformity Criteria.* London: British Standards Institution, 2006, 24 pp.

22. British Standards Institution. *BS EN 450-1:2012: Fly Ash for Concrete – Definition, Specifications, and Conformity Criteria.* London: British Standards Institution, 2012, 34 pp.

23. British Standards Institution. *BS EN 13263-1:2005: Silica Fume for Concrete – Part 1. Definitions, Requirements, and Conformity Criteria.* London: British Standards Institution, 2005, 28 pp.

24. British Standards Institution. *BS EN 12630:2013: Aggregates for Concrete.* London: British Standards Institution, 2013, 60 pp.

25. British Standards Institution. *PD 6682-1:2009: Aggregates: Aggregates for Concrete: Guidance on the use of BS EN 12620.* London: British Standards Institution, 2009, 34 pp.

26. British Standards Institution. *BS 8204-2:2003: Screeds, Bases, and In Situ Floorings – Part 2. Concrete Wearing Surfaces: Code of Practice.* London: British Standards Institution, 2003, 44 pp.

27. British Standards Institution. *BS EN 1097-2:2010: Tests for Mechanical and Physical Properties of Aggregates – Part 2. Methods for the Determination of Resistance to Fragmentation.* London: British Standards Institution, 2010, 38 pp.

28. British Standards Institution. *BS EN 1097-8:2009: Tests for Mechanical and Physical Properties of Aggregates – Part 8. Determination of the Polished Stone Value.* London: British Standards Institution, 2009, 34 pp.

29. Highways Agency. Pavement Design and Maintenance: Section 5. Pavement Materials: Part 1. Surfacing Materials for New and Maintenance Construction. In *Design Manual for Roads and Bridges*, v. 7, 2006, 20 pp.

30. British Standards Institution. *BS EN 1367-1:2007: Tests for Thermal and Weathering Properties of Aggregates – Part 1. Determination of Resistance to Freezing and Thawing.* London: British Standards Institution, 2007, 16 pp.

31. British Standards Institution. *BS EN 1367-2:2009: Tests for Thermal and Weathering Properties of Aggregates – Part 2. Magnesium Sulphate Test.* London: British Standards Institution, 2009, 18 pp.

32. British Standards Institution. *BS EN 1367-4:2008: Tests for Thermal and Weathering Properties of Aggregates – Part 4. Determination of Drying Shrinkage.* London: British Standards Institution, 2008, 18 pp.

33. Building Research Establishment. *BRE Digest 330: Part 2. Alkali–Silica Reaction in Concrete: Detailed Guidance for New Construction.* Watford: Building Research Establishment, 2004, 12 pp.

34. British Standards Institution. *BS 812-123:1999: Testing Aggregates – Method for Determination of Alkali–Silica Reactivity: Concrete Prism Method.* London: British Standards Institution, 1999, 18 pp.

35. British Cement Association. *Testing Protocol for Greywacke Aggregates: Protocol of the BSI B/517/1/20 Ad Hoc Group on ASR.* Crowthorne: British Cement Association, 1999, 8 pp.

36. British Standards Institution. *BS EN 1744-1:2009: Tests for Chemical Properties of Aggregates – Chemical Analysis.* London: British Standards Institution, 2009, 66 pp.

37. British Standards Institution. *BS EN 206:2013: Concrete – Part 1. Specification, Performance, Production, and Conformity.* London: British Standards Institution, 2013, 98 pp.

38. Juckes, L. M. Dicalcium silicate in blast-furnace slag: A critical review of the implications for aggregate stability. *Transactions of the Institution of Mining and Metallurgy C,* v. 111, 2002, pp. 120–128.

39. British Standards Institution. *BS EN 13055-1:2002: Lightweight Aggregates: Part 1. Lightweight Aggregates for Concrete, Mortar and Grout.* London: British Standards Institution, 2002, 40 pp.

40. Rixom, R. and N. Mailvaganam. *Chemical Admixtures for Concrete, 3rd ed.* London: Spon, 1999, 456 pp.

41. The Concrete Society. *Guidance on the Use of Macrosynthetic Fibre–Reinforced Concrete.* Technical Report Number 65. Camberley: The Concrete Society, 2007, 76 pp.

42. Bamforth, P. B. *Enhancing Reinforced Concrete Durability: Guidance on Selecting Measures for Minimising the Risk of Corrosion of Reinforcement in Concrete.* Camberley: The Concrete Society, 2004, 108 pp.

43. British Standards Institution. *BS EN 14889-1:2006: Fibres for Concrete – Part 1. Steel Fibres: Definitions, Specifications, and Conformity.* London: British Standards Institution, 2006, 30 pp.

44. British Standards Institution. *BS EN 14889-2:2006: Fibres for Concrete – Polymer Fibres: Definitions, Specifications, and Conformity.* London: British Standards Institution, 2006, 30 pp.

45. British Standards Institution. *BS 8500-1:2006: Concrete – Complementary British Standard to BS EN 206-1: Part 1. Method of Specifying and Guidance of the Specifier.* London: British Standards Institution, 2006, 66 pp.

46. Building Research Establishment. *BRE Digest 357: Shrinkage of Natural Aggregates in Concrete.* Watford: Building Research Establishment, 1991, 4 pp.

47. Teychenné, D. C., R. E. Franklin, and H. C. Erntroy. *Design of Normal Concrete Mixes, 2nd ed.* Watford: Building Research Establishment, 1997, 38 pp.

48. European Ready-Mixed Concrete Organisation. *The European Guidelines for Self-Compacting Concrete: Specification, Production, and Use.* Brussels, Belgium: European Ready-Mixed Concrete Organisation, 2005, 63 pp.

49. Concrete Society Working Party. *Design Guidance for High-Strength Concrete.* Concrete Society Technical Report 49. Slough: The Concrete Society, 1998, 168 pp.

50. Brady, K. C., G. R. A. Watts, and M. R. Jones. *Specification for Foamed Concrete.* TRL Application Guide AG39. Wokingham: Transport Research Laboratory, 2001, 60 pp.

51. The Institution of Structural Engineers. *Guide to the Structural Use of Lightweight Aggregate Concrete.* London: The Institution of Structural Engineers, 1987, 58 pp.

52. Fédération Internationale de la Précontrainte. *FIP Manual of Lightweight Aggregate Concrete, 2nd ed.* Glasgow, Norway: Surrey University Press, 1983, 259 pp.

53. American Concrete Institute. *ACI 211-2R-98: Standard Practice for Selecting Proportions for Structural Lightweight Concrete.* Farmington Hills, Michigan: American Concrete Institute, 1998, 20 pp.

54. American Concrete Institute. *ACI 211.1-91: Standard Practice for Selecting Proportions for Normal, Heavyweight, and Mass Concrete.* Farmington Hills, Michigan: American Concrete Institute, 1991, 38 pp.

Chapter 6

Construction of durable concrete structures

6.1 INTRODUCTION

Although the materials used in concrete structures and their design play an essential role in ensuring durability, a number of on-site construction activities may also be necessary to achieve appropriate durability performance. Some of these activities ensure the achievement of the potential properties of a concrete mix through good practice – such as following the use of appropriate curing procedures – whereas others involve the use of specific techniques that enhance performance. These aspects of the construction process are discussed in the next sections.

Since several of the techniques examined influence the surface characteristics of concrete, this chapter starts with a discussion of the characteristics of this zone. Then various approaches to controlling surface characteristics are examined, including the use of controlled permeability formwork, curing, surface finishing techniques and surface protection treatments. Finally, the use of cathodic protection systems – electrochemical techniques that protect steel reinforcement from corrosion – is discussed.

6.2 SURFACE OF CONCRETE

The protection afforded to concrete by its surface differs from that provided by the interior for a number of reasons. One major reason is that the wall effect (discussed in Chapter 5) causes a higher proportion of fine material, such as cement, to position itself at the surface that is confining the fresh concrete (for instance, formwork). This leads to the aggregate/cement ratio at the surface being significantly lower than elsewhere. Since the cement matrix of concrete is usually more porous than the aggregate, the effect of this is a higher porosity at the surface, as shown in Figure 6.1.

The segregation of water from the solid material in concrete leads to a layer of water forming at the surface. The process of segregation may reduce the water/cement (W/C) ratio of the bulk concrete slightly, with

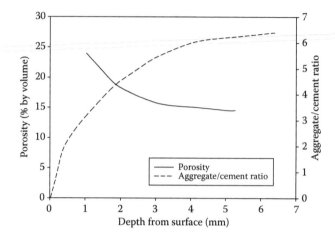

Figure 6.1 Changes in the porosity and aggregate/cement ratio at increasing depths into concrete. (From Kreijger, P. C., *Materials and Structures*, 17, 1984, 275–283.)

possible positive effects. However, where bleeding is significant, the material at the surface will tend to have a very high W/C ratio. This leads to the formation of surface 'laitance', which leaves a thin layer of weak, friable material at the surface. Concrete mixes that are prone to bleeding tend to have low cement contents and contain coarser cements. Cements with high tricalcium aluminate contents tend to bleed less.

Acting against the process of bleeding is the evaporation of water. If the rate of evaporation matches or exceeds that of the bleed rate, then laitance will not be formed.

6.2.1 Controlled permeability formwork

Conventional formwork is usually made from materials such as plywood or steel and will typically be either totally impermeable to water or of very limited permeability. More recently, controlled permeability formwork (CPF) has been developed, which limits the effects previously discussed.

CPF consists of two layers: a filter layer and a drainage layer. The filter layer is normally a textile with holes in it, which permit the movement of water through but prevent the movement of cement particles. Behind the filter layer is the drainage layer, which normally consists of an open network of plastic spacers that provide a means for the water to drain away.

By allowing water to drain from the concrete surface, the W/C ratio in this zone is reduced, leading to enhanced strength and permeation properties. Since air can also permeate the filter layer, bubbles trapped at the formwork surface, which would otherwise lead to the formation of surface defects known as 'blowholes', are also allowed to escape.

The effects of using CPF are illustrated in Figure 6.2, which shows the chloride contents of concrete made with and without CPF and subsequently exposed to a chloride-bearing environment. This is the result of the improved permeation properties at the surface, and similar results are obtained with respect to processes that involve the movement of substances other than chlorides, such as carbonation [2]. Moreover, resistance to physical deterioration processes is also enhanced, as shown in Table 6.1 for freeze–thaw attack.

The improvement of concrete surface quality potentially affords a reduction in cover depth, in situations where the protection of reinforcement from the effects of chloride ingress or carbonation define this depth. A method for calculating the equivalent cover thickness achieved using CPF has been proposed [3] using the equation

$$C_{eq} = 1.49 C_{CPF}$$

where C_{eq} is the equivalent cover thickness (mm), and C_{CPF} is the thickness of the outer layer of the concrete resulting from the use of CPF (mm).

From this, potential reductions in thickness can be deduced.

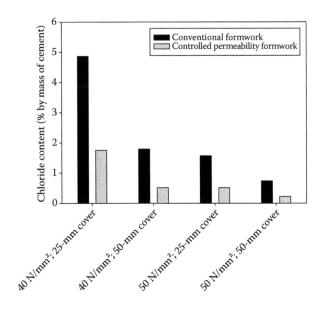

Figure 6.2 Total chloride content of concrete cast against conventional and controlled permeability formwork and subsequently exposed to a 3-h wetting/9-h drying regime using a 3.0-M NaCl solution. (From McCarthy, M. J. et al., *Materials and Structures*, 34, 2001, 566–576.)

Table 6.1 Average values of scaled surface area and mass loss resulting from scaling obtained from concrete specimens made with and without CPF after exposure to 20 freeze–thaw cycles

Formwork	Proportion of surface area suffering from scaling (%)	Weight loss (mg/mm²)
Normal	66	1.5
CPF	17	0.4

Source: Cairns, J., *Magazine of Concrete Research*, 51, 1999, 73–86.

6.2.2 Surface finish

Although laitance may not always present a problem, it clearly needs to be avoided where a good quality horizontal surface is being constructed, such as a pavement or a floor slab. In such cases, finishing of the surface is carried out. Finishing of an unhardened concrete surface can be done using several techniques and combinations of techniques, such as skip floating, brushing, power floating and power trowelling.

Once a concrete slab or a pavement has been placed and compacted, it must be struck off to the correct elevation and levelled using a skip (or 'bull') float, and its edges must be formed using an edger. At this point, the most basic surface finish that can be provided is either a brushed or tamped finish.

Brushing is used where skid resistance is required. A brushed (or 'broomed') surface is achieved by drawing a broom (ideally, one, specifically for this purpose) over the surface to create a roughed surface. Tamping involves imprinting ridges in the concrete surface using a tamping beam. Again, the main reason for tamping is to impart skid resistance to the surface.

If a smoother surface is required, power floating can be used. A power float consists of a rotating disk or an arrangement of rectangular blades joined at a rotating central axis, which is pushed over the concrete surface. The moving surface of the disk or blades acts to smooth the still-plastic concrete surface and close any openings.

Power floating of the concrete will alter the nature of the surface, since it will tend to bring finer particles of cement and sand to the surface at the expense of the coarse aggregate, with similar implications for surface characteristics as discussed previously for the wall effect.

Moreover, from a durability perspective, it is very important that power floating is carried out at the appropriate point in time. The correct procedure is to wait until all of the bleed water has evaporated and the concrete has stiffened to some extent (typically after ~3 h using conventional cements and under UK ambient conditions). Where power floating is carried out prematurely, the bleed water is driven below the surface and accumulates in a layer below the surface, which, as a result of the higher W/C ratio, is much weaker and more permeable.

Power floating leaves a surface with a gritty texture. If a still-smoother surface is required, power trowelling can be used after power floating. A power trowel has a similar configuration to a power float, with the exception that the blades are smaller and the resulting finish is much less textured than that resulting from power floating.

In the hardened state, concrete surfaces such as floors can also be ground or polished to provide a suitably even surface. Where concrete floors are required to have high levels of abrasion resistance, a dry shake/sprinkle finish may be applied to the surface. These are proprietary mixtures of abrasion-resistant aggregates and cement. They are sprinkled dry onto the fresh screed surface and trowelled in to create an abrasion-resistant surface.

6.3 CURING

In its freshly mixed condition, concrete will have adequate water for cement hydration down to a W/C ratio of 0.38 (at least where Portland cement is the only cement present). Below this, self-desiccation will occur – the free water will be reduced to such an extent that cement hydration is arrested. Regardless of the W/C ratio, where concrete surfaces are exposed to the atmosphere, water can evaporate, leaving the quantity of free water insufficient for complete cement reaction.

Factors that influence the rate of evaporation include ambient temperature, relative humidity, concrete temperature, wind speed and the exposed surface area. A reference chart has been produced by the American Concrete Institute (Figure 6.3), which allows the rate to be estimated based on most of these parameters. It should also be noted that sunlight incident on a concrete surface will act to increase its temperature.

The chart is based on an equation devised by Menzel [7], which has subsequently been simplified to

$$E = 5([T_c + 18]^{2.5} - r \cdot [T_a + 18]^{2.5})(V + 4) \times 10^{-6}$$

where E is the evaporation rate (kg/m²/h), T_c is the concrete temperature (°C), T_a is the air temperature (°C), R is the relative humidity (%) and V is the wind velocity (km/h) [8].

Inadequate curing affects both the mechanical properties and permeation characteristics of the concrete, particularly at the surface. Since all aspects of durability are influenced by one or both of these properties, durable concrete must be well cured. Figure 6.4 shows the development of compressive strength under different curing conditions, with both sealing

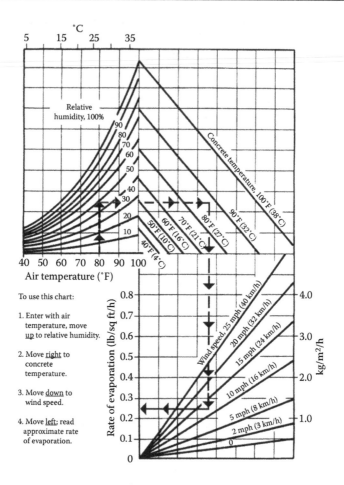

Figure 6.3 Chart for estimating the rate of evaporation of water from concrete. (From ACI Committee 308. *Standard Practice for Curing Concrete – ACI 308-92.* Farmington Hills, Michigan: American Concrete Institute, 1997, p. 11; National Ready-Mixed Concrete Association. *Plastic Cracking of Concrete.* Silver Spring, Maryland: NRMCA, 1960, p. 2; Menzel, C. A. Causes and prevention of crack development in plastic concrete. *Proceedings of the Portland Cement Association.* Shokie, Illinois: Portland Cement Association, 1954, pp. 130–136.)

of the concrete and water curing (submersion in water) producing significantly improved results relative to air curing.

Figure 6.5 shows water sorptivity profiles through concrete made with Portland cement (PC), ground granulated blast-furnace slag (GGBS) or fly ash (FA) after different curing periods. The plots show that the surface of the concrete is the most affected by inadequate curing (since this is the zone from which water is initially lost) and that the concrete containing FA and

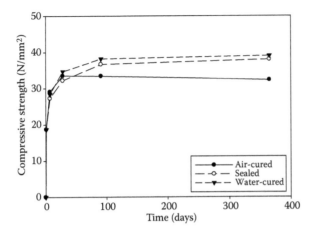

Figure 6.4 Strength development of 100-mm concrete cylinders under different curing conditions. Curing temperature = 20°C; relative humidity for air curing
and sealed curing = 50%. (From Aitcin, P.-C., *ACI Materials Journal*, 91, 1994,
349–354.)

GGBS is more sensitive to curing. This is the result of the slower reactions
of such materials – where water is able to evaporate, the pozzolanic and
latent hydraulic reactions of these materials will be occurring at times when
there is less moisture and thus less opportunity for the formation of hydration products, in comparison with the earlier reactions of PC.

The sensitivity of FA and GGBS to curing is also shown in Figure 6.6,
which shows oxygen permeability measurements for the same concrete
mixes characterised in Figure 6.5. The influence of curing on the diffusion
of chemical species through concrete is shown in Figure 6.7, in this case for
chloride ions.

Resistance to physical forms of deterioration is also compromised by
poor curing. Figures 6.8 and 6.9 illustrate the influence of curing on resistance to freeze–thaw attack and abrasion, respectively. The similarity of
the relationships shown in these various figures is notable.

Thus, under conditions where evaporation is likely to occur to any extent,
curing is essential. Curing refers to actions taken to prevent water loss from
occurring or to provide additional water to replace that which is lost.

The prevention of evaporation can be achieved through a number
of means. The most basic means is by striking formwork at later ages.
However, this may present problems in terms of preventing the use of formwork elsewhere and may be restricted by the need to strike early in certain
cases to avoid thermal contraction issues (see Chapter 2). Exposed concrete
surfaces can be covered by plastic sheeting or building paper. These coverings are normally effective but are far from 100% efficient and are prone

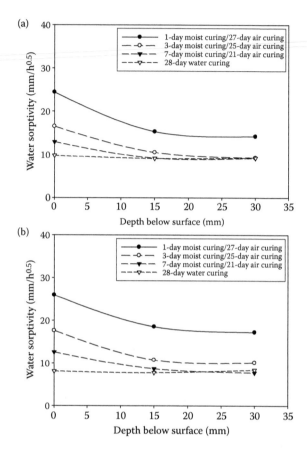

Figure 6.5 Water sorptivity values obtained from slices taken from concrete specimens made with different cement types under different curing regimes. (a) 100% PC (W/C ratio = 0.47); (b) 30% FA/70% PC (W/C ratio = 0.49); (c) 50% GGBS/50% PC (W/C ratio = 0.51). (From Ballim, Y., *Materials and Structures*, 26, 1993, 238–244.)

to disruption (for instance, by high winds). For this reason, it is essential that great care is taken in securing such coverings to the concrete surface.

Where curing is critical and only retention of water is required, a curing membrane (or 'curing compound') may be applied. Curing membranes are liquid formulations that are sprayed onto concrete surfaces and subsequently form a solid layer that prevents water vapour from passing through. A range of different products of this type are commercially available and include emulsions of waxes or bitumen, polymer solutions and solutions of inorganic silicate compounds such as sodium silicate. The emulsions and polymer solutions simply work by depositing a solid layer once the liquid

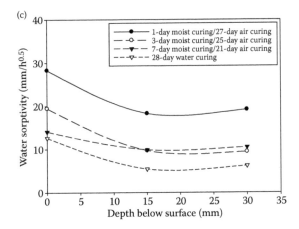

Figure 6.5 (Continued) Water sorptivity values obtained from slices taken from concrete specimens made with different cement types under different curing regimes. (a) 100% PC (W/C ratio = 0.47); (b) 30% FA/70% PC (W/C ratio = 0.49); (c) 50% GGBS/50% PC (W/C ratio = 0.51). (From Ballim, Y., *Materials and Structures*, 26, 1993, 238–244.)

component of the formulation has evaporated. The inorganic solutions work by reacting with soluble calcium from the concrete to form calcium silicate compounds.

The efficiency of curing membranes to retain water within concrete can be characterised using a method described in *BS 7542* [11]. In the United Kingdom, curing membranes must have an efficiency of greater than 75%. Some membrane formulations are available, which can achieve efficiencies in excess of 90%. Such formulations are referred to as 'super' grade, whereas between 75% and 90% efficiency is referred to as 'standard' grade.

For optimum water retention to be achieved, it is essential that a complete surface coverage of membrane is achieved. For this reason, many formulations contain a dye, which allows visual confirmation of this. The dye is fugitive, allowing the colour to rapidly fade after application.

In applications where large horizontal surfaces are exposed to sunlight, formulations containing reflective aluminised particles or a white pigment may be used. These membranes reflect solar radiation and prevent an increase in concrete temperature, which would otherwise accelerate evaporation.

One drawback of membranes is that, if they are used on construction joints or surfaces that are subsequently to be bonded to other materials, they need to be removed after an appropriate period, which typically requires some form of mechanical scarification.

Curing techniques that provide additional water include ponding and spraying. Spraying is self-explanatory, but ponding involves flooding

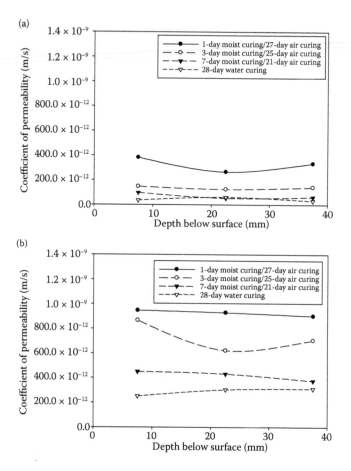

Figure 6.6 Oxygen permeability values obtained from slices taken from concrete specimens made with different cement types under different curing regimes. Mixes are the same as in Figure 6.5. (From Ballim, Y., *Materials and Structures*, 26, 1993, 238–244.)

horizontal surfaces with water. Additionally, absorbent materials saturated with water can be brought into contact with the concrete surface. Damp fabric, such as hessian, is most commonly used for this, often in combination with plastic sheeting to prevent evaporation from the fabric. It is also possible to place other absorbent, water-bearing materials against the concrete surface. UK guidance suggests the use of wet sand on horizontal surfaces, whereas guidance from the United States also suggests damp soil, sawdust, straw or hay [5].

EN 13670 [15] provides requirements for curing. It defines a series of Curing Classes. These are numbered 1 to 4, corresponding to increasing

(c)

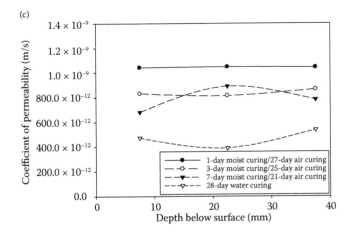

Figure 6.6 (Continued) Oxygen permeability values obtained from slices taken from concrete specimens made with different cement types under different curing regimes. Mixes are the same as in Figure 6.5. (From Ballim, Y., *Materials and Structures*, 26, 1993, 238–244.)

curing periods. Curing Class 1 involves curing for 12 h. Curing Classes 2 to 4 are not defined in terms of specific curing periods but in terms of the proportion of compressive strength, which is required to have developed before curing can be stopped. These are 35%, 50%, and 70% of the characteristic 28-day compressive strength for Classes 2, 3, and 4, respectively.

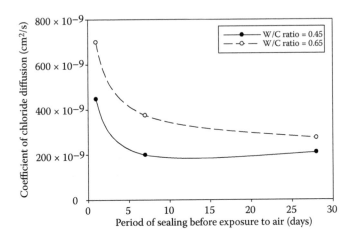

Figure 6.7 Coefficients of chloride diffusion obtained from concrete specimens after different periods of sealed curing, followed by exposure to air up to an age of 28 days. (From Hillier, S. R. et al., *Magazine of Concrete Research*, 52, 2000, 321–327.)

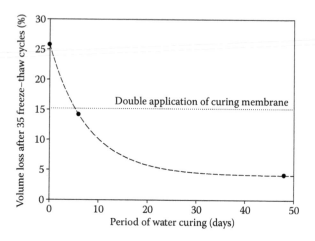

Figure 6.8 Influence of period of curing in water (and the application of a curing membrane) on the freeze–thaw durability of concrete specimens containing blast-
furnace slag. Specimens were tested at an age of 55 days, with the period of
water curing (20°C) occurring first, followed by a period of curing in air at
20°C/65% relative humidity. (From Gunter, M. et al., Effect of curing and type
of cement on the resistance of concrete to freezing in de-icing salt solutions.
In J. Scanlon, ed., *ACI Special Publication SP-100: Proceedings of the Katharine and
Bryant Mather International Symposium.* Detroit, Michigan: American Concrete
Institute, 1987, pp. 877–899.)

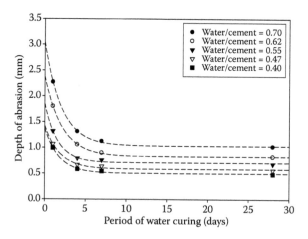

Figure 6.9 Influence of period of water curing (20°C) on the abrasion depth of concrete
specimens with various W/C ratios tested at an age of 28 days using a rolling steel wheel abrasion test machine. (From Dhir, R. K. et al., *Materials and
Structures,* 24, 1991, 122–128.)

6.4 SURFACE PROTECTION SYSTEMS

One of the most obvious means of preventing harmful substances from penetrating into concrete is to apply a protective layer at the concrete surface. Such treatments can take a number of forms. The simplest treatment is a coating that forms a continuous protective layer at the concrete surface (Figure 6.10a). However, there are potentially greater benefits if the substance used is capable of penetrating beneath the surface, thus blocking pores within a volume of concrete located at the surface (Figure 6.10b). Such materials are referred to as 'surface sealers' or 'impregnants'. Additionally, treatments can be applied to surfaces, which penetrate beneath the surface without altering the pore structure of the material in any way, but which impart a degree of hydrophobicity to the pore surfaces – hydrophobic impregnants (Figure 6.10c).

Surface protection systems are covered by a single standard – *BS EN 1504-2* [16] – that defines their required characteristics.

This section also looks briefly at the use of protective liner systems.

6.4.1 Surface coatings

Surface coating systems based on a range of materials are available commercially. They include both organic and inorganic formulations. Organic systems include those based on polymers, including polymers and elastomers delivered in the form of solutions in either water or organic solvents and deposited as the solvent evaporates; resin systems, which harden once applied to the surface; alkyds (fatty acid–modified polyesters); bitumen; and oleoresins (oil/resin mixtures obtained from various plants). Inorganic systems include cementitious formulations and alkali silicate products, similar to those used as the curing membranes previously discussed.

Coating formulations may also contain other substances, including pigments, plasticisers, rheology modifiers and fillers.

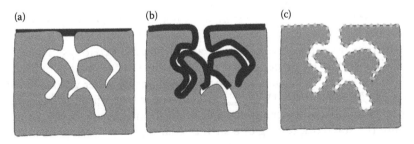

Figure 6.10 Different types of surface protection: (a) coating; (b) sealer; (c) hydrophobic impregnant.

BS EN 1504-2 identifies the key characteristics required of surface treatments based on the application for which they are intended. This is done by defining a series of 'principles' for the protection of concrete. These principles are as follows:

- Protection from ingress
- Moisture control
- Increasing physical resistance
- Resistance to chemicals
- Increasing resistivity

The standard identifies surface coatings as being potentially suitable for addressing all of these principles.

To establish whether a given surface treatment performs one of these functions, measures of performance are defined which the material must satisfy. These are known as 'compulsory performance requirements'. Additionally, other non-compulsory performance requirements may also be defined for certain applications where such characteristics may be useful. Regardless of the main purpose for using a surface coating, requires compulsory performance with regards to adhesion strength (measured using a pull-off test) and the capillary absorption (of water) and liquid water permeability characteristics of the resulting surface.

'Protection from ingress' means preventing the movement of substances potentially harmful to either concrete or steel reinforcement, including water, carbon dioxide and chloride ions. The compulsory performance requirements for coatings used for ingress protection in the standard are permeability to carbon dioxide and permeability to water vapour.

'Moisture control' primarily refers to actions that prevent the movement of liquid water into concrete while permitting it to 'breathe'. Although coatings have been traditionally designed to completely seal the surface, it is often desirable to allow water vapour to leave via the surface. The trapping of moisture beneath the concrete surface is normally considered to be undesirable, since it can exacerbate water-related durability problems. The need for water vapour permeability in certain applications has led to the introduction and development of anticarbonation coatings. These are typically acrylate elastomer–based formulations that are highly impermeable to the ingress of carbon dioxide while permitting water vapour to pass through. *BS EN 1504-9* [17] suggests that upper horizontal surfaces need not be permeable to water vapour, but vertical surfaces and soffits should. Within the standard, testing of a surface coating for water vapour permeability is required to classify the formulation as either 'permeable' or 'not permeable'.

Where the coating is applied to impart physical resistance, the aim is to improve performance with regards to impact, abrasion and freeze–thaw attack. Thus, the compulsory performance requirements are abrasion resistance and impact resistance.

'Chemical resistance' refers mainly to resistance to sulphate attack and acid attack. Thus, resistance to severe chemical attack is the additional compulsory performance requirement.

'Increasing resistivity' is similar to moisture control, as it refers to surface treatment to reduce moisture levels within the concrete during service. However, in this case, this is carried out to reduce the electrical conductivity of the concrete. This is a desirable condition where corrosion of reinforcement is to be prevented, since the electrochemical processes of steel corrosion can be slowed considerably if the resistivity of the concrete is very low. Thus, the requirements are identical with those for moisture control.

Other noncompulsory characteristics that may be desirable in certain applications are also included: linear shrinkage, compressive strength, coefficient of thermal expansion, resistance to thermal shock, crack bridging ability, reaction to fire, slip/skid resistance, resistance to weathering, antistatic behaviour and diffusion of chloride ions. Additionally, several additional adhesion performance requirements are included, as well as an additional means of determining resistance to chemical attack. In all cases, test methods for measuring each performance requirement are included, as well as criteria for passing such tests.

The quality of a coating is very much dependent on the manner in which it is applied. *BS 6150* [18] is a code of practice that covers the selection and application of coatings for a range of different surfaces, including concrete. In general, coatings are more successful when applied to dry concrete, and guidance is provided on measures that should be followed to ensure this. Efflorescence should be removed from surfaces, and painting should be ideally postponed until efflorescence has ceased. Additionally, the code of practice highlights the sensitivity of some organic coatings to the alkaline conditions at the concrete surface and suggests the application of alkali-resisting primers to the surface to enhance performance.

Service lives of surface coatings can be as long as 20 years, but periods of 10 to 15 years are more common, after which reapplication is necessary [3].

6.4.2 Surface sealers

Surface sealer formulations are often based around similar formulations to surface coatings, with the main distinction being a much lower viscosity to maximise the depth of penetration. They can include solvent solutions of acrylic polymers, polyurethane and epoxies. In some cases (specifically, epoxies and polyurethanes), an agent is required to curing the compounds, which is mixed with the solution prior to application. More recently, polyurethane formulations have been developed, which cure in contact with moisture. Inorganic formulations also exist, which are based on silicofluoride compounds and sodium or potassium silicate.

In terms of the principles of protection defined in *BS EN 1504-2*, surface sealers are effective at improving protection from ingress and improving physical resistance. Typically, surface sealers cannot be relied upon to improve chemical resistance.

Where the main objective is protection from ingress, the compulsory performance requirements are capillary absorption, permeability to water and depth of penetration. It should be pointed out that surface sealers will also protect against carbonation, although not to the extent that an anticarbonation coating can.

Inorganic sealers are, in most cases, unsuitable for improving protection against ingress. The main reason for using such formulations is improvement of physical resistance. Where this is the main objective, abrasion resistance, capillary absorption and permeability to liquid water are the compulsory performance requirements.

During and after application, sealants are not as sensitive to surface conditions, and so the preparation requirements are usually less demanding. However, it is important that, where protection from ingress is required, the application is sufficient to achieve this.

6.4.3 Hydrophobic impregnants

Hydrophobic pore-lining impregnants are liquids applied to concrete surfaces to render the surface and interior of pores near the surface hydrophobic. They have the benefit of having little effect on the surface appearance. Additionally, because they do not create a physical barrier between the concrete and the outside world, they allow the concrete to 'breathe'.

Most hydrophobic agents used on concrete are silane compounds, and the most common of these are alkyl trialkoxysilane monomers, such as isobutyl (trimethoxy) silane (Figure 6.11).

The basic structural configuration of most silane compounds used for concrete protection applications are three methoxy (CH_3-O-) or ethoxy (CH_3-CH_2-O-) groups bound to a silicon atom, along with an alkyl group

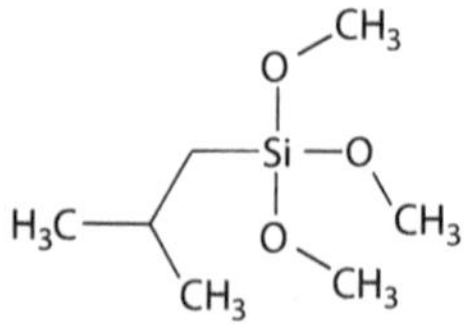

Figure 6.11 Chemical structure of isobutyl (trimethoxy) silane.

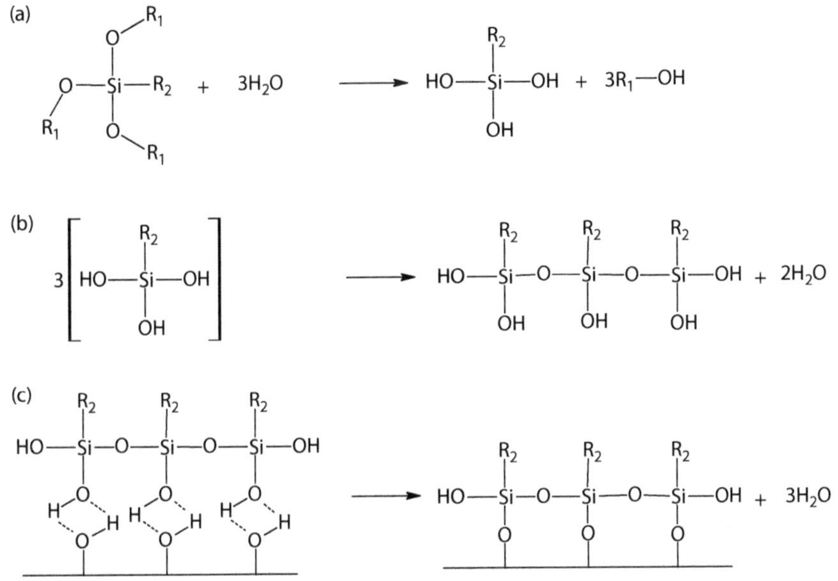

(a)

(b)

(c)

Figure 6.12 Reactions leading to the attachment of hydrophobic silane compounds to concrete surfaces: (a) hydrolysis; (b) condensation; (c) bonding to a surface.

(CH_3-, CH_3-C_2-, etc.). In water, the methoxy or ethoxy groups (R_1 in Figure 6.12a) undergo hydrolysis such that they become detached from the silicon atom. The hydrolysis reaction is base- and acid-catalyzed – it occurs at a faster rate under high- or low-pH conditions. This is ideal in the case of concrete, since the hydrolysis reaction is initiated by bringing the silane into contact with the alkaline, hardened cement. The hydrolysed molecules then undergo condensation reactions, leading to the formation of short chains (oligomers), as shown in Figure 6.12b. The oligomers become attached to hydroxide (OH) groups on the surface of cement hydration products via hydrogen bonds, with covalent bonds eventually forming to strongly bind the molecules to the surface (Figure 6.12c).

The alkyl group (R_2) is hydrophobic, and the resulting coating causes water droplets in contact with the concrete surface to display a high contact angle, typically up to approximately 120°. The benefits of this are two-fold. First, water falling onto the surface will make very little contact with it, reducing the extent to which it can enter the pores of the concrete. Second, where the silanes have penetrated further into the concrete and lined pore surfaces, capillary action (see Chapter 5) is essentially eliminated, since this relies on a low contact angle, which means that water is not drawn into the concrete's interior.

BS EN 1504-2 identifies hydrophobic impregnants as being suitable for use in applications where ingress protection, moisture control and increasing resistivity are required outcomes. In all cases, the compulsory performance requirements are depth of penetration, drying rate and testing for water absorption and alkali resistance. Depth of penetration is clearly important, since this will define the extent of protection offered. Where the penetration depth is <10 mm, hydrophobic impregnants are identified as being Class 1, whereas Class 2 formulations display greater penetration depths.

Hydrophobic pore-lining impregnants are applied by spraying. In the United Kingdom, the Highways Agency's *Design Manual for Roads and Bridges* [19] requires that the concrete surface has been allowed to naturally dry for 24 h to a surface-dry condition. The surface must also be clean and free of curing membrane residues, dust and debris. Spraying is carried out in two stages, with at least 6 h between applications and a coverage of 300 mL/m² at each application.

The *Design Manual for Roads and Bridges* also states that if application is carried out appropriately, surface hydrophobicity will remain effective for at least 15 years under UK conditions. However, where the surface is subject to degradation mechanisms, it is likely that this period will be reduced.

6.4.4 Screeds

Floors in a concrete structure can be located at ground level on a ground-supported slab or at higher levels as a suspended floor supported by a structural concrete slab, a combination of precast concrete units and a structural slab or a composite metal deck. In many cases, it may be desirable to directly finish the concrete surface (discussed in Section 6.2.2) such that the slab is appropriately level and abrasion resistant. However, in other cases, it may be necessary to lay a screed over the slab.

A screed is a layer of material – usually cement, sand and water – that is applied to a concrete base to provide a level surface (a 'levelling screed'), to act as a support for a subsequent layer of flooring or to act as a flooring surface in its own right. When the last of these three purposes is the main objective, the screed is referred to as a 'wearing screed', and durability becomes a significant issue, since the screed is required to resist abrasion (Chapter 2). In some cases, screeds may also be required to be resistant to chemical attack.

The requirements for the materials, design, execution, inspection and maintenance of screeds are detailed in the code of practice – *BS 8204-2* [20]. The basic requirement for cement is that it should have a strength class of 42.5 N or an additional amount of cement should be used to counteract low strength. The code requires that the cement used in a wearing screed should either be CEM I Portland cement, sulphate-resisting cement, calcium aluminate cement and various Portland-slag cements or combinations of Portland cement and GGBS. The standard also permits the use of rapid drying/hardening proprietary cements that do not have British Standards. In the case of calcium aluminate cements, the standard stresses that the advice of the manufacturer should be sought.

Combinations can also be produced in the mixer using Portland cement and additions of FA, blast-furnace slag, limestone fines, silica fume or metakaolin. The standard also allows other additions to be used if their suitability can be demonstrated from other applications.

Aggregates for wearing screeds are required to have a Los Angeles coefficient of less than 40 (Chapter 5). Furthermore, they are required to conform to *BS EN 12620 – Aggregates for Concrete* [21]. Fine aggregate should conform to *BS EN 13139* [22] or, if not conforming, there should be historical data demonstrating the successful use in wearing screeds.

Pigments and chemical admixtures may also be added. The standard requires that they should not compromise the durability of the screed.

Screeds can be used in a number of ways in terms of how they are attached to the underlying concrete base. 'Bonded' screeds are screeds that are laid onto the hardened concrete base, which has been treated in such a way as to maximise the bond between the two layers. This typically involves the removal of laitance and dirt from the concrete surface using a technique such as shot blasting, removal of loose debris resulting from this process, wetting of the surface and application of a layer of cement grout immediately prior to laying the screed. Sometimes, a bonding agent is applied to the surface or a bonding admixture is added to the grout layer.

A 'monolithic' screed is one that is laid onto the concrete layer while it is still in a plastic state (normally within 3 h of mixing, although this depends on ambient temperatures and on the cement and admixtures used). It is important that bleed water is not present at the surface, since this will produce a layer of high W/C ratio at the concrete–screed interface.

Bonded and monolithic screeds are the favoured design approaches in the code, since cracking and 'curling' resulting from the drying shrinkage of the screed tend to be reduced where there is a good bond with the substrate. Shrinkage can be reduced by using a screed with a low W/C ratio. To control shrinkage, the inclusion of joints in a screed design is necessary, with details provided in the standard.

In some cases, a good bond between screed and concrete may not be possible. This could include instances where the design requires that a damp-proofing membrane be placed between the concrete and the screed or where the concrete surface has been contaminated with a substance that will prevent bonding, for example, oil. In these circumstances, an 'unbonded' screed is necessary. In such cases, *BS 8204-2* requires that a concrete 'overslab' is placed over the membrane onto which the screed will then be applied. The overslab needs to have a thickness of between 100 and 150 mm, with a greater thickness providing greater resistance to curling. The overslab need not have a screed applied to it – it can be formulated and finished in such a manner that it is a directly finished surface with abrasion-resistant capabilities in its own right. The strength and minimum cement requirements of overslabs are defined in the standard.

Examples of configurations of cement-based wearing surfaces are shown in Figure 6.13.

When a bonded screed is used, a design thickness of 40 mm is recommended, with a minimum actual thickness (as a result of eccentricities in the underlying surface) of 20 mm. Thicknesses in excess of 40 mm are possible, but increase the risk of de-bonding from the underlying concrete as a result of drying shrinkage stresses. Monolithic screeds should have a thickness of 15 ± 5 mm.

Where high resistance to abrasion is needed, a dry shake/sprinkle finish may be applied (see Section 6.2). Impact resistance may also be improved through the use of steel or polymer fibres in the screed formulation.

Good finishing and curing of a wearing screed is essential. Surface finish techniques such as power trowelling are normally used to compact the screed and thus maximise its density. After laying and finishing, a screed may also have either an inorganic or organic surface sealer (surface hardener) applied, as discussed in Section 6.4.2.

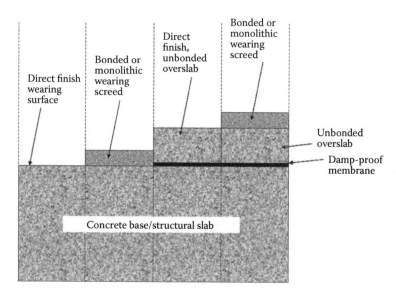

Figure 6.13 Different configurations for cement-based wearing surfaces on a ground-supported concrete slab.

Suspended floors provide an additional challenge to creating floors, since their deflection can produce stresses in conventional cement-based screeds, which can cause cracking. One option with such flooring is to use screeds based on polymer-modified cements or synthetic resins.

Polymer-modified cements are cements that are mixed with a water-based dispersion of polymer. In some cases, the cement may be supplied as a mixture of cement and polymer particles, to which water is added. As the free water is removed, both during the hydration of the cement and as a result of evaporation, polymer is deposited as a matrix among the conventional matrix of hydration products formed by the hydrating cement. Typically, polymers such as acrylates, styrene butadiene rubber and vinyl acetate are used.

The main reason for using polymer-modified cements is the higher flexural strengths that they display relative to conventional cement. In many cases, they may also enhance bond strength and render surfaces more resistant to impact and abrasion, which is clearly desirable for screeds.

Screeds based on polymer-modified cements are covered by the code of practice, *BS 8204-3* [23]. Materials for use in such screeds are essentially the same as those for conventional cement-based screeds. The code of practice takes the approach of identifying service conditions involving different levels of abrasion and defining mix proportions and screed thicknesses to suit such conditions. In terms of screed thickness, a greater thickness is required for more abrasive conditions (between 6 and 40 mm). In mix proportion terms, more abrasive conditions are countered by the use of lower ratios of cement to aggregate and lower levels of polymer/water dispersion (thus effectively reducing the W/C ratio).

Before laying a polymer-modified screed, the code of practice requires that a bonding agent is applied to the concrete base beneath. This can be an epoxy formulation but could also be a high-concentration dispersion of the polymer used in the screed or a cement paste modified with a high-concentration polymer dispersion.

Screeds made from synthetic resins are normally based around epoxy resins, methacrylates, and polyurethanes. The code of practice for resin-based screeds is *BS 8204-6* [24]. Recently, self-levelling resin-based screeds have grown in popularity. These are low-viscosity screeds that flow over the surface that they are being applied to, without the need for surface finishing.

Both synthetic resin- and polymer-modified cement-based screeds can provide good resistance to chemical attack, since they are less permeable. This is particularly true of resins. Alternatively, calcium aluminate cement–based screeds tend to be more resistant to chemical attack, rather than Portland cement–based products.

6.4.5 Protective liners

There are now many commercially available liner systems for concrete. Liners are normally manufactured from thermoplastic polymers, such as high-density polyethylene and polypropylene, although elastomeric products also exist. Most commonly, they take the form of sheets of polymer between approximately 2 to more than 10 mm in thickness, which can be installed against the formwork before placing and compaction. Most liner sheet systems have anchor studs on the concrete side to ensure a good bond.

Once the formwork is removed, adjacent sheets are welded together by extrusion welding – polymer is softened plastic is softened by heating and extruded using a hand-held extruder onto the joint. The edges of the joint are simultaneously softened by a jet of hot air from the extruder to enhance the bond between the extrudate and the edge.

Preformed liners may also be used in precast applications. For instance, preformed tubes can be located in the moulds for concrete pipes and filled with concrete, producing a pipe with a lined interior.

Because of their thickness and good chemical resistance, liners are normally used in applications where very aggressive chemical conditions are prevalent or where a high retention of liquid is necessary. The expansion of liners at higher ambient temperatures can present a problem, and some liners are sold in both light and dark forms. The reflectance of sunlight from the light-coloured material make it less prone to thermal expansion.

6.5 CATHODIC PROTECTION

In Chapter 4, approaches to designing concrete and concrete structures in such a manner as to minimise the corrosion of steel reinforcement were examined. These largely revolve around ensuring that the concrete is as impermeable as possible, exploiting the chemical composition of cement, controlling the electrochemical conditions within the concrete or using alternative reinforcement materials. However, another approach to minimising reinforcement corrosion is the installation of a cathodic protection system either during construction or retrospectively. Normally, when a cathodic protection system is fitted during construction, it is referred to as 'cathodic prevention'. However, the term 'protection' will be used throughout this section for simplicity, except where a distinction must be made.

As discussed in Chapter 4, the corrosion of steel reinforcement involves the establishment of an electrochemical cell in which different parts of the same steel article become the cathode and anode. The anodic regions of the steel corrode.

Cathodic protection involves placing the reinforcement in a concrete structure within a circuit in such a manner that the steel becomes polarised and acts as the cathode. As a result, it does not undergo corrosion at a significant rate. As is probably already evident, this is often not a trivial task, and the expense of cathodic protection means that it is normally limited to critical infrastructure. Two approaches to cathodic protection are possible: impressed current and galvanic. These techniques are discussed in the next sections.

6.5.1 Impressed current cathodic protection

Impressed current cathodic protection involves connecting the reinforcement that is to be protected to a metal or carbon component and applying a direct current between the reinforcement and the component such that the reinforcement becomes the cathode in an electrochemical cell and the other conductive material becomes the anode (Figure 6.14). Since it is the anode that will undergo corrosion, the steel remains intact (or, at least, loses material at a very slow rate). The anode is made from a conductive material that is inert and, thus, not vulnerable to corrosion.

The circuit is completed by the diffusion of ions through the concrete pore solution, with anions such as hydroxide and chloride ions (if present) migrating from the cathode to the anode, and cations (including calcium, sodium and potassium) moving toward the cathode. This can potentially have a positive effect, since it can reduce concentrations of chloride around the reinforcement (Figure 6.15), further reducing the probability of corrosion. Additional hydroxide ions are generated at the cathode through the reaction

$$2H_2O + O_2 + 4e^- \rightarrow 4OH^-$$

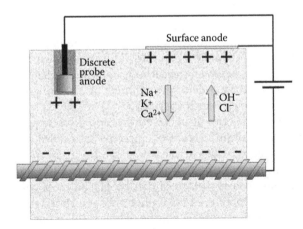

Figure 6.14 Impressed current cathodic protection.

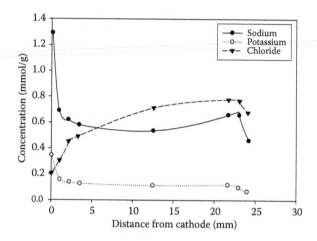

Figure 6.15 Concentrations of sodium, potassium and chloride ions in a concrete specimen subject to cathodic polarisation. (From Sergi, G. and C. L. Page. *The Effects of Cathodic Protection on Alkali–Silica Reaction in Reinforced Concrete.* Crowthorne, United Kingdom: Transport and Road Research Laboratory, 1992, p. 53.)

The changes in the chemistry around the reinforcement cause the anode to become polarised – even after the current is switched off, there is a potential difference between the electrodes, with the steel being negatively polarised.

6.5.1.1 Anodes

Where cathodic protection is to be used on steel reinforcement in a concrete structural element exposed to the atmosphere, materials suitable for the anode include titanium, certain zinc or aluminium alloys, conductive titanium-based ceramics and carbon [26]. The anode must be located such that a current can pass between the cathode and the anode. This is achieved by either embedding the anode in the concrete or attaching them to a relatively large area of the element's surface.

Embedding of anodes can be achieved through discrete probe electrodes (usually titanium or titania ceramics) inserted into holes in the concrete and grouted into place. Alternatively, slots can be cut in the concrete, and ribbons of titanium mesh can be inserted and grouted in a similar manner. Alternatively, similar layers of mesh can be placed on a concrete surface and overlaid with concrete, often applied through spraying.

Surface attachment of anodes can be achieved by painting or spraying a layer of the anode material onto the surface. This makes efficient use of the material, since a relatively small quantity is used to cover a large area and also ensures a good contact between the surface and the anode. Carbon-based paints are available for such applications. Alloy coatings can be applied using thermal spraying techniques, such as flame or arc spraying. Another approach is to apply a conductive cement-based carbon fibre composite overlay to the concrete surface.

It is not uncommon for combinations of different anode configurations to be used.

Where reinforced concrete elements are located beneath the ground or underwater, different conditions apply, because the water surrounding the concrete, either as the medium in which the element is submerged or as moisture in soil, will act to conduct the current between the cathode and the anode. Thus, anodes do not have to be physically in contact with the concrete. Moreover, a single anode can be used to protect a relatively large section of a structure.

In underwater and underground situations, the anodes can be made from a range of conducting materials, including platinum-coated titanium, magnetite and iron with a high silicon content. The use of titanium alone is usually not appropriate, since it has a layer of oxide at its surface, which increases its resistance to current flow. This leads to a situation where an excessively high voltage has to be applied across the electrodes to achieve the required current density, which ultimately leads to the oxide layer rupturing – the 'breakdown potential' of the titanium is exceeded – and the titanium undergoing local corrosion at a significant rate [27]. Instead, the titanium anode is coated in a layer of inert metal with good, maintainable conductivity. This has, in the past, been platinum. This approach is also economically viable, since platinum alone would be excessively expensive. More recently, titanium coated with a mixed metal oxide (MMO) layer has come to replace platinum-coated titanium. The metal oxides are normally based on mixtures of ruthenium and iridium oxide.

Where higher voltages are to be used, which would exceed the breakdown potential of the titanium (~8 V), niobium-coated tantalum anodes are used instead. Guidance produced on the subject has identified maximum operating voltages for many types of anode [26].

Where electrodes are buried underground, they are frequently backfilled with carbon-based granules that act to improve the flow of the current.

6.5.1.2 Operation

The current required to prevent corrosion in a structure equipped with a cathodic protection system is expressed in terms of current density – the current required per square meter of steel surface area to be protected. The British Standard for the cathodic protection of reinforced concrete (*BS EN ISO 12696* [28]) recommends a current density of 0.2 to 2 mA/m^2 for reinforced concrete that is free from chlorides. Where chloride contamination has occurred, either as a result of ingress from the environment or in structures where calcium chloride has been used as an accelerating admixture, a density of 2 to 20 mA/m² is recommended. The standard also states that normally cathodic prevention requires current densities in the lower of these two ranges, regardless of whether chlorides will ingress into the concrete.

6.5.1.3 Monitoring

To establish whether a cathodic protection system is having the desired effect, it is usually necessary to also equip a structure with a monitoring system. This involves either embedding reference electrodes (usually silver/silver chloride/potassium chloride double junction electrodes) in the concrete or taking readings using a surface-mounted electrode (usually a similar type of electrode or a manganese/manganese dioxide/sodium hydroxide electrode). The surface-mounted electrode can be a portable device that can be used on multiple locations.

Monitoring is usually conducted by measuring the potential difference between the reference electrode and the reinforcement immediately after switching off the cathodic protection current and again some time after switching it off. Switching off the current causes the steel to become depolarised – the potential relative to the reference electrode becomes less negative – but this occurs gradually. Thus, these two measurements give an indication of the corrosion potential if the cathodic protection was not there and an indication of the extent to which the system is improving matters.

A potential of less than (i.e., more negative than) –150 mV measured using a Ag/AgCl/0.5-M–KCl electrode is likely to mean that conditions at the reinforcement surface are noncorroding. The type of electrode and the concentration of electrolyte (potassium chloride [KCl] in the electrode described above) define the magnitude of the potential measured and, where a different electrode is used, a conversion factor is required. The standard also provides guidance on the other sensors that can be used as a means of monitoring performance with respect to corrosion protection.

BS EN ISO 12696 defines stricter criteria for the adequacy of cathodic protection. These are any one of the following:

1. A potential measurement more negative than –720 mV (using a Ag/AgCl/0.5-M–KCl reference electrode) immediately after switching off the current
2. A potential decay of >100 mV over a period of 24 h
3. A potential decay of >150 mV over a longer period

Where a system is found to not be performing adequately, it may be necessary to use a higher current density. Although this is possible, a cathodic protection system should only be operated under such conditions for a relatively short period of time (a matter of months), and the current density should not be excessively high. High current densities can lead to a number of problems, including hydrogen embrittlement of the reinforcement, loss of bond between the reinforcement and the concrete and alkali–aggregate reaction (Chapter 3).

Hydrogen embrittlement is a process that occurs when high concentrations of hydrogen gas are present at the surface of steel. Such atoms are sufficiently small that they can diffuse through solid steel with relative ease and tend to become trapped in grain boundaries. The accumulation of hydrogen in these locations gradually leads to a loss in strength. Hydrogen can be formed if the potential of the steel reinforcement exceeds a level known as its 'hydrogen evolution potential'. Hydrogen embrittlement is usually only an issue for high tensile steels used for pre- and posttensioned reinforcement. Where such reinforcement is present, *BS EN ISO 12696* states that it is necessary to maintain operating conditions that do not exceed a potential of –900 mV measured using a Ag/AgCl/0.5-M–KCl electrode. The standard also places a limit of 1100 mV for plain reinforcing steel.

The movement of alkali ions towards the anode during the cathodic protection of reinforced concrete leads to the accumulation of alkali ions at the surface. Where aggregates are present, which are vulnerable to alkali–aggregate reaction (see Chapter 3), there exists the possibility that the threshold concentration above which this reaction will occur will be exceeded to an extent that is harmful. Experiments have demonstrated that this is the case [25]. However, the current densities required to obtain the high potentials needed to cause the reaction were, at least for the initial period of the experiment, far in excess of those recommended in *BS EN ISO 12696* – in most cases, approximately 400 mA/m^2. Thus, it is considered unlikely that cathodic protection presents a risk with respect to alkali–aggregate reaction as long as operating conditions dictated in the standard are maintained.

Another related phenomenon is loss of bond. This also results from the accumulation of alkali ions at the cathodic reinforcement, which leads to the decalcification of calcium silicate hydrate gel in the hardened cement paste. This causes a weakening of the cement matrix (Figure 6.16), which produces a loss in strength of the bond between the reinforcement and the concrete. Reductions in bond strength of almost 35% have been observed in laboratory experiments [29]. However, these magnitudes of bond loss were achieved using current densities in excess of 500 mA/m^2, and it can again be concluded that such effects are unlikely if the operating conditions are maintained within the limits suggested.

The diffusion of hydroxide and chloride ions leads to the following reaction once the ions reach the anode:

$$4OH^- \rightarrow 2H_2O + O_2 + 4e^-$$

$$2Cl^- \rightarrow Cl_2 + 2e^-$$

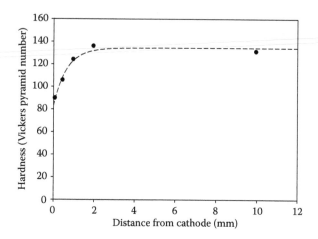

Figure 6.16 Hardness measurements taken over a range of distances from the cathode in a concrete specimen having undergone cathodic polarisation. (Rasheeduzzafar et al., *ACI Materials Journal*, v. 90, 1993, pp. 8–15.)

The overall effect is the evolution of oxygen and chlorine gas. The evolution of gas can present a serious problem where its escape is prevented, either by a surface coating or simply as the result of concrete pores being saturated with water. Additionally, the oxidation of hydroxide ions to water and oxygen at the anode can lead to a drop of pH to acidic conditions, which can attack the concrete. These problems are generally avoided by limiting the current density. For instance, *BS EN ISO 12696* recommends limiting the long-term current density to below 110 mA/m², where MMO-coated titanium is used. Higher densities may still be used for short periods of time – for instance, to set up conditions of polarisation – but the duration should be limited to short periods.

6.5.2 Galvanic cathodic protection

Galvanic cathodic protection is a passive form of protection, in the sense that a potential difference is not required to be actively applied. Instead, the potential difference is produced by connecting the reinforcing steel to an electrode made from a metal that is more anodic (Figure 6.17), leading to galvanic corrosion (as described in Chapter 4), in which the more anodic electrode corrodes sacrificially and the steel remains largely unaffected. The absence of a need for an applied electrical current means that such systems are usually more economical when compared with impressed current systems.

As for impressed current cathodic protection, the technique can be applied to structures exposed to the atmosphere and those that are buried

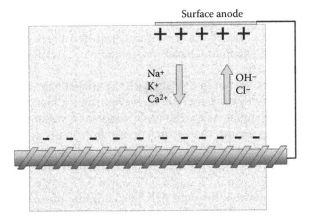

Figure 6.17 Galvanic cathodic protection.

or submerged, and similar approaches to the location of the anode are required. The anodic metal is almost always zinc, although aluminium–zinc–indium alloys are sometimes used.

Where structures are exposed to the atmosphere, the anode must be in contact with the concrete and can take the form of sheeting attached to the surface, mesh applied to the surface and held in place by permanent formwork, thin layers applied using thermal spraying or blocks held in shallow cavities in the concrete.

In the case of zinc, the products of corrosion are zinc oxide and hydroxide, whose formation leads to a net increase in volume. Therefore, where an embedded block is used, it is important that enough space is provided for these products. This is usually achieved by surrounding the blocks with a volume of conductive backfill material that is porous and thus permits product formation without damaging the concrete. The backfill is normally carbon based to permit it to efficiently conduct current from the anode to the cavity surfaces.

To optimise the protection imparted by a galvanic cathodic protection system, an activator must also be present around the anode. These are compounds that cause soluble forms of corrosion product to be formed. These can be compounds that increase the surrounding pH to very high levels (>14), where the hydroxide becomes more soluble, or sulphate or halide compounds, whose presence will lead to the formation of soluble chloride and sulphate salts of zinc. The use of halides has durability implications, since the corrosive nature of chloride ions (discussed in Chapter 4) is clearly an issue. As covered in Chapter 3, sulphates may also present problems. However, cathodic anodes are usually supplied as a single product with appropriate backfill and activator formulations, with guidance on appropriate use.

During service, the anode is consumed as part of the corrosion process. In such applications, although the intention is usually for anodes to last for the intended service life of the structure, sheet and block anodes can be replaced as part of the maintenance programme of a structure if they are consumed excessively.

Where structures are not in contact with the atmosphere, the anode need not be in contact with the concrete. However, since replacement is potentially a more challenging task, the type, shape and volume of the anode must be selected to ensure that sufficient anode remains at the end of the design life, in an appropriate shape, to provide adequate protection. The materials used for these anodes are typically zinc, aluminium and magnesium alloys.

The voltage achievable through galvanic cathodic protection is around 1 V. This is less than what would typically be applied across an impressed current system, and the overall effect is that the level of protection is limited. Under chloride-free conditions, this is unlikely to be a major issue, but, where chlorides are present, it is normally the case that galvanic protection will merely act as a means of slowing corrosion rather than preventing it.

Monitoring the performance of galvanic cathodic protection systems is usually done by measuring the current between the reinforcement and the anode or anodes. In some cases, the corrosion potential of the steel may also be measured, although this involves connecting parts of the reinforcement to a half-cell electrode, which may not be practical.

REFERENCES

1. Kreijger, P. C. The skin of concrete: Composition and properties. *Materials and Structures*, v. 17, 1984, pp. 275–283.
2. Cairns, J. Enhancements in surface quality of concrete through the use of controlled permeability formwork liners. *Magazine of Concrete Research*, v. 51, 1999, pp. 73–86.
3. Bamforth, P. B. *Concrete Society Technical Report 61: Enhancing Reinforced Concrete Durability*. Camberley, United Kingdom: Concrete Society, 2004, 161 pp.
4. McCarthy, M. J., A. Giannakou, and M. R. Jones. Specifying concrete for chloride environments using controlled permeability formwork. *Materials and Structures* v. 34, 2001, pp. 566–576.
5. ACI Committee 308. *Standard Practice for Curing Concrete – ACI 308-92*. Farmington Hills, Michigan: American Concrete Institute, 1997, 11 pp.
6. National Ready-Mixed Concrete Association. *Plastic Cracking of Concrete*. Silver Spring, Maryland: NRMCA, 1960, 2 pp.
7. Menzel, C. A. Causes and prevention of crack development in plastic concrete. *Proceedings of the Portland Cement Association*. Shokie, Illinois: Portland Cement Association, 1954, pp. 130–136.

8. Uno, P. J. Plastic shrinkage cracking and evaporation formulas. *ACI Materials Journal*, v. 95, 1998, pp. 365–375.

9. Aitcin, P.-C. Effect of size and curing on cylinder compressive strength of normal and high-strength concretes. *ACI Materials Journal*, v. 91, 1994, pp. 349–354.

10. Ballim, Y. Curing and the durability of OPC, fly ash, and blast-furnace slag concretes. *Materials and Structures*, v. 26, 1993, pp. 238–244.

11. British Standards Institution. *BS 7542:1992: Method of Test for Curing Compounds for Concrete*. London: British Standards Institution, 1992, 12 pp.

12. Hillier, S. R., C. M. Sangha, B. A. Plunkett, and P. J. Walden. Effect of concrete curing on chloride ion ingress. *Magazine of Concrete Research*, v. 52, 2000, pp. 321–327.

13. Gunter, M., T. Bier, H. Hilsdorf. Effect of curing and type of cement on the resistance of concrete to freezing in de-icing salt solutions. In J. Scanlon, ed., *ACI Special Publication SP-100: Proceedings of Katharine and Bryant Mather International Symposium*. Detroit, Michigan: American Concrete Institute, 1987, pp. 877–899.

14. Dhir, R. K., P. C. Hewlett, and Y. N. Chan. Near-surface characteristics of concrete: Abrasion resistance. *Materials and Structures*, v. 24, 1991, pp. 122–128.

15. British Standards Institution. *BS EN 13670:2009: Execution of Concrete Structures*. London: British Standards Institution, 2009, 76 pp.

16. British Standards Institution. *BS EN 1504-2:2004: Products and Systems for the Protection and Repair of Concrete Structures – Definitions, Requirements, Quality Control, and Evaluation of Conformity: Surface Protection Systems for Concrete*. London: British Standards Institution, 2004, 50 pp.

17. British Standards Institution. *BS EN 1504-9:2008: Products and Systems for the Protection and Repair of Concrete Structures – Definitions, Requirements, Quality Control, and Evaluation of Conformity: General Principles for Use of Products and Systems*. London: British Standards Institution, 2008, 32 pp.

18. British Standards Institution. *BS 6150:2006: Painting of Buildings: Code of Practice*. London: British Standards Institution, 2006, 174 pp.

19. The Highways Agency. *Design Manual for Roads and Bridges, Vol. 2 Highway Structures: Design (Substructures and Special Structures) Materials, Part 2. BD43/03: The Impregnation of Reinforced and Prestressed Concrete Highway Structures using Hydrophobic Pore-Lining Impregnants – BD43/03*. Norwich: HMSO, 2003, 14 pp.

20. British Standards Institution. *BS 8204-2:2003: Screeds, Bases, and In Situ Floorings – Part 2. Concrete Wearing Surfaces: Code of Practice*. London: British Standards Institution, 2003, 44 pp.

21. British Standards Institution. *BS EN 12620:2002: Aggregates for Concrete*. London: British Standards Institution, 2002, 60 pp.

22. British Standards Institution. *BS EN 13139:2002: Aggregates for Mortar*. London: British Standards Institution, 2002, 44 pp.

23. British Standards Institution. *BS 8204-3:2004: Screeds, Bases, and In Situ Floorings – Part 3. Polymer-Modified Cementitious Levelling Screeds and Wearing Screeds: Code of Practice*. London: British Standards Institution, 2004, 36 pp.

24. British Standards Institution. *BS 8204-6:2008: Screeds, Bases, and In Situ Floorings – Part 6. Synthetic Resin Floorings: Code of Practice.* London: British Standards Institution, 2008, 52 pp.

25. Sergi, G. and C. L. Page. *The Effects of Cathodic Protection on Alkali–Silica Reaction in Reinforced Concrete.* Crowthorne, United Kingdom: Transport and Road Research Laboratory, 1992, 53 pp.

26. The Concrete Society. *Technical Report 73: Cathodic Protection of Steel in Concrete.* Camberley, United Kingdom: The Concrete Society, 2011, p. 105.

27. Cotton, J. B. Platinum-faced titanium for electrochemical anodes. *Platinum Metals Review*, v. 2, 1958, pp. 45–47.

28. British Standards Institution. *BS EN ISO 12696:2012: Cathodic Protection of Steel in Concrete.* London: British Standards Institution, 2012, 56 pp.

29. Rasheeduzzafar Ali, M. G. and G. J. Al-Sulaimani. Degradation of bond between reinforcing steel and concrete due to cathodic protection current. *ACI Materials Journal*, v. 90, 1993, pp. 8–15.

Chapter 7

Serviceability, repair and maintenance of concrete structures

7.1 INTRODUCTION

In previous chapters, the deterioration mechanisms that act to challenge concrete durability have been examined, along with the ways in which concrete structural elements can be designed and fabricated to provide suitable protection. However, adequate protection was not necessarily used in concrete structures built in the past, and many of the processes that compromise durability were unknown before their construction and have only come to light during the service of such structures. Moreover, protective measures can fail in ways that were not anticipated.

It should be recognised that structures will inevitably deteriorate with time, and what is most important is that a structure remains able to carry out its function for the intended service life. However, the function of a structure may change in a manner that could not be anticipated by its designers, or it may be desirable to continue using a structure beyond the intended service life for economic or practical reasons. In such cases, it is essential to establish the nature and extent of deterioration to allow a judgement to be made with regards to a structure's suitability and whether anything can be done to rectify problems or improve performance.

This chapter initially discusses the manner in which a structure's ability to satisfy its function changes with time and at what point it is deemed to have ceased to do this. It then outlines the way in which a structure can be appraised in terms of performance and how attempts can be made to predict future performance. Test methods that can be used *in situ* and in the laboratory to assist in this appraisal are examined, as well as some of the measures available to the engineer to rectify concrete durability problems.

7.2 SERVICEABILITY OF STRUCTURES

It is clearly the case that the collapse of a structure is an event that must be avoided at all cost and that structures must be designed with this in mind. Despite this, structural collapses still occur, and such events are well-publicised as a result of loss of lives or the dramatic images that result. However, a much more common form of failure occurs when structures reach a point of deterioration where they can no longer perform the function that they were designed to serve. Although such events are much less widely publicised and the consequences are less traumatic, where this loss of 'serviceability' occurs within the intended service life of a structure, a failure of the design and/or execution of a structure has evidently occurred.

7.2.1 Limit states

The 'limit state' approach is now commonly used in the design of structures and is one of the fundamental philosophies behind the design process described in *Eurocode 2* [1]. Under such a system, an event that has implications for the safety of people or the structure, such as a collapse, signifies that an ultimate limit state (ULS) has been exceeded. *EN 1990* [2] identifies the following ULSs:

- 'Loss of equilibrium of the structure or any part of it, considered as a rigid body'
- 'Failure by excessive deformation; transformation of the structure or any part of it into a mechanism; rupture; loss of stability of the structure or any part of it, including supports and foundations'
- 'Failure caused by fatigue or other time-dependent effects'

The points beyond which a structure fails to satisfy its function are referred to as serviceability limit states (SLSs). *EN 1990* identifies a number of SLSs. The first of these are 'deformations that affect the appearance, the comfort of users, the functioning of the structure (including the functioning of machines or services) or that cause damage to finishes or nonstructural members'. In addition, the standard includes 'vibrations that cause discomfort to people or that limit the functional effectiveness of the structure', as well as 'damage that is likely to adversely affect the appearance, the durability, or the functioning of the structure'.

Eurocode 2 defines three types of reinforced concrete structure behaviour that define the end of serviceability: stress limitation, crack control, and deflection control.

'Stress limitation' refers to maintaining stresses in structural elements below defined levels, with the aim of reducing the likelihood of cracking and creep. Longitudinal cracking of elements is undesirable on the grounds that such cracks are likely to reduce the protection to steel reinforcement provided by concrete and create other durability problems. In environments where chlorides are present or where freeze–thaw attack is a possibility, the standard recommends that the stress carried by a structural element be below 0.6 f_{ck} to avoid such cracking. f_{ck} is the characteristic cylinder strength of the concrete. The standard also points out that, alternatively, longitudinal cracking can be avoided through confining transverse reinforcement, or the depth of cover can be increased in the compression zones of an element to counteract the effect of cracking on durability.

The risk of cracking is also controlled by a recommended maximum tensile stress in the reinforcement of 0.8 f_{yk}. f_{yk} is the characteristic yield strength of the reinforcement. Where reinforcement takes the form of prestressed tendons, the mean value of tensile stress should not exceed 0.75 f_{yk}. *Eurocode 2* also recommends defining serviceability criteria to keep creep within acceptable magnitudes.

'Crack control' is the limitation of crack widths whose presence would 'impair the proper functioning or durability of the structure or cause its appearance to be unacceptable'. For reinforced members and prestressed members with unbonded tendons, the limit state crack width is 0.3 mm in the UK annex to *Eurocode 2* [2]. For bonded prestressed tendons, which are potentially more sensitive to the effects of corrosion, the crack width is limited to 0.2 mm. In addition, the design of prestressed members containing bonded tendons must be checked for 'decompression' under the quasipermanent combination of loads: it must be established whether the loads that the member will experience during service will be sufficiently high to exceed the tensile loads of any of the tendons, thus putting the concrete in tension. Where regions of the concrete are found to be 'decompressed', it must be ensured that no bonded tendons lie in such regions and that they are at least 25 mm within concrete in compression. This is to ensure that tensile loads do not extend to depths that would threaten the durability of the tendons.

Eurocode 2 includes a method for calculating the minimum area of reinforcement necessary to achieve appropriate crack control. The standard also includes a method for estimating crack widths in members. These methods are outlined in Chapter 2. The method can be used to determine the minimum bar diameter required to achieve crack widths below limit state values. To avoid the need for calculation, a table of minimum reinforcement bar diameters to achieve sufficiently narrow crack widths for given levels of stress carried by the steel is provided. The alternative option of limiting the spacing between bars is also presented, with a similar table provided to determine this for a given stress level.

The standard makes the point that, where no detrimental effect is realised by the presence of a crack, regardless of its width, such a crack is permitted.

Deflection is limited on the grounds that it will impair the function and appearance of a structure. The standard provides general guidance on limit values for deflection. It states that the general utility and appearance of a structure is likely to be compromised when the amount of sag in a beam, slab, or cantilever exceeds 1/250 of the span under quasipermanent loads.

Furthermore, it suggests that deflection beyond 1/500 of the span has the potential to damage the adjacent parts of a structure.

The standard stresses that appropriate limits for deflection may need to be less than those previously discussed, where movement could damage the features of a building, such as surface finishes and glazing. It also points out that sagging of flat roofs may lead to the formation of depressions that allow ponding of rainwater.

The limit states covered by *Eurocode 2* need not be the only ones defined for a structure. The designer of a structure will specify a series of serviceability criteria that will then be agreed with the client. Other criteria might include vibration and abrasion of concrete floors.

7.2.2 Aspects of durability influencing serviceability

It should be noted that all of the SLSs defined by *Eurocode 2* are in some ways, related to durability. Stress limit states are partly defined to limit cracking that would otherwise compromise the effectiveness of cover in protecting steel. Crack widths are primarily limited for the same reason. Moreover, although excessive deflection or excessively high stresses need not compromise durability, they may be the result of the deterioration of reinforcement.

Table 7.1 identifies the manner in which the deterioration mechanisms of concrete have a direct influence on compromising serviceability, along with some of the key secondary effects. It should be stressed that the listed secondary effects are very much an oversimplification–for instance, the formation and growth of cracks resulting from alkali–aggregate reaction (AAR) is, in reality, likely to have an exacerbating effect on all deterioration mechanisms acting on part of a structure to some degree.

7.2.3 Durability and performance

Although the deterioration of a structure tends to become visually apparent later in life, it should be stressed that the process of deterioration will have started potentially while it was still being built. Figure 7.1a illustrates how the performance of a structure may decline with time relative to any one of its serviceability limits. The first curve shows a scenario in which the design and/or workmanship is inadequate, leading to a rate of deterioration that leads to the serviceability limit being exceeded before the intended service life has been reached.

The ideal scenario is one in which deterioration proceeds at a rate such that the performance of the structure does not fall below the serviceability limit until after the structure has reached the end of its intended service life. The most desirable means of doing this is to ensure that both the design and quality of construction are such that performance above the serviceability limit is

Table 7.1 Direct and secondary influences of concrete deterioration mechanisms on serviceability

Deterioration mechanism	Direct influence on serviceability	Secondary influence on serviceability
Plastic shrinkage	Crack formation	Possible increased rate of chloride ingress and/or carbonation Possible increased rate of sulphate attack and/or acid attack
Drying/autogenous shrinkage	Crack formation Deflection	Possible increased rate of chloride ingress and/or carbonation Possible increased rate of sulphate attack and/or acid attack
Thermal contraction	Crack formation Deflection	Possible increased rate of chloride ingress and/or carbonation Possible increased rate of sulphate attack and/or acid attack
Freeze–thaw attack	Increase in stress levels resulting from loss of section	Loss of cover leading to higher stress levels in structural members Loss of cover leading to possible reduced period for chlorides and carbonation front to reach reinforcement
Abrasion/erosion	Increase in stress levels resulting from loss of section	Loss of cover leading to higher stress levels in structural members Loss of cover leading to possible reduced period for chlorides and carbonation front to reach reinforcement
Sulphate attack	Increase in stress levels resulting from loss of section	Loss of cover leading to higher stress levels in structural members Loss of cover leading to possible reduced period for chlorides and carbonation front to reach reinforcement
AAR	Crack formation Deflection	Possible increased rate of chloride ingress and/or carbonation Possible increased rate of sulphate attack and/or acid attack
Acid attack	Increase in stress levels resulting from loss of section	Loss of cover leading to higher stress levels in structural members Loss of cover leading to possible reduced period for chlorides and carbonation front to reach reinforcement

(continued)

Table 7.1 (Continued) Direct and secondary influences of concrete deterioration mechanisms on serviceability

Deterioration mechanism	Direct influence on serviceability	Secondary influence on serviceability
Chloride ingress	Cracking from corrosion product formation	Increased rate of chloride ingress and/or carbonation
	Deflection through corrosion of reinforcement	Possible increased rate of sulphate attack and/or acid attack
Carbonation	Cracking from corrosion product formation	Increased rate of chloride ingress and/or carbonation
	Deflection through corrosion of reinforcement	Possible increased rate of sulphate attack and/or acid attack
		Reduced chloride binding

maintained throughout the service life from an initial level of performance that is both cost-effective and resource-efficient.

Maintaining adequate performance can of course also be achieved through overdesign, such that the starting level of performance far exceeds that which is required. Such an approach is unlikely to be cost-effective and would usually be viewed as being wasteful. Nonetheless, this approach has led to much of the world's infrastructure remaining serviceable far beyond intended periods of service.

7.2.4 Repair to maintain serviceability

Figure 7.1b illustrates how repair of a structure can permit it to remain serviceable despite an excessive rate of deterioration. Repair may take the form of frequent small-scale repairs or infrequent major rehabilitation, but the additional expense is likely to be comparable, regardless of the strategy adopted.

Realistically, however, it is likely that repair of some form is likely to be necessary during the service life of a structure. The following sections examine assessment procedures and some of the repair methods available to the engineer.

7.3 APPRAISAL OF STRUCTURES

The need to evaluate a structure can arise for a number of reasons. These can include concerns regarding the deterioration of the structure as a result of exposure to its normal environment or from its service, concerns regarding inadequate design or workmanship, a change in the use of a structure,

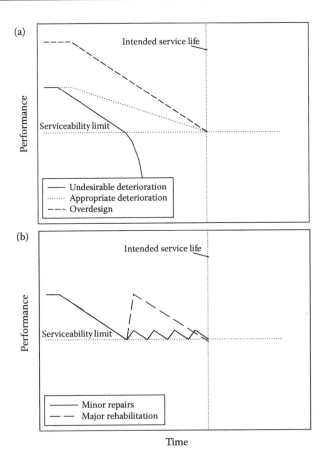

Figure 7.1 Changes in performance over the service life of a structure without repair (a) and with repair (b).

alterations made to the structure itself or changes in environmental conditions. There are also many instances where an evaluation will be a requirement of purchase, provision of insurance or for various legal reasons.

7.3.1 Appraisal process

The approach to the evaluation outlined in this section is largely based around the Institution of Structural Engineers (ISE) guidance [3], with some additional components taken from other pieces of UK-based guidance. The general procedure that should be followed in carrying out an appraisal is examined, before some of the specific aspects of appraising concrete in a structure are discussed.

The ISE guidance states that an evaluation must establish whether a structure is adequately safe and whether it will remain that way in the future. In addition, it must ascertain whether the structure satisfies the requirements for serviceability. The tasks that make up a typical appraisal are

- Desk study
- Reconnaissance and site inspection
- Testing and monitoring
- Appraisal
- Report

although this sequence is not necessarily followed, since testing and monitoring may be required at various stages.

The initiation of the desk study involves gathering as much documented information regarding the structure as possible. This information will include

- Whether the structure is a listed building or scheduled ancient monument
- Drawings
- Calculations
- Specifications
- History of structure
- Explanatory information

Although it is not necessarily the case that all this information will be available, the more recent a structure is, the more likely that relevant documents can be located. This should be particularly true of structures built after 1994, since that year saw the introduction of the Construction (Design and Management) Regulations in the United Kingdom, which require a Health and Safety File of information, providing information relevant to ensuring the health and safety of those involved in future construction and maintenance activities, to be compiled throughout the lifetime of a project and handed to the client once completed.

'Explanatory information' is any information that, although not being specific to the structure in question, provides useful background information. Such information can include contemporary standards and codes of practice, manufacturers' product descriptions and papers and textbooks relating to the materials or techniques that may have been used.

The desk study then needs to use the information gathered to identify hazards that may arise during an inspection of the structure with consideration of those carrying out the inspection, users of the structure, and members of the public. An evaluation of the risk posed by each of these hazards should be conducted.

Reconnaissance of a site involves establishing

- Site conditions
- How access may be gained to the site
- Whether materials need to be removed to expose the structural features covered by finishes

In addition, a systematic plan of how the inspection should be carried out must be devised. This plan should be informed by the risk assessment previously conducted and may include requirements for appropriate clothing and personal protective equipment.

Inspection will definitely involve visual inspection of the structure, but will almost certainly also involve obtaining measurements of the parts of the structure. Dimensions may be provided in the drawings and plans of the structure obtained during the desk study, but there may be a significant variation between the designs and the actual structure, and so these cannot be relied upon exclusively. The ISE guidance also makes the point that it is usually impractical to obtain a full set of dimensions, and so, through studying the information available prior to inspection, dimensions critical to the performance of the structure should be identified for verification during inspection.

The combination of the inspection and desk study should address the following additional aspects:

- Structural arrangements
- Materials of construction
- Structure condition
- Actions and loadings
- Lateral stability
- Soil pressures and ground movement
- Aggressive ground conditions
- Thermal effects
- Changes of humidity
- Creep
- Moisture ingress
- Deleterious materials
- Fungal and insect infestation
- Atmospheric conditions
- Abrasion and erosion
- Vandalism

Detailed discussions of all these factors are provided in the guidance. However, from a concrete durability perspective, it is likely that if deterioration has occurred in a manner that compromises structural performance and/or serviceability, it will manifest itself visually in some manner. Figures 7.2, 7.3, and 7.4 are flowcharts indicating how information from visual inspection relating to concrete durability can be interpreted and what additional investigation is required to confirm the probable mechanism and establish the extent of the problem.

These flow diagrams have been adapted from the tables in the ISE guidance. It should be noted that they concentrate exclusively on durability-related problems, and the document also provides guidance on the interpretation of problems relating to the aspects of structural performance, including detailing of reinforcement and possible overloading of structures. Guidance is also provided for problems unique to specific types of structural element, which are not included in the flow diagrams.

Although these diagrams are potentially useful, it should be stressed that such methods of diagnosis are, at best, only likely to point the appraiser in the right direction in terms of identifying problems. This is partly because of the close interrelationship between the different mechanisms of deterioration. For instance, investigation of cracking may indicate corrosion of the reinforcement. However, this interpretation may miss underlying mechanisms that have led to the concrete cover being ineffective–cracking resulting from alkali-silica reaction (ASR), for example.

Visual inspection will also involve examination of structural elements for signs of SLSs being, or close to being, exceeded.

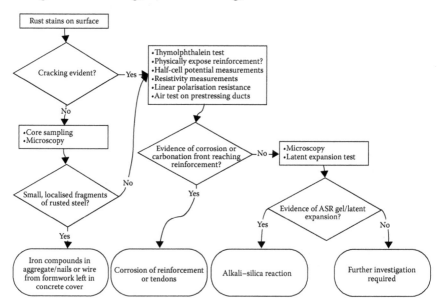

Figure 7.2 Flow diagram for the diagnosis of rust staining, based on ISE guidance. (From The Institution of Structural Engineers. *Appraisal of Existing Structures, 3rd ed.* London: The Institution of Structural Engineers, 2010.)

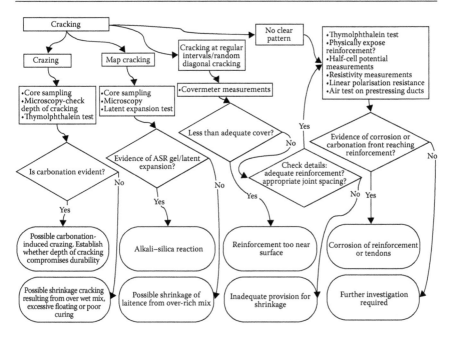

Figure 7.3 Flow diagram for the diagnosis of cracking, based on ISE guidance. (From The Institution of Structural Engineers. *Appraisal of Existing Structures, 3rd ed.* London: The Institution of Structural Engineers, 2010.)

The actual process of appraisal is divided into four stages in the ISE guidance. These are not discussed here in any great detail, but it is useful to outline the roles of the different stages.

The first stage of appraisal establishes whether any parts of the structure show visible signs of deterioration, whether structural details are in good condition and robust and whether any new use that a structure is intended to be put to will impose heavier loads than those originally designed for. Where a structure is not likely to be loaded beyond the originally intended loads and deterioration can be remedied through repair, this stage may be the only appraisal stage required and recommendations for repair can be made. However, where loads are likely to exceed those originally anticipated or repair is problematic, the appraiser is required to move to the second stage.

The second appraisal stage involves structural analysis to establish the overall robustness of the structure and to evaluate the adequacy of each individual structural element. Where the structure is found to have an adequate factor of safety with respect to the relevant standard, code of practice, or guidance, and the structure shows no visual indications of deterioration, the structure can normally be pronounced safe after checking calculations. If the factor of safety is calculated to be one or less, it is possible that the

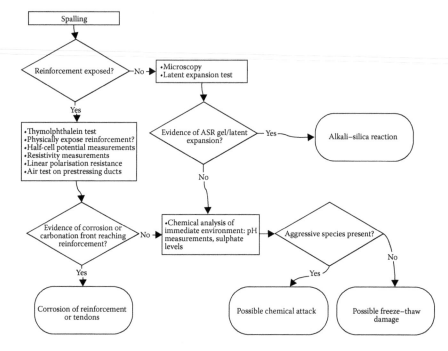

Figure 7.4 Flow diagram for the diagnosis of spalling, based on ISE guidance. (From The Institution of Structural Engineers. *Appraisal of Existing Structures*, 3rd ed. London: The Institution of Structural Engineers, 2010.)

structure is grossly overloaded. However, particularly where there is little visual evidence of deterioration, it is necessary to check calculations prior to devising remedial action.

The third possible outcome of evaluating robustness is that the factor of safety exceeds a value of one, but is less than that specified in the relevant codes. In such a case, further investigation is required to establish whether this presents a problem. There are various reasons why a safety factor may spuriously be found to be sitting between unity and the requirements of the design codes.

First, it is possible that the assumptions made and simplifications adopted during structural analysis are yielding an inaccurate result, and it may be possible to refine the mathematical model used in the calculations and revise the assumptions to obtain more accurate results.

Second, it is highly likely that the strengths of the materials in the structure are not the same as those used in the original design process. This is particularly true of concrete, where strength at a given age (typically, 28 days) will have been specified by the designer. However, concrete is likely to continue to develop strength beyond this age, which means that the concrete may be outperforming the performance assumed for design purposes. Thus, it is advisable to determine the actual strengths of the materials in a structure, as well as the actual dead loads, where possible.

Thus, the third stage of appraisal, if required, involves revising the structural model and measuring *in situ* strengths (and, possibly, other characteristics) with the view of revising the structure's safety factors. It should be stressed that, although this may lead to a reduction in safety factors – and, potentially, to the conclusion being reached that a structure's calculated capacity is adequate – where the quality of workmanship or materials is substandard, deterioration of materials has occurred or more accurate calculations indicate lower safety factors, the opposite outcome is entirely possible.

If the structure is to be put to a new use, the third stage may also involve examining whether the new loads are higher and whether the structure has the capacity for these. This evaluation may also analyse the risks presented by the structure's environment, in the form of snow and wind loads, which could cause load-bearing capacity to be exceeded.

The process of appraisal is shown in Figure 7.5, which summarises considerably more detailed decision trees provided in the ISE guidance.

7.3.2 Predicting future deterioration

A final stage of appraisal may also be requested of the engineer – prediction of future performance. This involves projecting how deterioration mechanisms acting on the structure will advance with time. Prediction requires three components if it is to be done successfully: adequate data about the current state of the deterioration of the structure, adequate historical data relating to how the structure has been used and maintained, and suitable models for projecting deterioration forward. Although an appropriately detailed inspection and testing programme is likely to adequately deal with the first requirement, historical data may be much harder to obtain and may be incomplete. However, the greatest problem posed is that, even when an adequate model is available for predicting rates of deterioration, with deterioration mechanisms that are nonlinear with respect to time, meaningful extrapolation is often only possible with three or more data points. In the case of an appraisal of a structure, it is likely that only two data points are available: the initial condition and the condition determined by inspection or testing. Moreover, in some cases, the first of these points is theoretical, since it will be based on the intended state of the structure rather than a real one.

In some instances, additional data from previous appraisals may exist, in which case prediction of a nonlinear process may be possible. However, care must be taken to ensure that previously obtained information is compatible with that obtained during the present appraisal. For instance, techniques

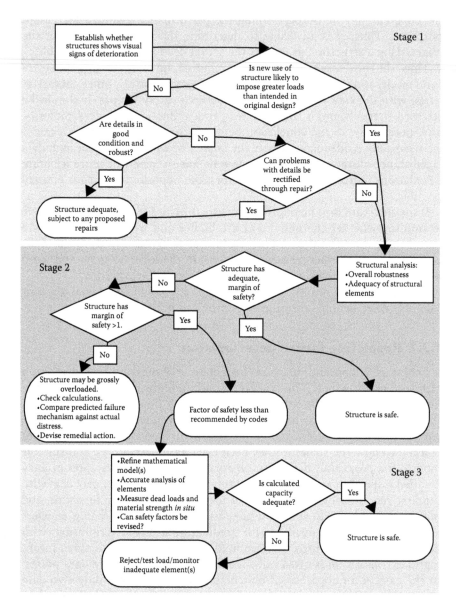

Figure 7.5 Structure evaluation decision tree.

used to measure permeability-related characteristics may produce widely different results, and it is up to the engineer to establish whether two different data sets are compatible or whether one set can be converted into a form that renders them compatible.

In some cases, the nonlinear nature of deterioration mechanisms can be overcome through appropriate testing and analysis. For instance, an estimate of the diffusion coefficient of a chemical substance (such as sulphate or chloride) moving through the cover of concrete can be obtained by conducting a chemical analysis on samples taken at progressive depths into the concrete to obtain concentration profiles (see Section 7.5.1.1). However, in other cases – carbonation is a good example – this approach cannot be adopted.

The actual approach taken by an engineer toward predicting the decline in the performance of a structure will depend on how much data are available and what testing of the structure can be conducted and on the importance of structural elements and the deterioration mechanisms acting on them. Table 7.2 provides some pointers toward possible models that may be used in the modelling of deterioration, with references to sections in this book, where relevant.

It should also be stressed that, where multiple mechanisms may be acting on a structure, secondary impacts on durability may have to be considered, as outlined previously in Table 7.1. For instance, where ASR has been found to be occurring, loss of strength and stiffness resulting from alkali–silica reaction should be modelled. However, the cracking that this reaction produces will reduce the ability of the concrete cover to resist chloride ingress, with possibly much larger implications for load bearing capacity. Thus, the increase in the coefficient of chloride diffusion will also require estimation to predict its impact on the rate of chloride ingress and its subsequent influence on reinforcement corrosion. A number of software packages have emerged in recent years, which attempt to assist the engineer in making such predictions, particularly where so-called 'multiaggressive' environments are present.

It is typical in projecting the longer term behaviour of materials to use statistical methods to indicate the confidence range within which predicted values lie. The small number of data points along the timeline of service life resulting from a typical appraisal normally means that such an option is not available. Given the uncertainties surrounding projections of rates of deterioration, any prediction that is made by an engineer should be qualified as an estimate in the documentation in which it is presented.

With the exception of drying shrinkage and thermal contraction, an aspect of concrete deterioration mechanisms, which has so far not been addressed, is the rate at which cracks grow as processes that cause cracking progress. In particular, the processes of AAR, cracking from the formation of corrosion products on steel reinforcement and the formation of cracks during sulphate and freeze–thaw attacks require further discussion.

The development of cracks from AAR is initially a process of crack development and widening. This development is shown in Figure 7.6, which plots the maximum crack width and the surface crack density (expressed as the

Table 7.2 Possible means of predicting rates of deterioration and information required

Deterioration mechanism	Possible model(s)	Information required	
Crack development			
Drying shrinkage	*Eurocode 2* equations for average crack width and crack spacing prediction.	Tensile stress acting on element Age and history of structure Reinforcement ratio Modulus of elasticity of concrete Cover depth Reinforcement diameter	Chapter 2, Section 2.2.2
Cracking through AAR	Crack width typically increases linearly with time.	Crack widths Age of structure	This chapter
Cracking from sulphate attack	See subsequent text.	Characteristic compressive strength of concrete specified in design Current strength of concrete	This chapter
Cracking from corrosion of reinforcement	See subsequent text.	Reinforcement detailing Current strength of concrete Rate of corrosion from corrosion current measurements Chloride penetration/ carbonation data if cracking has not initiated	This chapter
Freeze–thaw attack	Loss of stiffness and strength is approximately linear with respect to the freeze–thaw cycles experienced. Longer term loss of mass from scaling normally tends toward a linear relationship with respect to the number of freeze–thaw cycles.	Ambient temperature data over life of structure Characteristic compressive strength of concrete specified in design Strength/elastic modulus of *in situ* concrete	Chapter 2, Section 2.4.2

Table 7.2 (Continued) Possible means of predicting rates of deterioration and information required

Deterioration mechanism	Possible model(s)	Information required	
Abrasion	Rates of abrasion are typically constant, assuming that abrasive action remains constant.	Depth of abrasion Age of structure History of abrasive action	Chapter 2, Section 2.5.1
Chemical degradation			
Sulphate attack	Fick's second law Loss of strength is approximately linear with respect to time for conventional sulphate attack.	Sulphate concentration profile Age of structure/ history of sulphate exposure Characteristic compressive strength of concrete specified in design Current strength of concrete	Chapter 3, Section 3.2.3; this chapter
AAR	Loss of strength and stiffness typically follows the relationship $y = (a + bt)/(c + t)$	Characteristic compressive strength of concrete specified in design Current strength	Chapter 3, Section 3.3.4
Chemical attack	Because loss of surface is caused by mechanical attrition, a similar approach to abrasion may be suitable.	Depth of abrasion Age of structure History of abrasive action	Chapter 3
Chloride ingress			
Chloride diffusion	Fick's second law	Chloride concentration profile Depth of cover Age of structure/ history of chloride exposure	Chapter 5, Section 5.2.5
Chloride ingress via flow	Darcy's law	Fluid pressure acting on the surface Permeability measurements Depth of cover	Chapter 5, Section 5.2.4

(continued)

Table 7.2 (Continued) Possible means of predicting rates of deterioration and information required

Deterioration mechanism	Possible model(s)	Information required	
Chloride ingress via absorption	Capillary action equation	Pore size distribution of concrete Contact angle between water and concrete Depth of cover Absorption measurements	Chapter 5, Section 5.2.3
Carbonation	Modified Fick's law equation	Local atmospheric concentration of CO_2 Cement type and, hence, CO_2 required for complete reaction If possible, CO_2 diffusion coefficient Depth of cover	Chapter 4, Section 4.4.4
	Influence of cracks on mass transport	Crack width Crack density Depth of cracking	Chapter 4, Section 4.3.2; Chapter 5, Sections 5.2.3 and 5.2.4
Reinforcement corrosion		Chloride concentration at reinforcement and change with time Carbonation depth and change with time Likely chloride threshold Corrosion current and other electrochemical parameters	This chapter; Chapter 4

total length of cracks per unit area of surface). The plot shows an initially rapid increase in crack density, which rapidly levels off, with crack widening continuing in a linear manner. These results are also used to calculate a crack spacing factor (see Chapter 4), which indicates an initial rapid decline in spacing factor, which levels out to a value of approximately 100. Indeed, it has been observed that the crack spacing factor seldom falls below 100, regardless of the mechanism of cracking and the degree of deterioration [4].

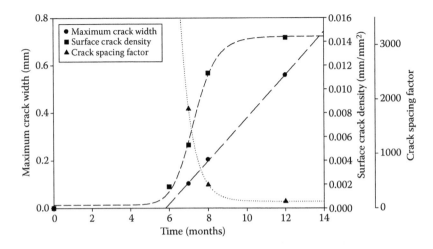

Figure 7.6 Increase in maximum crack width and crack density (determined using image analysis on published crack maps) for reinforced concrete beams exposed to 0.5-N mixed alkali solution at 38°C. Crack spacing factor is also estimated using these two characteristics. (From Fan, S. and J. M. Hanson, *ACI Structural Journal*, 95, 1998).

Two key points can be deduced from these results. First, where a concrete surface is found to be cracked as a result of AAR, it is more likely to be at a stage where the formation of new cracks has slowed down or ceased, and linear increase in crack width is the dominant mode of deterioration. Perhaps, more importantly, the fact that the spacing factor tends toward a lower limit, where there is uncertainty regarding how the development of cracks will progress, it may be appropriate to take a conservative approach and assume a crack spacing factor of 100.

The development of cracks resulting from freeze–thaw attack is not dissimilar to that for AAR, in the sense that the rate at which new cracks are formed is initially high and gradually decreases, whereas there is a steady growth in crack width. The process is illustrated in Figure 7.7, which plots maximum crack width and average distance between cracks against the number of freeze–thaw cycles experienced by concrete specimens. An increase in the number of cracks results in a decline in the average distance between cracks. The effect of the changes in these parameters on the crack spacing factor is also plotted, and it is evident that, as for AAR, this tends toward a value of approximately 100.

It should be stressed that these relationships are plotted against freeze–thaw cycles rather than time, and this means that prediction of damage resulting from freeze–thaw action requires historical ambient temperature

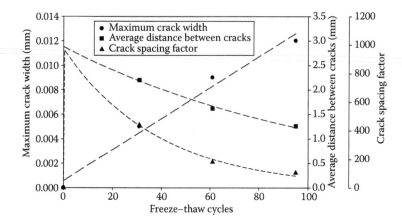

Figure 7.7 Increase in crack width, decrease in average distance between cracks, and the influence on crack spacing factor for concrete exposed to freeze–thaw cycles. (From Gérard, B. and J. Marchand, *Cement and Concrete Research*, 30, 2000, 37–43.)

data. Even when such data are available, this will only be of limited relevance to a given structure, since it will reflect conditions in the macroclimate in which the structure resides, rather than the microclimate in its immediate vicinity. This means that freeze–thaw events in the lifetime of the structure may be under- or overcounted.

The formation of cracks at concrete surfaces resulting from the formation of corrosion products on steel reinforcement is, in some respects, a simpler process than the previously discussed mechanisms. This is because, since the expansive forces derive from a localised area (a reinforcement bar), the damage normally takes the form of a single or possibly double crack – as shown in Figure 7.8 – which subsequently widens.

Although corrosion crack development follows a relatively simple course, the process of predicting corrosion cracking is made more complex by the fact that it may be possible to predict events prior to cracking. Deterioration resulting from AAR and freeze-thaw attack will usually only be identified once cracking has started. In contrast, the onset of corrosion cracking can potentially be predicted either via electrochemical measurements or through the measurement of chloride ingress profiles and/or carbonation fronts.

Many models for predicting this 'time to corrosion cracking' – of varying complexity – have been devised. The key parameters that influence the duration are the rate of corrosion, the mechanical properties of the concrete and the configuration of reinforcement with respect to the concrete surface – in particular, the depth of cover and the thickness of the porous. From the

(a)   (b)

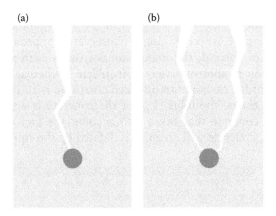

Figure 7.8 Crack formation resulting from corrosion of steel reinforcement.

perspective of an engineer attempting to predict the future performance of a structure, the following equation has the benefit of using characteristics that are likely to be obtainable from on-site measurements that could be conducted as part of an appraisal [6]:

$$t_{cr} = \left[\frac{7117.5(D+2\delta_0)(1+v+\Psi)}{iE_{ef}} \right] \left[\frac{2Cf_{ct}}{D} + \frac{2\delta_0 E_{ef}}{(1+v+\Psi)(D+2\delta_0)} \right]$$

where t_{cr} is the time from corrosion initiation to corrosion cracking (h), D is the diameter of steel reinforcing bar (mm), C is the concrete cover (mm), δ_0 is the thickness of 'porous zone' (μm), v is the Poisson's ratio of concrete (typically, 0.18), i is the corrosion current (μA/cm²), E_{ef} is the effective elastic modulus of the concrete (N/mm²), f_{ctm} is the mean tensile strength of the concrete (N/mm²), and Ψ is a constant dependent on D, C, and δ_0.

Some of these parameters require further explanation. The 'porous zone' is the volume of concrete around a reinforcement bar that is sufficiently porous to accommodate the initial formation of corrosion products without the development of stress. The thickness of this zone (δ_0) has been estimated as being between 10 and 20 μm [7]. Since there is no simple means of determining this either from *in situ* measurements or laboratory analysis, the approach advised by the developers of the model is to use both these values in calculations to obtain a time range in which cracking is estimated to occur.

If corrosion has already commenced, the corrosion current (i) can be obtained from linear polarisation resistance measurements, as discussed in Section 7.4.5. Where corrosion has not yet initiated, the corrosion current once corrosion has started will need to be estimated. This is not a straight-forward task, since values can vary considerably, depending on environ-mental conditions. However, taking a conservative approach, it has been suggested that, in general, the maximum corrosion current for chloride-induced corrosion is approximately 100 µA/cm², whereas the maximum corrosion current for carbonation-induced corrosion is 10 µA/cm² [8].

The effective elastic modulus (E_{ef}) of the concrete is a measure of the stiffness of the concrete that takes into account the fact that, over long loading periods, concrete will creep. It is defined by the equation

$$E_{ef} = \frac{E_c}{1+\varphi}$$

where E_c is the modulus of elasticity of the concrete (N/mm²) and φ is the creep coefficient (see Chapter 2, Section 2.2.2).

Eurocode 2 [1] provides guidance on estimating the modulus of elasticity from (*in situ* or laboratory-based) measurements of compressive strength using the equation

$$E_{cm} = 22\left[\frac{f_{cm}}{10}\right]^{0.3}$$

where f_{cm} is the mean cylinder compressive strength of the concrete (N/mm²).

Similarly, *Eurocode 2* provides guidance on the estimation of tensile strength (f_{ctm}). When the mean compressive cylinder strength (f_{cm}) of the concrete is 50 N/mm² or less, the mean tensile strength can be estimated using the equation

$$f_{ctm} = 0.30 f_{cm}^{2/3}$$

Where the strength is higher, a different equation should be used:

$$f_{ctm} = 2.12\ln\left(1+\left(\frac{f_{cm}}{10}\right)\right)$$

Ψ is defined by the equation

$$\Psi\, \frac{D^2}{2C(C+D+\delta_0)}$$

Once a crack has appeared at the surface, its width will typically increase in a linear manner with respect to time, as shown in Figure 7.9. As for time to cracking, many models have been developed to describe crack width growth resulting from reinforcement corrosion, using techniques including finite-element analysis and fracture mechanics, with the majority showing good parity with reality. Again, reinforcement configuration and rate of corrosion are the most influential factors. The equation cited here is again chosen primarily on the grounds that its parameters can be obtained from site measurements [9]:

$$w = -(D+2C)\frac{2m_s(1-r_v)+2\pi\rho_s\delta_0 D+\pi\varepsilon_{cr}\rho_s(D^2+2DC+2C^2)}{(D^2+2DC+2C^2)\rho_s}$$

where w is the crack width (m), m_s is the mass of steel consumed by corrosion (kg/m of reinforcement bar), r_v is the volumetric expansion ratio, ρ_s is the density of steel (7800 kg/m^3), and ε_t is the tensile strain capacity of the concrete. The other parameters have been previously defined, but in this equation, all distances are in meters.

If corrosion has already started, the mass of steel consumed may be approximately estimated using corrosion current measurements and Faraday's law (see Section 7.4.5), as long as an estimate of when corrosion had been initiated can also be made.

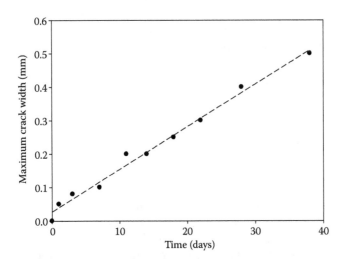

Figure 7.9 Crack growth resulting from corrosion of reinforcement induced through the application of an electrical current. (From Andrade, C. et al., *Materials and Structures*, 26, 1993, 453–464.)

The volumetric expansion ratio is the ratio of the volume of corrosion products to the volume of steel that formed them. This can vary but is typically approximately 3 [10].

The tensile strain capacity is the tensile deformation required to cause cracking. This can be determined from the elastic modulus and tensile strength of the concrete, which can, in turn, be estimated from *in situ* or laboratory measurement of compressive strength (see Chapter 2, Section 2.2.2).

All of the above cracking mechanisms produce damage that can be described in terms of discrete cracks or at least in terms of their distribution (such as via the crack spacing factor). However, in the case of sulphate attack, deterioration takes the form of the generation of large numbers of microcracks whose identification and measurement are potentially much more problematic, some of which only later become macrocracks. For this reason, a 'continuum damage mechanics approach' is usually adopted – in other words, the cracking is treated as a general deterioration of the concrete affected by sulphate attack, leading to a decrease in strength and an increase in rate at which mass transport occurs.

The approach to modelling this type of damage selected for this chapter is the one described by Sarkar et al. [12]. This is based around the estimation of a nucleated crack density (C_d), a measure of how much microcracking has been sustained by concrete in a structural element undergoing expansion as a result of sulphate attack [13]. The methodology described by the researchers uses curve fitting to the stress–strain curve obtained from initially undamaged concrete to determine constants that describe the manner in which microcracks begin to form under loading. However, the opportunity to test undamaged concrete in a manner that permits a full stress–strain curve to be obtained may not be an option available to an engineer. The approach taken here is to combine the relationships described with others described in *Eurocode 2* to give an equation that relates the strength of concrete at the beginning of service and the strength at the time of appraisal to the nucleated crack density:

$$C_d = 0.56\left(1 - \frac{f_{cm}^{10/3}}{f_{cm0}^{10/3}}\right)$$

where f_{cm0} is the mean cylinder compressive strength of the concrete at the beginning of a structure's service life (N/mm²).

This relationship is valid over periods of deterioration where only microcracks have formed. As shown in Chapter 2, Section 3.2.2, loss of strength is approximately linear with respect to time when deterioration occurs through conventional sulphate attack, which assists prediction of future deterioration. Decline in strength for magnesium sulphate attack is usually not a linear process, which makes prediction of future damage problematic.

It will be necessary to estimate the effect that sulphate attack has on the movement of sulphate ions (or other harmful chemical species) into a concrete element. Where macrocracks have not yet formed, this is described by the equation [12]

$$D_i = D_{i0}\left[\left(1+\frac{32}{9}C_d\right)\right]$$

where D_i is the diffusion coefficient of a given chemical species through the damaged concrete (cm²/s) and D_{i0} is the diffusion coefficient of the chemical species through the undamaged concrete (cm²/s).

Obtaining a value for D_{i0} presents some problems. It may be possible to conduct diffusion tests on cores taken from undamaged parts of an element or other elements made from the same concrete. However, this may not be present, coring may not be permissible or budgetary constraints may prohibit this. In such cases, it may be possible to locate diffusion coefficients measured on similar concrete mixes in the literature.

Where macrocracks have formed, the equation is modified to

$$D_i = D_{i0}\left[\left(1+\frac{32}{9}C_d\right)-\frac{(C_d-C_{dc})^2}{(C_{dec}-C_d)}\right]$$

where C_{dc} is the 'conduction percolation threshold' and C_{dec} is the 'rigidity percolation threshold'.

Values for C_{dc} and C_{dec} have been estimated as 0.182 and 0.712, respectively [14,15].

7.4 IN SITU TESTING

Testing of concrete elements in a structure may be required for many different reasons. It may be required simply to establish the current characteristics of concrete and other materials in a structure as part of an appraisal. It may also be required to confirm the possible deterioration mechanisms identified as part of the processes previously shown in Figures 7.2, 7.3, and 7.4 – a process sometimes referred to as 'building pathology'. Finally, as indicated in Table 7.2, information may be required pertaining to a deterioration process that will be necessary in attempting to predict the progress of deterioration in the future.

Testing can be conducted on site – *in situ* testing – or in a laboratory. *In situ* testing has a number of advantages over laboratory testing, the main one being that the material need not be removed from a structure (in the form of cores, etc.) and, in the case of many test methods, little or no damage is done. Moreover, where laboratory facilities are not readily available or transport of samples to a laboratory presents practical problems, *in situ* testing may be a solution.

The *in situ* methods described below are methods with direct relevance to concrete durability. There is a wide range of other techniques, in particular, those related with the destructive and nondestructive measurements of *in situ* strength. Such measurements are likely to also play a role in the appraisal of a structure and may indeed be necessary to predict the likely course of future deterioration. Moreover, a number of techniques including radiographic and short-pulse radar techniques may provide insight into the nature of cracking in concrete elements. The reader is directed elsewhere for further details on these tests [16].

7.4.1 Cover measurement

Where processes that lead to the corrosion of reinforcement are under investigation, it is usually necessary to know the depth of concrete cover present between the external environment and the steel, as well as the configuration of reinforcement bars within a structural element. This configuration and the nominal depth of cover may well be available to an engineer in the form of design drawings and specifications. However, this information may not be available, and it may not be prudent to assume parity between the design and the actual structure, and so further investigation will usually be required.

Determining the depth of cover can be achieved through drilling into the concrete until the steel is reached, and an indication of the configuration of reinforcement and the dimensions of bars used can be obtained by more extensive 'opening up'. However, such methods are clearly destructive, and such damage may not be acceptable and rectifying the damage will be expensive.

Various 'covermeter' devices are commercially available, which use electromagnetic pulse induction to estimate the depth of cover. Pulse induction involves passing an alternating current through a coil to create an alternating magnetic field. Where a metallic object lies inside this magnetic field, eddy currents are generated in the object that induces a magnetic field in the opposite direction to the first magnetic field, which is measured in the form of a change in voltage in another coil. The change in voltage can be used as a means of estimating the distance of the object from the coil.

The depth to which a covermeter can detect metal varies from model to model, but machines with a range of up to 200 mm are now not uncommon. However, any other metallic or magnetic materials can interfere with covermeter measurements. Such materials can include steel fibres and magnetite in aggregate or fly ash. Modern covermeters contain multiple coils arranged in configurations that allow not only measurement of depth but

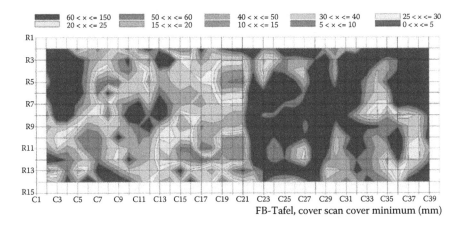

60 < x <= 150	50 < x <= 60	40 < x <= 50	30 < x <= 40	25 < x <= 30
20 < x <= 25	15 < x <= 20	10 < x <= 15	5 < x <= 10	0 < x <= 5

FB-Tafel, cover scan cover minimum (mm)

Figure 7.10 Example of a map of cover depth obtained using a covermeter. (From Reichling, K. et al., *Materials and Corrosion*, 64, 2013, in press.)

also estimation of the diameter of reinforcement bars. Where bar diameter estimation is required, the maximum cover depth over which the meter can operate is usually somewhat curtailed. More advanced systems are able to build up a map of the reinforcement as it is scanned (Figure 7.10).

Most covermeters consist of a display unit attached by a cable to a hand-held probe device in which the coils are housed. The probe is placed at the concrete surface, and the display unit processes the signal from the sensor to give a depth measurement. Units will also normally contain a system for temperature sensing and compensation, since temperature can otherwise influence the result obtained.

It should be stressed that the results from covermeters should be viewed as estimates, and where cover depth is critical, physical investigation may still be necessary to some extent.

7.4.2 Surface absorption

Where it is necessary to establish the quality of a concrete surface, with particular emphasis on its ability to resist the ingress of water, the Initial Surface Absorption Test (ISAT), described in *BS 1881-208* [18], is one of the methods that can be used to conduct such measurements.

The ISAT consists of a cap, fastened to the concrete surface, with a minimum surface area of 5000 mm², which is attached to a water reservoir positioned such that a head of 200 mm of water is maintained against the concrete surface (Figure 7.11). A seal between the cap and the concrete surface is achieved with an elastomeric gasket or a 'knife edge' that is subsequently sealed to the surface using modified modelling clay. After water is introduced into the cap, the supply of water from the reservoir is

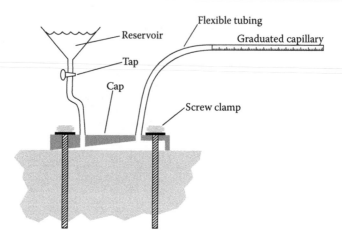

Figure 7.11 Schematic diagram of the ISAT apparatus.

periodically shut off using a tap, and water in a capillary tube becomes the source of water whose rate of absorption is measured by measuring the rate of movement of the water meniscus using graduations marked on the tube. Once the equipment has been calibrated, the rate of movement of water along the capillary can be related to the rate of absorption (in millilitres per square meter per second).

Because surface absorption declines with time, as the concrete approaches a state of local saturation, the rate of absorption should ideally be measured at three (and sometimes four) intervals from the beginning of the test: 10, 30, 60, and 120 min. However, it is often the case that, because a larger number of results from different points on a structure are considered of primary importance, only 10-min readings will be taken.

It has been concluded that ISAT measurements are not readily related to the sorptivity of concrete (see Chapter 5, Section 5.2.3) [19]. Guidance has been developed to provide an approximate means of judging the quality of a concrete surface (shown in Table 7.3), with high absorption rates equating to low quality. However, quality can really only be evaluated meaningfully through comparison with other concrete surfaces.

Table 7.3 Guidance for interpreting ISAT results

10-min is at value (ml/m²/s)	Concrete quality
>1.00	Very poor
0.85–1.00	Poor
0.70–0.85	Moderate
0.65–0.70	Good
<0.65	Very good

The ISAT test presents a number of practical problems when conducted *in situ*. The first is how to secure the cap to the concrete surface. This is best done using bolts, but this is clearly destructive and not always practical. The standard suggests the use of modelling clay softened with grease as an alternative. The second key problem is ensuring that the external surfaces tested are all in a comparably dry condition. The standard states that concrete should be protected from water for 48 h prior to testing and from direct sunlight for at least 12 h.

The test can also be carried out in the laboratory, in which case specimens can be dried in much more controlled conditions. The standard allows both oven drying at 105°C and drying in air under normal interior conditions at 20°C. Temperatures of 105°C will lead to the partial decomposition of certain cement hydration products, and so air drying is recommended.

Although the description of the ISAT procedure seems relatively simple, the reality of actually carrying it out is often less so for a novice. Therefore, practice in carrying out the test prior to conducting it on site is highly recommended.

A simple means of measuring water absorption is through the use of a standpipe test. The most commonly encountered of these is the Karsten (or RILEM) tube test, which is intended for use on masonry, but which can be used on concrete. The apparatus is simple and consists of a clear tube that can be fixed onto the surface to be tested using putty (Figure 7.12). The tube is filled with water to give an initial head of 100 mm, and the level of water is measured at times of 5 and 15 min after the start of the test [20], with the head declining as the level drops. The volume absorbed between these two times is called the 'water absorption coefficient' (Δ_{5-15}), which can be used to estimate the sorptivity (S) of the concrete surface [21]:

Figure 7.12 Karsten tubes for horizontal and vertical surfaces.

$$S = \frac{-\pi R^2 \left(\sqrt{15} - \sqrt{5}\right) + \sqrt{D}}{20 \dfrac{\pi R \gamma}{\theta_{cap}}}$$

where R is the radius of the contact zone (cm), γ is a constant (0.75), and θ_{cap} is the water content of the concrete in a saturated condition (m³/m³).

D is obtained from the equation

$$D = \pi^2 R^4 \left(\sqrt{15} - \sqrt{5}\right) + 40 \frac{\pi R \gamma}{\theta_{cap}} \Delta_{5-15}$$

This approach requires an estimate of θ_{cap}. This can be established from water sorption measurements, although these would require analysis of samples in the laboratory, which somewhat defeats the purpose of using an *in situ* test. However, the use of values typical for the concrete under investigation will yield results that are likely to remain useful where sorptivity values are required for subsequent calculations.

Two other techniques that can be used to measure the surface absorption characteristics of concrete are the Figg and Autoclam tests. The Figg test initially involves drilling a hole into the concrete surface to a depth of 40 mm. A silicone seal is placed at the opening of the hole, and a hypodermic needle attached to a syringe is used to fill the hole with water (Figure 7.13). The syringe is shut off with a tap, making the only source of water a graduated capillary positioned in such a manner as to subject the interior of the hole to a head of 100 mm. In a manner similar to the ISAT test, the rate of absorption is determined by measuring the time taken for the meniscus in the capillary to travel 50 mm.

The Figg test is relatively easily conducted, although the hole needs to be subsequently repaired. The test, in its basic form, clearly does not measure the characteristics of the true surface of the concrete, although a surface chamber is now supplied as part of the Figg kit, which allows measurements to be made directly at the surface.

A proposed system for the interpretation of Figg water absorption values is shown in Table 7.4.

The Autoclam involves a somewhat more technically complex device. It consists of a 'clam' – a chamber that is fixed to the concrete via a ring that is attached with epoxy resin (Figure 7.14). Water can be introduced and pressurised within the clam via a piston, and the movement of this piston can be automatically measured. The pressure within the clam is measured with a pressure gauge or a pressure transducer connected to a digital display.

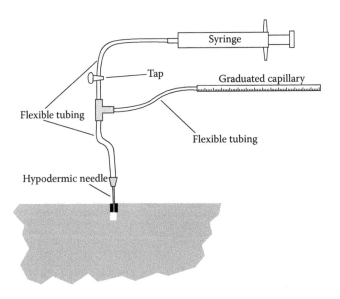

Figure 7.13 Schematic diagram of the apparatus used for the water absorption Figg test.

Table 7.4 Guidance on relating the measured time from the
Figg water absorption and air permeability tests to
concrete protective quality

Measured time (s)		
Water absorption	Air permeability	Protective quality
<20	<30	Poor
20–50	30–100	Not very good
50–100	100–300	Fair
100–500	300–1000	Good
>500	>1000	Excellent

Source: Basheer, P.A.M., *Proceedings of the ICE: Structures and Buildings*, 99, 1993, 74–83.

The test is conducted by filling the clam with water and applying a constant pressure using the automated piston. Two levels of pressure can be applied – 1 kPa for a sorptivity test and 150 kPa for a 'water permeability' test. As water moves into the concrete, the piston moves down to maintain pressure, and the movement of the piston is monitored over a 15-min period. The results of the test are expressed as a 'permeation index' in cubic metres per minute$^{0.5}$.

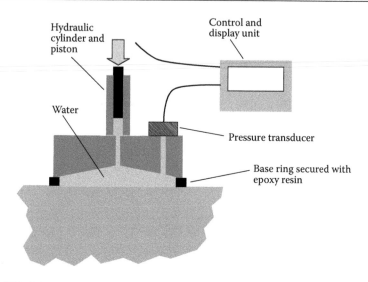

Figure 7.14 Schematic diagram of the apparatus used for the Autoclam test.

Table 7.5 Guidance on relating the permeation index obtained from the Autoclam water tests to concrete protective quality

	Test type		
Water sorptivity test (m³ × 10⁻⁷/min)	Water permeability test (m³ × 10⁻⁷/min)	Air permeability test (ln(pressure)/min)	Protective quality
≤1.30	≤1.30	≤0.10	Very good
>1.30 ≤2.60	>1.30 ≤2.60	>0.10 ≤0.50	Good
>2.60 ≤3.40	>2.60 ≤3.40	>0.50 ≤0.90	Poor
>3.40	>3.40	>0.90	Very poor

Source: Basheer, P.A.M., *Proceedings of the ICE: Structures and Buildings*, 99, 1993, 74–83.

It has been pointed out that the change in the rate of water take-up with time at both test pressures is, in fact, characteristic of absorption [23]. Suggested guidance on interpreting results is provided in Table 7.5.

7.4.3 Permeability

Both the Figg and Autoclam tests discussed above can be configured for measuring air permeability. In the case of the Figg test, the fundamental changes are that the syringe is replaced with a vacuum pump and the graduated capillary is replaced with a manometer. The vacuum pump is used to reduce the pressure in the hole to −55 kPa before being isolated from the system and switched off. The pressure measured on the manometer is monitored until the vacuum reduces to −50 kPa as the result of air permeating the concrete pores and infiltrating the hole. The time taken for the decay in vacuum is called the 'air permeability index'. Table 7.4 provides details of a proposed system for interpreting results.

The Autoclam apparatus can be run with air rather than water. The air is pressurised by the piston, and the decay of pressure from 50 kPa is monitored over a maximum period of 15 min. The drop in pressure observed is expressed as another 'air permeability index' – in this case, the natural log of the pressure drop divided by the time taken. Guidance on the interpretation of results is provided in Table 7.5.

A number of other air permeability tests exist, which operate on very similar principles [24]. It should be stressed that none of the tests actually measure permeability but rather characterise how air moves through concrete under a pressure differential. Laboratory testing of cores to obtain actual air permeability measurements is possible.

The point has already been made that the water permeability test mode of the Autoclam is, in reality, measuring absorption. Even in the laboratory, water permeability is a challenging task, since permeability can only be measured once a material is wholly saturated with water, and achieving this condition is problematic.

7.4.4 Half-cell potential

In the description of the corrosion of steel reinforcement in Chapter 4, it was shown that, when corrosion occurs, a galvanic cell is set up where part of the reinforcement will become anodic in nature and corrode, whereas another adjacent part will become cathodic and hence will not corrode. Where such a cell is established, a potential difference will exist between the two regions of the steel. Half-cell potential measurements determine the magnitude of this potential difference, which can be used to identify areas within a reinforced concrete member where corrosion is most likely to be a problem.

The key components used for half-cell corrosion potential measurements are a reference electrode and a voltmeter. Traditionally, copper/copper sulphate electrodes have been used for this purpose, but more recently, the development of low-maintenance silver/silver chloride electrodes has seen a shift towards these products [25]. The reference electrode requires regular calibration prior to use on site, and this is normally done with a calomel (mercury/mercury chloride) electrode under laboratory conditions.

The configuration for half-cell corrosion measurement is shown in Figure 7.15. A small length of reinforcement is exposed by drilling, having

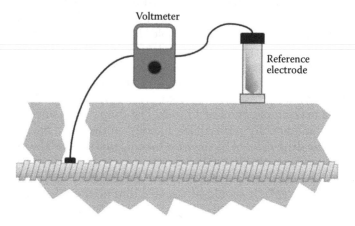

Figure 7.15 Half-cell potential circuit.

identified the location of bars using a covermeter. The bar is then drilled, and a self-tapping screw is driven into the hole to act as a connection to the voltmeter cable. It is usually beneficial to expose two relatively distant lengths of reinforcement to check that they are electrically connected, using the continuity setting on a multimeter.

The circuit is completed by connecting the reference electrode to the voltmeter and bringing it into contact with the concrete surface. To ensure contact between the electrode and the reinforcement, the surface must be thoroughly wetted with water. A detergent solution is sometimes used for this purpose because it has a lower surface tension and is likely to penetrate the pores of the concrete with greater ease.

Normally, half-cell measurements are taken by marking out a grid on the concrete surface, placing the electrode at each intersection on the grid and taking a reading from the voltmeter. These can be used to construct a map of half-cell potentials measured at the surface (Figure 7.16).

Guidance on the interpretation of half-cell potentials has been developed in *ASTM C876* [26]. This is summarised in Table 7.6.

Where exposure of reinforcement is not permitted, it is also possible to use two reference electrodes – one that remains static and one that is moved to each grid point.

7.4.5 Linear polarisation resistance

In Chapter 4, it was stated that the current passing through steel reinforcement, which is undergoing corrosion, gives a measure of the rate of corrosion. The actual mass (in grams) of a metal consumed by corrosion (m) at

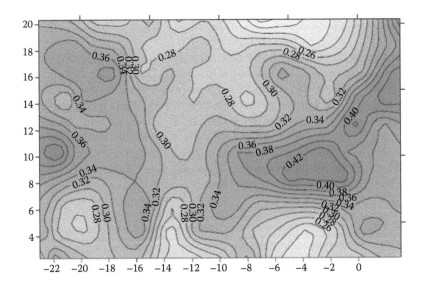

Figure 7.16 Example of a half-cell potential map. (From U.S. Department of Transportation, Federal Highway Administration. *Highway Concrete Pavement Technology Development and Testing, Vol. 5: Field Evaluation of Strategic Highway Research Program (SHRP) C-206 Test Sites (Bridge Deck Overlays).* McLean, Virginia, 2006, p. 41.)

time *t* (in seconds) after corrosion has initiated can be determined using Faraday's law of electrolysis:

$$m = \frac{MIt}{zF}$$

where M is the molar mass of metal (g), I is the corrosion current (A), z is the ionic charge of an ion of the metal (electrons removed during ionisation, e^-) and F is the Faraday constant (96,485 C/mol).

Table 7.6 Relationship between half-cell potential and risk of corrosion

Half-cell potential (mV)	Corrosion risk (%)
>−100	<10
>250 ≤−100	10–90
≤−250	>90

Source: American Society for Testing and Materials. ASTM C876-09: Standard Test Method for Corrosion Potentials of Uncoated Reinforcing Steel in Concrete. West Conshohocken, Pennsylvania: American Society for Testing and Materials, 2009, 7 pp.

In the case of conventional mild steel, it is valid to treat the material as being iron, which means that M is equal to 56 g and that z is equal to 2 ($Fe \rightarrow Fe^{2+} + 2e^-$; see Chapter 4, Section 4.2.2).

The corrosion current can be determined using linear polarisation resistance measurements. This technique involves a hand-held electrode that is held at the concrete surface and used to apply an external current that acts to change the current in the reinforcement by a small quantity, ΔI [25]. This acts to shift the corrosion potential, E_{corr} (see Section 7.4.4), by an amount ΔE. As a result of this polarisation process, the steel at the surface of the reinforcement will display a change in resistance (R_p) in response to the change in potential difference. The corrosion current in the reinforcement bar (I_{corr}) is given by the equation

$$I_{corr} = \frac{B\Delta E}{\Delta I}$$

where B is a constant (mV).

This is the Stern–Geary equation [28].

The configuration of the electrode is shown in Figure 7.17 – it consists of a ring-shaped auxiliary electrode with a reference electrode positioned in the middle of the ring. The reference electrode is usually a copper/copper sulphate electrode. This electrode is used in a half-cell configuration with the reinforcement (see Section 7.4.4) to measure ΔE and is thus connected,

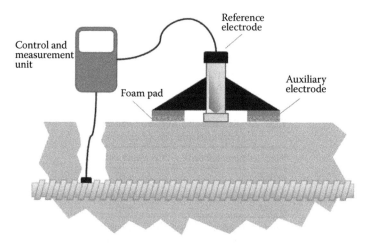

Figure 7.17 Schematic diagram of apparatus for linear polarisation resistance measurements.

via an instrumentation unit, to the reinforcement that must be uncovered at localised points to allow this.

During measurement, the instrumentation unit is used to apply a known potential difference (typically ±10 mV), which is controlled using a measurement of ΔE from the reference electrode. The resulting current in the circuit between the auxiliary electrode and the reinforcement is also measured by the instrumentation, which then uses the values obtained to determine I_{corr} using the above equation. B is dependent on whether the steel is undergoing corrosion, with a high value indicating passivation. When steel is undergoing corrosion, the value is approximately 26 mV [29]. The normal approach is to err on the side of caution and assume this value.

The rate of corrosion of the steel can be predicted if the area of the reinforcement can be estimated using the equation

$$\Delta x = \frac{31,557,600 \cdot MI_{corr}}{zF\rho A}$$

where Δx is the rate of loss of diameter of the steel bar (mm/year), ρ is the density of steel (7652 kg/m³), and A is the surface area of the reinforcement (m²).

This assumes that corrosion is occurring at a uniform rate over the entire surface of the reinforcement. In Chapter 2, the possibility of this not occurring is discussed – the process of localised pitting (often promoted by the presence of chloride ions) will lead to localised corrosion at a rate far higher than that obtained from the above equation. Thus, it should be used with care, and other techniques should be used to explore whether pitting may be occurring, such as chemical analysis for chlorides.

Table 7.7 also provides a means of interpreting linear polarisation resistance values using corrosion current density – the corrosion current per unit surface area of reinforcement.

Table 7.7 Guidance on interpreting linear polarisation resistance values

Corrosion current density (μA/cm²)	Rate of corrosion
<0.1	Negligible
0.1–0.5	Weak
0.5–1.0	Moderate
>1.0	High

Source: Andrade, C. and C. Alonso, *Materials and Structures*, 37, 2004, 623–643.

7.4.6 Resistivity

In Chapter 4, the dependency of corrosion on good electrical conductivity within concrete was discussed. This relationship has led to the adaption of the Wenner technique [31], originally used in soil engineering applications, to concrete. The Wenner technique uses four spring-loaded metal probes mounted in a holder such that they are spaced an equal distance apart. This distance is typically 50 mm. This apparatus is used to measure the resistivity of the concrete, which can be used as an indicator of the rate of corrosion, but is usually best used after half-cell corrosion potential measurements have established that corrosion is likely to be occurring.

The tips of the probes are pushed against the surface of the concrete, and an alternating current of known magnitude is passed between the two outer probes through the concrete. The two inner probes are connected through a voltmeter, which is used to measure the potential difference (V), in volts, between them. The resistivity (ρ) in kilo-ohm centimeters, of the concrete is determined using the equation:

$$\rho = 2\pi\alpha\frac{V}{I}$$

where α is the distance between the probes (cm), V is the potential difference (mV), and I is the current applied (μA).

Typically, a current of approximately 250 μA is used.

It should be stressed that results from resistivity measurements are very much dependent on the nature of the structural element being tested. Since the measurement is of a volume of concrete close to the surface, the usual surface effects (see Chapter 6), plus the effects of finishing and curing (or a lack of it), may give resistivity values that are not representative of the bulk concrete [32]. Defects that lead to higher levels of porosity, such as cracks and air voids, will have a similar effect. The depth of cover may also play a role in influencing the values obtained.

Guidance has been developed for interpreting resistivity measurements in terms of the relative corrosion rate (Table 7.8).

Table 7.8 Guidance for relating resistivity with corrosion rate

Resistivity (kΩcm)	Corrosion rate
>20	Low
10–20	Low to moderate
5–10	High
<5	Very high

Source: Langford, P. and J. Broomfield, *Construction Repair*, 1, 1987, 32–36.

7.4.7 Abrasion resistance

Where abrasion of a concrete surface is evident in a structure, *in situ* abrasion testing is probably not necessary – measurements of depths of abrasion will probably suffice to evaluate the scale of the problem and may be useful to predict how this will progress in the future. However, abrasion testing may be necessary where a structure is to undergo a change of use.

The British Standard method for the abrasion testing of screeds and directly finished concrete floors is *BS EN 13892-4* [34]. It uses a device sometimes referred to as a 'British Cement Association accelerated wear machine'. The machine applies a load of 65 kg to the floor surface through three 75-mm hardened steel wheels with individual fixed axles. These wheels travel in a fixed circle for 2850 revolutions, and the depth of abrasion is measured using a depth gauge at eight predetermined points on the circle.

Other abrasion tests exist, many of which can also be conducted *in situ*. Some of these have been discussed in further detail in Section 2.5.2 of Chapter 2. By their nature, abrasion tests are destructive, and repairs will normally be necessary to fill the abraded test zone.

7.5 LABORATORY TESTING

7.5.1 Chemical analysis

The techniques available for the chemical analysis of concrete are extremely varied, ranging from simple wet chemical methods to analysis using various forms of spectrometry, chromatography and gravimetry. Despite this wide variety, it is necessary, in some cases, to adhere to standard methodologies to ensure compatibility of results with others previously obtained. This is often important, since different analytical methodologies, regardless of their relative accuracy, will in many cases be measuring different things.

An example of this would be the measurement of chlorides in hardened concrete. Wet chemical methods can be used for this but require digestion of the sample such that the chlorides in the solid sample are released into aqueous solution. The extent to which chloride ions are released will depend on the digestion method used – exposure to water will only release chloride in readily soluble form, whereas acid digestion will certainly dissolve the majority of cement hydration products, leading to higher concentrations of chloride in solution. The type of acid is also significant – dilute nitric acid will dissolve hydration products and calcareous aggregates but will not dissolve a proportion of chlorides present in aggregate, whereas hydrofluoric acid is likely to dissolve all of the solid constituents. This may be important, since it may be the case that aggregate contains quantities of otherwise unavailable chlorides, which would normally be of no relevance from a durability perspective. In a similar way, spectrometry techniques that can be used to analyse solid concrete specimens without digestion, such as X-ray fluorescence, will give total chloride concentrations.

In the United Kingdom, *BS 1881-124: Methods for Analysis of Hardened Concrete* [35] describes a number of standard methods for the chemical analysis of concrete. The methods are outlined in Table 7.9.

The standard provides guidance on the appropriate number of samples and their minimum dimensions and mass to ensure that sampling is representative. The minimum sample dimension should be at least five times the maximum aggregate size, whereas the minimum sample mass is 1 kg, except for certain tests where a larger mass is required (see comments in Table 7.9). The number of samples taken depends on the volume of concrete under investigation: for a volume of less than 10 m³, two to four samples are required, which must be analysed separately. Larger volumes require more than 10 samples.

In many cases, it is useful to obtain other information relating to the original constituent materials used in the production of concrete in a structure. In particular, if data relating to the chemical composition of these materials are available, it is likely to enhance the accuracy of the deductions made from chemical analyses.

A number of test methods described in the standard require the material to be in the form of a ground powder. This must be of a quantity greater than 20 g, all of which passes a 150-μm sieve. A procedure for achieving this is provided in the standard, as outlined in Figure 7.18, although it should be stressed that if the same results can be achieved with fewer stages, that approach is also acceptable.

As discussed in Chapter 4, the cement matrix of concrete undergoes carbonation reactions when in contact with an atmosphere containing carbon dioxide. Although carbonation of intact concrete takes tens of years, carbonation of powdered concrete can occur much more rapidly because of the higher surface area. For this reason, the standard recommends conducting grinding as quickly as possible to minimise carbonation. In addition, specimens should be stored in such a way as to minimise the extent to which carbonation can occur – storage in air-tight bags or containers in an environment where atmospheric moisture is low (such as a desiccator containing anhydrous silica gel), ideally with reduced concentrations of carbon dioxide (achieved through the use of soda lime or even a pure nitrogen atmosphere).

Although all of the measurements described in *BS 1881-124* may play a part in investigating the aspects of deterioration in a structure, the following subsections concentrate on the aspects of concrete chemistry of particular relevance to concrete durability.

Table 7.9 Analysis methods described in BS 1881-124

Characteristics	Techniques	Comments
Cement and aggregate content	• Determination of insoluble residue: digestion with dilute hydrochloric acid, followed by washing with ammonium chloride, hydrochloric acid, and water. Gravimetric determination of residue. • Determination of soluble silica: analysis of the mixture of filtrate and washings from determination of insoluble residue using gravimetric analysis after precipitation of silica from solution or atomic absorption spectrometry. • Determination of CaO: analysis of the mixture of filtrate and washings from determination of insoluble residue by titration or atomic absorption spectrometry. • Calculation of aggregate and cement content.	Conducted on powdered specimen. Knowledge of composition of original materials is beneficial to accuracy. Additional methods for determining sulphide content, loss on ignition, and carbon dioxide content are also provided to provide further information, such as ground granulated blastfurnace slag content.
Aggregate grading	• Break concrete samples into less than 50-mm particles. • Separate into coarse and fine fractions using 5-mm sieve. • Clean fractions in acid solutions to remove cement. • Conduct a sieve analysis of fine fraction and calculate relative coarse and fine fractions.	Minimum sample mass: 4 kg.
Original water content	• Determination of the capillary porosity of concrete and aggregate (when available) by measuring the absorption of 1,1,1-trichlorethane under vacuum. • Determination of the combined water content of concrete and aggregate (when available) by gravimetric methods. • Determination of cement content as described above.	Initially requires a block of concrete cut from core, or similar, for capillary porosity measurements. The sample is then powdered. Minimum sample mass: 2 kg.

(continued)

Table 7.9 (Continued) Analysis methods described in BS 1881-124

Characteristics	Techniques	Comments
Type of cement		
Analysis of matrix	• Obtain fine fraction by crushing. • Determine soluble SiO_2, CaO, Al_2O_3 and FeO. • Correct for insoluble residue (aggregate) and combined water and carbon dioxide from carbonation (measured using loss-on-ignition). • Compare analysis against typical cement compositions.	
Microscope examination	• Surface of sawn concrete surface polished and treated with either potassium hydroxide solution or hydrofluoric acid vapour to yield differences in colour between anhydrous cement phases in unhydrated 'relics' of cement grains. • Examination of surface under reflected-light microscope.	Requires a block of concrete cut from core, or similar. Requires the presence of unhydrated 'relics' of cement grains – may not be possible with mature concrete. Hydrofluoric acid is extremely dangerous.
Type of aggregate	• Examination of aggregates exposed at broken or sawn surface, possibly with low-power microscope. • Treatment of exposed aggregate particles with dilute hydrochloric acid – effervescence indicates the presence of carbonate minerals.	Requires a block of concrete cut from core or similar.
Chloride content	• Digestion in boiling dilute nitric acid. • Titration of resulting solution.	
Sulphate content	• Digestion in boiling dilute hydrochloric acid. • Gravimetric measurement of sulphate via barium sulphate precipitation.	
Sodium and potassium oxide content	• Digestion in boiling dilute nitric acid. • Flame photometer analysis for sodium and potassium.	

7.5.1.1 Chloride

Measuring the quantity of chloride in concrete may be necessary for a number of reasons. Analysis of bulk concrete may be needed to establish whether a chloride-based accelerator was used during the construction process. Analysis of the concrete in contact with steel reinforcement

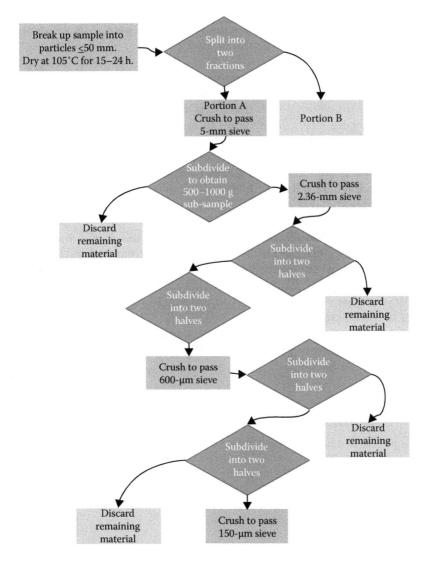

Figure 7.18 Procedure from *BS 1881-124* for obtaining a ground subsample of material for chemical analysis from a concrete sample.

will provide an engineer with an indication of whether chloride from the external environment has ingressed to an extent that is likely to have initiated corrosion. One of the more powerful ways in which chloride analysis can assist in the appraisal of a structure is through the construction of chloride profiles, which can be used to estimate the coefficient of chloride diffusion. This is particularly useful in situations where chlorides have not yet reached steel reinforcement but are known to be present on the exterior of a structural element, because it permits prediction of when corrosion is likely to initiate.

Obtaining chloride profiles requires the incremental removal of quantities of concrete from the surface, with the resulting specimens undergoing analysis for chloride such that a plot of concentration versus depth is obtained (Figure 7.19). The removal of material can be conducted *in situ* or on cores obtained from a structure. This is best achieved using a profile grinding apparatus specifically designed for the task, where a motorised abrasive disk is used to remove material and the resulting dust is collected by the machine. This has the benefit of allowing the removal of material from a relatively large area of surface and ensuring that the majority of abraded material is captured. However, reasonable results are possible by using a hammer drill to bore to progressively greater depths and, at each increment, collecting the resulting dust as it emerges and removing any remaining dust in the hole with a brush. Sawing thin slices from cores using a suitably configured diamond saw is also a possible option, but it should be stressed that contact with coolant water is likely to leach soluble chlorides to some extent.

Once each sample has been analysed, a process of curve fitting can be used to fit the following equation (which is a form of Fick's second law) to the chloride profile plot:

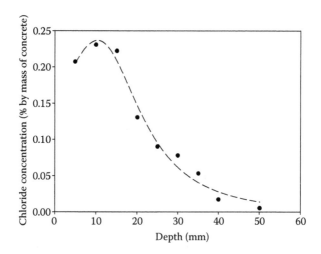

Figure 7.19 Chloride concentration profile obtained from a concrete in a dockyard exposed periodically to chloride-bearing estuary water. (From Costa, A. and J. Appleton, *Materials and Structures*, 32, 1999, 252–259.)

$$c = erf\left(\sqrt{\frac{x}{4Dt}}\right)$$

where c is the concentration of chloride at a depth x (m), D is the apparent chloride diffusion coefficient (m^2/s), and t is the period that the structure has been exposed to a chloride-bearing environment (s).

A value of t can normally be relatively easily estimated. In many cases, it may simply be the period of service of a structural element, although in some cases, exposure to chlorides may have been periodic or may have started at a point in time between the beginning of service and the present day, which may complicate matters. Thus, during curve fitting, a fixed value of t is used and D is refined to obtain the best fit. D can then be used to calculate the time necessary for the concentration of chlorides to reach the probable threshold concentration for corrosion at the reinforcement surface, using the same equation.

It should be noted that, as seen in Figure 7.10, lower concentrations of chlorides may be observed at depths close to the surface compared with some distance further into the concrete. As discussed in Chapter 4, Section 4.3.3, this is partly the result of a drop in pH at the surface, which occurs because of carbonation at the surface, the leaching of OH ions, or a combination of both. Some caution is required in using chloride profiles of this type, and it may be advisable to ignore results from the immediate surface.

The concentration term in the equation can be expressed in any form but is normally either given as a percentage by mass of concrete or as a percentage by mass of cement, where the cement content in each sample is estimated from the calcium content and knowledge of the cement used (or most likely to have been used). The latter approach is compatible with the normal means of defining corrosion thresholds for chlorides but is also cited as providing a more accurate value of D. This is because chlorides will, in most cases, be predominantly diffusing through the cement matrix and, as seen in Chapter 6, differences in the relative volumes of cement matrix and aggregate occur between the surface of the concrete and its interior.

In Chapter 4, the binding of chloride ions by the cement matrix was seen to be an important factor in extending the period before corrosion initiates. There is, therefore, some value in establishing how much chloride is actually in a bound state throughout the chloride profile (and thus more significantly how much is free). Many different approaches have been proposed for measuring the proportion of free chlorides, most of which involve exposing the sample to water and analysing the concentrations of chloride ions in the solution. Although there is nothing wrong with this approach, it should be stressed that the conditions of extraction will influence the result, and so, differences in the temperature of extraction and the ratio of water to solids between methodologies will lead to different free chloride values being obtained.

The most commonly used method for estimating free chlorides is the American Society for Testing and Materials test, which involves refluxing a mixture of water and powdered concrete (1:5 solid-to-water ratio) for 5 min before cooling and filtering. The filtrate has a quantity of nitric acid and hydrogen peroxide added and is briefly boiled and cooled before analysis of the resulting solution [37]. This approach can be used to calculate a value of D more closely related to the unbound chlorides – the free chloride diffusivity, which is arguably more useful in predicting when corrosion is likely to initiate. However, it should be stressed that the results of extraction of chlorides in this way do not represent an absolute measure of free chloride concentrations.

Although the *BS 1881-124* method for chloride analysis of the solution resulting from digestion of concrete uses titration, a number of other techniques are equally suitable for such analysis for solutions including atomic absorption spectrometry, ion-exchange chromatography, colorimetric methods and inductively coupled plasma emission spectrometry. A number of kits for chloride analysis also exist, most of which involve a strip or stick that changes colour in proportion to the concentration of chloride ions that it is exposed to. In addition, X-ray fluorescence spectrometry can be used to determine the total chloride content of solid powdered concrete specimens, although this will include all chlorides present in aggregate. This may not be an issue where aggregates can be demonstrated to have contained low concentrations of chloride prior to service.

7.5.1.2 Sulphate

It is possible to measure sulphate profiles in concrete in the same manner as for chlorides. However, because Portland cement contains calcium sulphate, unlike a chloride profile, a sulphate profile will not tend towards zero, but toward a baseline corresponding to the normal level in the concrete (Figure 7.20). This means that prior to curve fitting, to determine the apparent sulphate diffusion coefficient, it is necessary to subtract this baseline concentration from the profile data. It should also be stressed that, because concrete exposed to sulphates will become damaged, the diffusion coefficient obtained will be very much an apparent diffusion coefficient. Nonetheless, for the purpose of estimating rates of ingress, it is likely to yield usable values.

BS 1881-124 uses the gravimetric determination of sulphate by precipitating barium sulphate from a solution obtained by digestion in dilute hydrochloric acid. Similar alternative techniques to chlorides may be

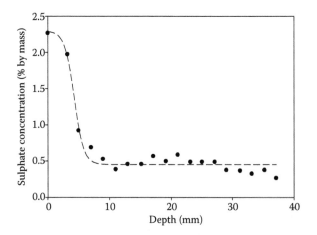

Figure 7.20 Concentration profile of sulphate in concrete exposed to water in a sewage treatment plant. (From Weritz, F. et al., *Construction and Building Materials*, 23, 2009, 275–283.)

used for analysis. The results shown in Figure 7.20 were obtained using laser-induced breakdown spectroscopy [38].

7.5.1.3 Alkalis (sodium and potassium)

From a durability perspective, the main reason for analysis for alkalis is to establish whether levels of sodium and potassium are likely to be sufficiently high to induce AAR. Only soluble alkalis will play a role in this reaction, and so analytical techniques that determine total alkalis (such as X-ray fluorescence) are unlikely to be useful on their own. Aggregates may contain relatively high levels of alkali in the form of feldspars, mica and clay minerals, which would mean that a total alkali analysis might significantly overestimate their availability.

Therefore, determination of alkalis should involve a digestion that is likely to remove as much of the alkali in the cement matrix while leaving as much of the insoluble alkali in the aggregates. In *BS 1881-124*, this is attempted through a digestion in boiling dilute nitric acid. It should be stressed that this approach will still dissolve some of the alkalis in aggregate minerals. This is possibly not entirely undesirable – in Chapter 3, it was seen that alkalis in aggregates prone to AAR will gradually be dissolved as the aggregate particles succumb to AAR. Thus, the overestimation of soluble alkalis given by acid digestion will give an indication of the future availability of alkalis.

BS 1881-124 uses flame photometry as the analytical technique for both sodium and potassium. Similar alternative techniques to those suitable for chlorides and sulphates may also be used to analyse extracted solutions.

7.5.1.4 Carbonation

The extent of carbonation in a concrete core can, in theory, be investigated through chemical analysis of samples taken from progressive depths, as discussed above. In this case, the analysis techniques used would be to determine concentrations of either portlandite, calcium carbonate or CO_2 present as carbonate. Techniques for such analysis might include thermogravimetry, X-ray diffraction or the method for determining carbon dioxide in *BS 1881-124*. However, there is a much more straightforward means of determining the depth of carbonation: the application of a solution of 1% thymolphthalein by mass in a 70% (by mass) ethanol in water solution [39] to a fractured concrete surface will turn blue if the pH exceeds around 10, but will remain colourless below this pH value. Thus, a very clear demarcation between carbonated and uncarbonated concrete can be obtained.

The thymolphthalein solution is best applied as a spray. It is necessary to use several measurements of depth, since there is typically some variation of carbonation depth at different points along the carbonation front, as a result of inhomogeneity within the concrete.

Historically, phenolphthalein has been used for the same purpose, but concerns about its safety, coupled with the fact that it changes colour around a pH of 9.5 (which could potentially lead to the underestimation of the risk of depassivation of steel, which occurs at pH levels somewhat higher than this), has led to a switch to thymolphthalein by most practitioners.

7.5.2 Air-void characteristics

The use of air-entraining agents is highly effective at protecting concrete from freeze–thaw damage. As discussed in Chapter 2, the distribution of bubbles as measured using the air bubble 'spacing factor' is a crucial aspect of the effectiveness of entrained air. In certain circumstances, where deterioration as a result of freeze–thaw attack appears to have occurred, it may be necessary to determine the spacing factor and, thus, establish whether air entrainment was adequate.

The spacing factor, and other air-void characteristics, can be determined using a method described in *BS EN 480-11* [40]. The method uses a microscope mounted on a linear traverse platform to count bubbles in the cement matrix of a polished concrete surface. The procedure involves moving the microscope along a series of regularly spaced lines and recording the total distance traversed through aggregate, through cement and through air, as well as the number of air bubbles encountered on each traverse line.

The method used to calculate the spacing factor (L) is dependent on the volumetric ratio of cement paste to aggregate (p/A), which is calculated using the equation

$$\frac{p}{A} = \frac{T_p}{T_a}$$

where T_p is the total traverse length through cement paste and T_a is the total traverse length through aggregate.

Where the p/A ratio is less than 4.342, the spacing factor is calculated using the equation

$$L = \frac{T_p}{4N}$$

where N is the total number of air voids encountered.

Where the p/A ratio is higher, the equation used is

$$L = \frac{3T_a}{4N}\left[1.4\left(1 + \frac{p}{A}\right)^{1/3} - 1\right]$$

Various other characteristics can be calculated from test results, including air content, void frequency and specific surface.

7.5.3 Latent expansion test

In cases where AAR is suspected but is identified in the very early stages, or where aggregate known to be potentially susceptible to the reaction is present, it may be useful to establish the extent to which the reaction will progress. This requires testing of cores taken from a structure under conditions that are likely to promote reaction and measuring dimensional changes. AAR testing on specimens taken from a structure where the reaction may have already initiated is referred to as 'latent expansion testing', since it is carried out to establish how much more expansion is likely. Although there is no British Standard for latent expansion as a result of AAR, the ASR test for establishing the reactivity of aggregate, BS 812-123 [41], can just as easily be applied to cores taken from a structure, with some minor changes.

The test involves the preparation of a standard concrete mix containing the aggregate under investigation – and usually additions of potassium sulphate to obtain a standard alkali content – and casting series prisms of dimensions 75 × 75 × 300 mm with studs in each end to allow measurement of dimension change using a comparator. Once removed from their moulds at an age of 28 days, the prisms are stored in individual airtight specimen holders wrapped in water-saturated fabric within a polymer bag at a temperature of 20°C for 7 days before being transferred to a temperature of 38°C. The length of the specimens is periodically measured for a period of at least 12 months.

Adapting the method to concrete taken from a structure is straightforward – cores with a diameter of 75 mm can be fitted with comparator studs by drilling holes at each end and embedding comparator studs using epoxy resin. The concrete in a core would clearly not be the same as the standard mix used in the standard, which has a notably high cement content. Moreover, the levels of alkali in the concrete would be different from a standard mix. This is likely to mean that rates of expansion would be slower than for a standard mix (which expand relatively slowly themselves). Thus, latent expansion testing can only be realistically initiated where there is sufficient time available and sufficient patience from the party requiring the appraisal.

7.6 CONCRETE REPAIR PRODUCTS

Where it is necessary to repair a structure as a result of deterioration, an appropriate repair method must be devised, and suitable repair products must be selected. Concrete repair products are covered in the United Kingdom by the *EN 1504* series of standards. In these documents, products are divided into the following categories:

- Surface protection systems for concrete
- Structural and nonstructural repair materials
- Structural bonding
- Concrete injection
- Anchoring of reinforcing steel bar
- Reinforcement corrosion protection

'Surface protection systems' are the same as the systems described in Chapter 6 and so are not discussed further in this chapter.

'Structural repair' refers to the removal and replacement of concrete in an element that is subsequently required to carry loads. 'Nonstructural repair' materials are used to fill holes or cutouts from concrete where the repair material will not be subjected to loads and will not be in contact with reinforcement – in other words, the repair material is playing a purely cosmetic role. In some cases, 'semistructural' repair may be required – missing concrete is replaced with material that is not subject to loads but is in contact with reinforcement and must provide adequate protection from corrosion to the steel [43].

A range of different products are available for both structural and nonstructural repair, including cementitious repair products containing cements normally encountered in conventional concrete and polymer-modified cementitious products in which polymeric additives have been included to improve adhesion to the concrete surface that it is placed against and to impart characteristics such as shrinkage, which will better match the mature concrete being repaired. These additives include styrene butadiene rubber (SBR) emulsions, which not only improve bond strength, but also renders the repair material relatively resistant to water permeation and chemical attack. In some cases, repair materials based on polymeric resins such as polyesters and epoxies can be used, although problems with compatibility of properties have often been found to be a problem.

EN 1504-3 [43] defines the performance requirements that repair materials must satisfy. These characteristics include compressive strength, adhesive bond, restrained shrinkage/expansion and coefficient of thermal expansion. Various durability requirements are also defined. The crucial difference between materials suitable for structural and nonstructural repair materials is that higher performance is demanded for structural repair.

It is often the case that, before a repair material is used, a bonding layer is applied first to enhance the bond between the repair material and the underlying surface. These 'structural bonding' materials can take many different forms. In the simplest form, the layer may simply be a cement slurry. However, superior bonding is often achieved using a cement slurry modified with a latex (SBR, polyvinyl acetate, epoxy or acrylic) or an epoxy resin [44].

The performance requirements for structural bonding materials are defined in EN 1504-4 [45]. The requirements depend on the application that the bonding agent is used for – the materials can also be used for bonding plate reinforcement for strengthening operations or bonding precast concrete repair materials. However, where the bonding agent is used for bonding mortar or concrete, the performance requirements include modulus of elasticity in flexure and compression, compressive and shear strength, and coefficient of thermal expansion.

'Concrete injection' involves the introduction of mortar into cracks and voids under pressure such that the space is completely infiltrated. Concrete injection materials are normally epoxy- or polyurethane-based grouts. The performance requirements (in EN 1504-5 [46]) depend on how the filling material is to be used. If it is intended to carry stress across the original crack – 'force transmitting' – the key characteristics are adhesion (under both tensile and shear loading) and volume stability. In some cases, where further movement is envisaged, it may be desirable for the material used to possess ductile characteristics. In such cases, the requirements are adhesion, elongation and watertightness. Watertightness is also required of the third type of injection material – swelling materials that expand once injected.

'Anchoring' products are used for the installation of additional reinforcement bars to strengthen a structure. The standard requires them to form an adequately strong bond, which is defined in terms of a pull-out test value [47]. There is also a limit in the amount of creep that the materials will undergo. As for structural repair materials, materials for anchoring reinforcement will usually be based on hydraulic cements, resins, or mixtures of cement and resin.

'Reinforcement corrosion protection' products are primer coatings applied to the surface of exposed steel reinforcement prior to repair of a concrete element with structural repair products. Depending on their nature, they may be applied with a brush, roller or spray. They fall into two categories: barrier coatings, which act as an impermeable layer at the steel surface, and active coatings, which offer some form of chemical protection [48]. Barrier coatings tend to be polymeric resins. Active coatings usually work by one of three mechanisms. First, they may simply contain alkaline constituents, including Portland cement, which help to maintain passivation at the steel surface. This may be important where a polymeric repair material is to be placed against the steel – such materials are not alkaline and, so, offer no protection from corrosion (other than physical protection from ingress of corrosive substances). Second, they may contain corrosion inhibitors (see Chapter 4). Third, they may provide galvanic protection. The most common examples of such a coating are zinc-containing primers – zinc particles dispersed in an epoxy resin. The zinc particles perform the same role as the zinc coating on galvanised steel – acting as a sacrificial layer that corrodes instead of the steel itself (see Chapter 4).

7.7 REPAIR METHODS

Where concrete must be repaired, careful consideration of the mechanisms of deterioration is necessary to ensure that the repair is wholly effective. A key part of a repair procedure should therefore be to ensure that deterioration processes are stopped prior to the repair being made. It is important that the material used will not exacerbate the same durability problem that is being addressed by the repair. In addition, the repair material must be compatible with the original materials that it is being placed against, both chemically and in terms of its physical characteristics – particularly those associated with change in dimension.

In many cases, it may be necessary to remove concrete prior to repair. This may be because the concrete is damaged to such a large extent that keeping it in place is not an option or because it is sufficiently contaminated with aggressive substances that it will continue to present a durability problem. This is particularly true of chloride-contaminated concrete: leaving highly contaminated concrete in place and repairing a cracked or spalled area will lead to volumes of concrete with very different concentrations adjacent to each other. This can lead to the 'incipient anode' effect, where the contaminated parts become anodic and an aggressive corrosion cell is established [42]. Where the removal of concrete will compromise structural integrity, appropriately designed support will be required until the repair is complete and the repair material is able to carry the required load.

EN 1504-9 [49] outlines the different approaches that can be taken toward using repair materials and prevention and rehabilitation methods to deal with durability problems in concrete structures. The standard takes the approach of defining a series of principles towards repairing, rehabilitating or protecting a structural element and identifies means by which these aims can be achieved. Each of these principles are discussed below.

7.7.1 Protection against ingress

Preventing substances in the external environment from entering concrete can be achieved in a number of ways. The most obvious is the application of the various types of surface protection – surface sealers, hydrophobic impregnants and surface coatings – which has been discussed in detail in Chapter 6. In a similar manner, a concrete surface can be rendered impermeable through the installation of protective liners and even external panels.

Since cracks offer the easiest route for ingress, they must also be addressed. The method used largely depends on whether a crack is live – whether it undergoes movement under the influence of loads and temperature changes. Where this is the case, the above approaches are unlikely to be appropriate, and cracks will probably require filling. They may be filled by hand using structural repair materials or by injection. EN 1504-9 also suggests the use of bandaging. Bandages, in this context, are strips of polymer membrane, most commonly used for sealing joints where significant movement is anticipated. They can be used to cover surface cracks using an adhesive to secure them to the concrete.

Where cracking is the result of an ongoing process such as reinforcement corrosion, filling of cracks, although probably still being necessary, will not solve the problem, and additional measures will also be necessary to arrest the expansive reaction. Measures for corrosion are discussed in a later section.

Finally, the possibility of converting a crack into a joint is included. The appearance of a crack may well be a sign that insufficient joints are present. Thus, by transforming a crack into a joint and sealing it appropriately, this problem may be resolved.

7.7.2 Moisture control

Given the crucial role that water plays in many deterioration mechanisms, reducing the amount of moisture present within concrete may provide an effective means of arresting degradation. The surface protection methods discussed above are likely to be effective, along with the installation of panels.

7.7.3 Concrete restoration

The restoration of concrete where it has broken away from a structural element or where concrete has been removed can be achieved using structural or nonstructural repair materials, as appropriate. Repair materials can take the form of concrete or mortar. Mortars are often applied by hand but can also be sprayed on. Concrete can be recast within formwork in a manner similar to the original element, or spraying may be possible.

The standard also includes the option of replacement of entire structural elements, which may be necessary if damage is too extensive.

Another means of achieving concrete restoration, which is not mentioned in the standard, is the use of precast panel repairs that are secured using mortar and the space around the perimeter injected with grout [50].

7.7.4 Structural strengthening

Structural strengthening can involve adding embedded bars, bars anchored in drilled holes, external bars or externally bonded plates. The use of external prestressed tendons is also an option.

Another means of strengthening a structural element is through the recasting of concrete or mortar to increase the cross-section. The standard also includes the filling of cracks by hand or by means of injection. This is likely to only have a slight influence on strength in most instances, but the standard emphasises the need to return a cracked strengthened structure back to its original condition.

Where reinforcement bars have corroded excessively, it may be necessary to remove these and replace them.

7.7.5 Increasing physical and chemical resistance

'Physical resistance', in this context, means resistance to impact, abrasion and, in some cases, freeze–thaw attack. As previously discussed in Chapter 6, hydrophobic impregnants are not suitable for protecting against such processes, but other forms of surface protection are. The provision of an additional protective layer of concrete or mortar is also suggested, such as a screed on a floor subject to abrasion.

Chemical resistance is enhanced through the same means.

7.7.6 Preserving or restoring passivity

Realkalisation and electrochemical chloride extraction are techniques that offer a means of preserving and possibly restoring passivity around steel reinforcement. These methods are discussed in detail in Section 7.8.

In cases where chloride contamination is severe or carbonation is complete, the best means of restoring a high-pH environment around steel reinforcement may be to remove the overlying concrete and replace it with new material. When this is done, the option of increasing the depth of cover exists.

Additional cover may also be added to carbonating concrete where the carbonation front has not yet reached the reinforcement. This will provide an additional barrier before the existing carbonation front can continue to progress. The same approach is less advisable where chloride ingress is occurring, since chlorides already in the concrete will continue to diffuse towards the reinforcement.

7.7.7 Increasing resistivity

On the assumption that changing the pore structure of mature concrete is difficult, the main available means of reducing the resistivity of concrete is through the reduction of moisture. All of the options identified as suitable for moisture control are thus suitable for increasing resistivity.

7.7.8 Cathodic control and cathodic protection

Cathodic control refers to any measure that limits the concentration of oxygen at the reinforcement, thus reducing the risk of corrosion. Approaches include the use of surface protection (excluding hydrophobic impregnation, which will permit the ingress of oxygen). The standard also includes ensuring that the concrete is permanently saturated with water, which may be possible in certain cases.

Cathodic protection refers to electrochemical techniques used to ensure that steel reinforcement remains permanently cathodic and thus unlikely to corrode. Approaches to achieving this are discussed in detail in Chapter 6.

7.7.9 Control of anodic areas

The need for the control of anodic areas arises from situations where it is not possible to remove chloride-contaminated concrete. As discussed earlier, this can lead to incipient anode corrosion. A solution is to treat the exposed steel with active reinforcement corrosion protection prior to restoring the concrete. Another solution is to coat the bar with barrier reinforcement corrosion protection. However, the standard points out that if this approach is adopted, the entire reinforcement bar must be exposed and fully coated, as it has been shown in Chapter 4 that incomplete coating will still lead to corrosion. Moreover, the bars must be completely cleaned of corrosion products.

The use of corrosion inhibitors is also an option (Chapter 4). These can be added to the repair material used for restoration, or migrating inhibitors can be applied to the concrete surface after restoration (Section 7.8).

7.7.10 Devising strategies for repair and rehabilitation

Using the principles proposed in *EN 1504* and an understanding of the mechanisms of deterioration, which have been examined throughout this book, a possible means of repairing and rehabilitating concrete structural elements that have suffered from durability problems can be devised. The possible options are outlined in Table 7.10. It should be stressed that not all of the options would be necessary for a given repair. For instance, it would probably be unlikely to require both the introduction of corrosion inhibitors in a repair material and at the concrete surface after repair.

7.8 REHABILITATION OF CONCRETE STRUCTURES

A number of techniques exist that can be used to reduce the potential threat of chloride-induced and carbonation-induced corrosions processes before they have occurred. It should be stressed that the value of applying these techniques after corrosion has initiated is questionable.

7.8.1 Electrochemical chloride extraction

Electrochemical chloride extraction is a technique that reverses the movement of chloride ions into concrete through the application of a potential

Table 7.10 Approaches to the repair of damaged concrete

Deterioration mechanism	Action	Reason for action
Freeze–thaw attack	Remove excessively damaged concrete	
	Replacement of concrete and/ or filling of cracks	Concrete restoration
	Surface protection	Moisture control: protection will limit the quantity of water entering the concrete after repair
		Increasing physical resistance (where surface protection is surface sealer or surface coating)
Abrasion	Replacement of concrete	Concrete restoration
	Install screed or apply surface protection (surface sealer or surface coating)	Increasing physical resistance
Sulphate attack	Remove excessively damaged concrete	
	Replacement of concrete	Concrete restoration
		Extend period before deterioration becomes unacceptable
	Surface protection	Moisture control: protection will limit the quantity of sulphate ions entering the concrete after repair
		Increasing chemical resistance (where surface protection is surface sealer or surface coating)
AAR	Remove excessively damaged concrete	
	Replacement of concrete and/ or filling of cracks	Concrete restoration
	Surface protection	Moisture control: protection will limit the quantity of water entering the concrete after repair, which would lead to swelling of AAR products

(continued)

Table 7.10 (Continued) Approaches to the repair of damaged concrete

Deterioration mechanism	Action	Reason for action
Chloride-induced corrosion	Remove excessively contaminated concrete	Attempt to reduce chloride concentrations at the reinforcement to levels that will not induce corrosion
	Replace corroded reinforcement	Restore load-bearing capacity to structural element
	Apply reinforcement corrosion protection to steel	Control anodic areas
	Replacement of concrete, possibly with inclusion of corrosion inhibitor	Concrete restoration Inhibitor may prevent or slow down future corrosion
	Apply migrating corrosion inhibitor	Inhibitor may prevent or slow down future corrosion
	Surface protection	Moisture control: limit ingress of chlorides after repair Increasing resistivity: reduced moisture levels in concrete will limit rates of corrosion Cathodic control (only for surface coating or surface sealer)
	Cathodic protection	Limit corrosion rates
Chloride ingress (corrosion has not yet initiated)	Electrochemical chloride extraction	Reduce levels of chloride in the concrete cover
	Apply migrating corrosion inhibitor	Inhibitor may prevent or slow down future corrosion
	Surface protection	Limit ingress of chloride ions into concrete Increasing resistivity: reduced moisture levels in concrete will limit rates of corrosion Cathodic control (only for surface coating or surface sealer)
	Cathodic protection	Limit corrosion rates
Carbonation-induced corrosion	Remove carbonated concrete	Remove low-pH concrete in contact with reinforcement
	Replace corroded reinforcement	Restore load-bearing capacity to structural element
	Replacement of concrete, possibly with inclusion of corrosion inhibitor	Concrete restoration Inhibitor may prevent or slow down future corrosion

Table 7.10 (Continued) Approaches to the repair of damaged concrete

Deterioration mechanism	Action	Reason for action
Carbonation-induced corrosion	Apply reinforcement corrosion protection to steel	Control anodic areas
	Apply migrating corrosion inhibitor	Inhibitor may prevent or slow down future corrosion
	Surface protection (surface sealer or surface coating)	Limit diffusion of carbon dioxide into concrete
		Moisture control: reduce relative humidity in concrete to below optimum levels for carbonation
		Increasing resistivity: reduced moisture levels in concrete will limit rates of corrosion
		Cathodic control
	Cathodic protection	Limit corrosion rates
Partial carbonation (corrosion has not initiated)	Realkalisation	Increase pH around reinforcement
	Apply migrating corrosion inhibitor	Inhibitor may prevent or slow down future corrosion
	Recast concrete with additional cover	
	Surface protection (surface sealer or surface coating)	Limit diffusion of carbon dioxide into concrete
		Moisture control: reduce relative humidity in concrete to below optimum levels for carbonation
		Increasing resistivity: reduced moisture levels in concrete will limit rates of corrosion
		Cathodic control
	Cathodic protection	Limit corrosion rates

difference between an electrode at the concrete surface and the reinforcing steel. The configuration required to do this is not unlike that for impressed current cathodic protection, as seen in Chapter 6, although a temporary external electrode is used and the voltage used is somewhat higher. Typically, a voltage of approximately 10 to 40 V is suitable, compared with potential differences of a few hundred millivolts for cathodic protection [51]. A current density between 1 and 2 A/m^2 is usual [52].

The principle is very simple. When a potential difference is established between a surface electrode and the reinforcing steel, negative ions will migrate towards the anode. By making the external electrode the anode, the negative ions will move towards the concrete surface. In concrete containing chloride ions, the migrating ions will be mostly chloride ions and hydroxide ions. As these ions make their way towards the surface, hydroxide ions are formed at the cathode.

The basic configuration for electrochemical chloride extraction is shown in Figure 7.21. A tank is attached to the concrete surface. A titanium or steel mesh is supported within the tank, and the tank is filled with an electrolyte solution (usually calcium hydroxide). In some cases, a mass of cellulose gel containing a solution of electrolyte is used instead. A reinforcement bar is uncovered at a point along its length, and a potential difference is applied between this and the mesh with a direct current power source. As discussed in Chapter 6, when chloride ions reach the anode during cathodic protection, they form chlorine gas. However, where the electrolyte is present, calcium chloride is formed instead, which can be removed through draining and refreshing of the tank solution.

The effects of chloride extraction are shown in Figure 7.22, which plots the concentration of free chloride at different depths from a reinforcement bar with time. According to these results, by 120 h of treatment, the chloride concentration throughout most of the concrete is significantly reduced. It is important to note that free chloride levels increase initially, which means that an adequate treatment time is essential to ensure that the situation is not worsened. Typically, a period of 2 to 6 weeks is required to fully treat concrete [52].

Concerns have been raised about the possible side effects of the high current densities used during electrochemical chloride extraction. These concerns are the same as for cathodic protection: potential exacerbation

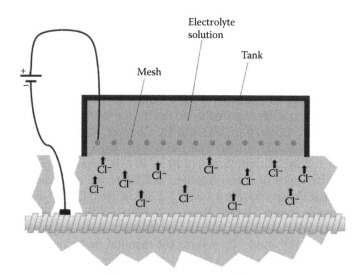

Figure 7.21 Schematic diagram of the configuration for electrolytic chloride extraction.

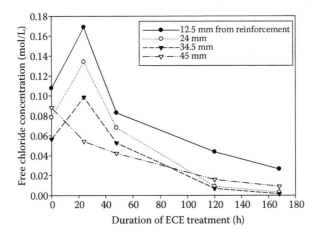

Figure 7.22 Free chloride concentrations at different distances from a reinforcement bar versus time of electrochemical chloride extraction (ECE). (From Concrete Repair Association. *Electrochemical Rehabilitation of Steel-Reinforced Concrete Structures.* Advice Note 4. Aldershot, United Kingdom: Concrete Repair Association, 2009.)

of AAR resulting from the production of hydroxide ions at the cathode and deterioration due to the evolution of hydrogen gas at the cathode in the form of hydrogen embrittlement of steel and possible microcracking of the concrete. Experiments designed to investigate AAR found alkali–silica gel only in specimens exposed to elevated temperatures (60°C) after chloride extraction and concluded that the risk under normal conditions was minimal [53]. Despite this, it has been proposed that, where AAR is a risk, lithium electrolytes (such as lithium borate) should be used as a result of lithium's ability to control alkali–silica reaction (see Chapter 3) [54].

As for cathodic protection, the potential difference applied across the concrete leads to the polarisation of the steel. This means that measurements of half-cell corrosion potential immediately after treatment cannot be relied on [55], and the concrete must be left for a number of months to allow the corrosion potentials to stabilise through the diffusion of hydroxide ions in the concrete and oxygen gas from the exterior [56].

7.8.2 Realkalisation

Realkalisation is another electrochemical technique that is used to produce hydroxide ions at the reinforcement (as occurs during cathodic protection and chloride extraction) and cause the migration of alkali metal ions toward the steel to passivate its surface. It is usually used on carbonated concrete, where pH levels within a structural element have dropped, or threaten to drop, below levels that would normally protect steel reinforcement.

The difference in operation is that the electrolyte used is normally potassium carbonate (K_2CO_3) or sodium carbonate (Na_2CO_3), which hopefully will permeate the concrete and further increase levels of alkalinity in the concrete pore solution. Furthermore, a lower current density to chloride extraction is used (typically, 0.5–1.0A/m²) over a shorter period of time (3–10 days) [52].

Rather than use a power supply to drive a current between the reinforcement and the concrete surface, a sacrificial external electrode can be used in a configuration similar to that used for galvanic cathodic protection (see Chapter 6) [57]. The sacrificial mesh is normally made from aluminium alloy. The process is considerably slower than the conventional technique.

Figure 7.23 shows the change in the concentration of various chemical species at the reinforcement over a period of treatment. The increase in hydroxide ions at the surface is the result of their production at the reinforcement surface. However, the increase in sodium is the result of the migration of ions from within the concrete and from the electrolyte solution used in the treatment. Evidence for the Na_2CO_3 electrolyte having reached the reinforcement is the (relatively small) increase in CO_3^{2-} ion concentration.

Figure 7.24 shows the change in pH produced through realkalisation (using the sacrificial anode technique) at different depths through a concrete

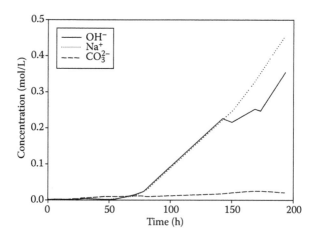

Figure 7.23 Change in the concentration of chemical species at the reinforcement of a carbonated concrete specimen undergoing realkalisation using a Na_2CO_3 electrolyte. (From Andrade, C. M. et al., *Materials and Structures*, 32, 1999, 427–436.)

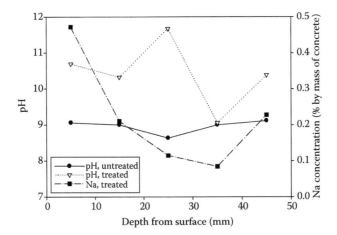

Figure 7.24 Profiles of pH in an untreated carbonated concrete specimen and a realka-lised specimen after treatment for 33 days using Na_2CO_3 as the electrolyte using the sacrificial anode technique. Na concentrations are also plotted. (From Tong, Y. Y. et al., *Cement and Concrete Research*, 42, 2012, 84–94.)

specimen. After treatment, the pH is considerably higher in most parts of the specimen and highest around the reinforcement, principally due to the production of hydroxide ions in this zone. Also plotted in Figure 7.24 is the sodium concentration, showing clear evidence of its migration into the concrete from the electrolyte solution.

The pH around the reinforcement will typically decrease with time after treatment, as a result of the diffusion of ions caused by the differences in concentrations of species at different locations in the concrete. Nonetheless, it appears that, where treatment is carried out correctly, pH should stabilise at around 10.5, which should be adequate for passivation [58].

As for chloride removal, polarisation of the reinforcing steel can initially give misleading measurements when electrochemical techniques are used to estimate corrosion rates immediately after treatment. The point has been made that, where corrosion has already initiated, realkalisation is unlikely to repassivate steel reinforcement. This is illustrated by the results shown in Figure 7.25 of corrosion current density measurements made on corroded and clean steel bars cast into uncarbonated concrete – the corrosion current density (I_{corr}) of the corroded steel remains high, despite being surrounded by an alkaline environment. Thus, realkalisation is probably only of value in structural elements where corrosion has not yet been initiated.

Realkalisation can also be carried out without the use of a potential difference. A tank of realkalising solution (in most cases, Na_2CO_3 or K_2CO_3) is secured at the surface and left to diffuse into the concrete. Permeation is clearly a slower process under these conditions.

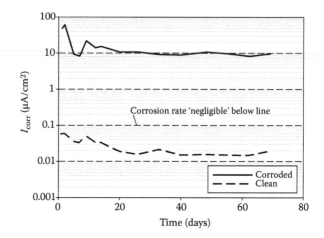

Figure 7.25 Corrosion current density (I_{corr}) of corroded and clean steel bars cast into uncarbonated concrete. (From González, J. A. et al., *Materials and Corrosion*, 51, 2000, 97–103.)

7.8.3 Migrating corrosion inhibitors

Corrosion inhibitors have already been encountered in Chapter 4, where the mechanisms by which these additives work have been explored. However, a new development in corrosion inhibitor technology has been their use as migrating agents applied to the surface of hardened concrete, which may already contain chlorides or have undergone carbonation, and which permeate the concrete to reach the reinforcement. These agents are referred to as 'migrating' corrosion inhibitors.

Inhibitors can be applied to the concrete surface with a brush or spray. Migration can be accelerated through the application of a potential difference between the reinforcement and the surface on which the inhibitor is present. This is achieved in much the same way as for chloride extraction and realkalisation – the inhibitor is held at the surface in a tank, gel, or saturated sponge along with a metallic mesh, which acts as the anode. The process is known as 'electromigration'.

Figure 7.26 shows concentration profiles of a migrating corrosion inhibitor through concrete. The profile shape is characteristic of diffusion and presents an important point – the application of even a large quantity of inhibitor at the surface will lead to only a relatively small quantity reaching the reinforcement. The best results are therefore normally achieved with multiple applications. The effectiveness of the procedure is also dependent on the depth of cover and the water/cement (W/C) ratio of the concrete, with a shallow depth of cover and a higher W/C ratio yielding greater improvements in corrosion resistance [60].

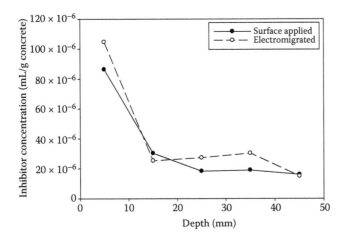

Figure 7.26 Concentration profiles through concrete specimens treated with an amine-based corrosion inhibitor. Surface application was conducted by painting the top surface of the concrete with inhibitor every 3 weeks for 300 days. Electromigration was carried out using a current density of 7.4 µA/cm² for 2 weeks. (From Holloway, L. et al., *Cement and Concrete Research*, 34, 2004, 1435–1440.)

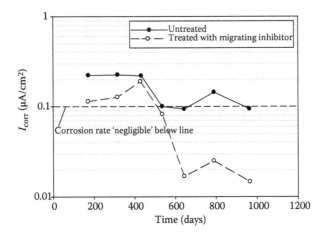

Figure 7.27 Corrosion current density (I_{corr}) of a reinforced concrete sample held in a marine environment after treatment with a migrating corrosion inhibitor, compared with an untreated specimen. (From Morris, W. and M. Vázquez, *Cement and Concrete Research*, 32, 2002, 259–267.)

An example of the influence that migrating corrosion inhibitors can have is shown in Figure 7.27, which compares the corrosion current density of treated and untreated specimens in an aggressive environment. Longer-term monitoring of the performance of such inhibitors has shown that their inhibiting effect decreases with time, which means that such treatments are probably best used as an ongoing process of maintenance [61].

REFERENCES

1. British Standards Institution. *BS EN 1992-1-1:2004: Design of Concrete Structures—General Rules and Rules for Buildings*. London: British Standards Institution, 2004, 230 pp.
2. British Standards Institution. *BS EN 1990:2002: Eurocode—Basis of Structural Design*. London: British Standards Institution, 2002, 120 pp.
3. The Institution of Structural Engineers. *Appraisal of Existing Structures, 3rd ed.* London: The Institution of Structural Engineers, 2010, 187 pp.
4. Gérard, B. and J. Marchand. Influence of cracking on the diffusion properties of cement-based materials: Part 1. Influence of continuous cracks on the steady-state regime. *Cement and Concrete Research*, v. 30, 2000, pp. 37–43.
5. Fan, S. and J. M. Hanson. Effect of alkali–silica reaction expansion and cracking on structural behaviour of reinforced concrete beams. *ACI Structural Journal*, v. 95, 1998, pp. 498–505.
6. Bhargava, K., A. K. Ghosh, Y. Mori, and S. Ramanujam. Modelling of time to corrosion-induced cover cracking in reinforced concrete structures. *Cement and Concrete Research*, v. 35, 2005, pp. 2203–2218.
7. Thoft-Christensen, P. Stochastic modelling of the crack initiation time for reinforced concrete structures. In M. Elgaaly, ed., *ASCE Structures Congress: Advanced Technology in Structural Engineering*. Reston, Virginia: American Society of Civil Engineers, 2000, pp. 1–8.
8. Andrade, C., C. Alonso, J. A. Gonzalez, and J. Rodriguez. Remaining service life of corroded structures. *Proceedings of IABSE Symposium on Durability of Structures*, pp. 359–363.
9. Cao, C., M. M. S. Cheung, and B. Y. B. Chan. Modelling of interaction between corrosion-induced concrete cover crack and steel corrosion rate. *Corrosion Science*, v. 69, 2013, pp. 97–109.
10. Suda, K., S. Misra, and K. Motohashi. Corrosion products of reinforcing bars embedded in concrete. *Corrosion Science*, v. 35, 1993, pp. 1543–1549.
11. Andrade, C., C. Alonso, and F. J. Molina. Cover cracking as a function of bar corrosion: Part 1. Experimental test. *Materials and Structures*, v. 26, 1993, pp. 453–464.
12. Sarkar, S., S. Mahadevan, J. C. L. Meeussen, H. van der Sloot, and D. S. Kosson. Numerical simulation of cementitious materials degradation under external sulphate attack. *Cement and Concrete Composites*, v. 32, 2010, pp. 241–252.

13. Karihaloo, B. L. *Fracture mechanics and structural concrete*. Harlow, United Kingdom: Longman, 1995, 346 pp.
14. Charlaix, E. Percolation threshold of a random array of discs: A numerical simulation. *Journal of Physics A: Mathematical and General*, v. 19, 1986, pp. L533–L536.
15. Sornette, D. Critical transport and failure in continuum crack percolation. *Journal Physique*, v. 49, 1988, pp. 1365–1377.
16. Bungey, J. H., S. G. Millard, and M. G. Grantham. *Testing of Concrete in Structures, 4th ed*. Boca Raton, Florida: CRC Press, 2006, 352 pp.
17. Reichling, K., M. Raupach, J. Broomfield, J. Gulikers, V. L'Hostis, S. Kessler, K. Osterminski, I. Pepenar et al. Full surface inspection methods regarding reinforcement corrosion of concrete structures. *Materials and Corrosion*, v. 64, 2013 (in press).
18. British Standards Institution. *BS 1881-208:1996: Testing Concrete—Part 208: Recommendations for the Determination of the Initial Surface Absorption of Concrete*. London: British Standards Institution, 1996, 14 pp.
19. Hall, C. Water movement in porous building materials: Part 4. The initial surface absorption and sorptivity. *Building and Environment*, v. 16, 1981, pp. 201–207.
20. RILEM TC 25-PEM. Recommended tests to measure the deterioration of stone and to assess the effectiveness of treatment methods. *Materials and Structures*, v. 13, 1980, pp. 175–253.
21. Hendrickx, R. Using the Karsten tube to estimate water transport parameters of porous building materials. *Materials and Structures* (in press) v. 46, 2013, pp. 1309–1320.
22. Basheer, P. A. M. A brief review of methods for measuring the permeation properties of concrete *in situ*. *Proceedings of the ICE: Structures and Buildings*, v. 99, 1993, pp. 74–83.
23. Leeming, M. B. *Standard Tests for Repair Materials and Coatings for Concrete—Part 2. Permeability Tests*. Technical Note 140. London: CIRIA, 1993, 63 pp.
24. The Concrete Society. *Permeability Testing of Site Concrete*. Technical Report 31. Camberley, United Kingdom: The Concrete Society, 2008, 80 pp.
25. The Concrete Society/Institute of Corrosion. *Electrochemical Tests for Reinforcement Corrosion*. Technical Report 60. Camberley, United Kingdom: The Concrete Society, 2004, 32 pp.
26. American Society for Testing and Materials. ASTM C876-09: *Standard Test Method for Corrosion Potentials of Uncoated Reinforcing Steel in Concrete*. West Conshohocken, Pennsylvania: American Society for Testing and Materials, 2009, 7 pp.
27. U.S. Department of Transportation, Federal Highway Administration. *Highway Concrete Pavement Technology Development and Testing, Vol. 5: Field Evaluation of Strategic Highway Research Program (SHRP) C-206 Test Sites (Bridge Deck Overlays)*. McLean, Virginia, 2006, 41 pp.
28. Stern, M. and A. L. Geary. Electrochemical polarization: A theoretical analysis of the shape of polarization curves. *Journal of the Electrochemical Society*, v. 104, 1957, pp. 56–63.

29. Dhir, R. K., M. R. Jones, and M. J. McCarthy. Quantifying chloride-induced corrosion from half-cell potential. *Cement and Concrete Research*, v. 23, 1993, pp. 1443–1454.

30. Andrade, C. and C. Alonso. Test methods for on-site corrosion rate measurement of steel reinforcement in concrete by means of the polarization resistance method. *Materials and Structures*, v. 37, 2004, pp. 623–643.

31. Wenner, F. A method of measuring earth resistivity. *Bulletin of the Bureau of Standards*, v. 12, 1915, pp. 469–478.

32. Gowers, K. R. and S. G. Millard. Measurement of concrete resistivity for assessment of corrosion severity of steel using Wenner technique. *ACI Materials Journal*, v. 96, 1999, pp. 536–541.

33. Langford, P. and J. Broomfield. Monitoring the corrosion of reinforcing steel. *Construction Repair*, v. 1, 1987, pp. 32–36.

34. British Standards Institution. *BS EN 13892-4:2002: Methods of Test for Screed Materials—Part 4. Determination of Wear Resistance: BCA.* London: British Standards Institution, 2002, 12 pp.

35. British Standards Institution. *BS 1881-124:1988: Testing Concrete—Part 124: Methods for Analysis of Hardened Concrete.* London: British Standards Institution, 1988, 24 pp.

36. Costa, A. and J. Appleton. Chloride penetration into concrete in marine environment: Part 1. Main parameters affecting chloride penetration. *Materials and Structures*, v. 32, 1999, pp. 252–259.

37. American Society for Testing and Materials. *ASTM C1218/C1218M-99: Standard Test Method for Water-Soluble Chloride in Mortar and Concrete.* West Conshohocken, Pennsylvania: ASTM, 2008, 3 pp.

38. Weritz, F., A. Taffe, D. Schaurich, and G. Wilsch. Detailed depth profiles of sulphate ingress into concrete measured with laser-induced breakdown spectroscopy. *Construction and Building Materials*, v. 23, 2009, pp. 275–283.

39. Vennesland, O. Documentation of electrochemical maintenance methods. In R. K. Dhir, M. R. Jones, and L. Zheng, eds., *Repair, Rejuvenation, and Enhancement of Concrete: Proceedings of the International Seminar.* London: Thomas Telford, 2002, pp. 191–198.

40. British Standards Institution. *BS EN 480-11:2005: Admixtures for Concrete, Mortar, and Grout—Test Methods: Determination of Air-Void Characteristics in Hardened Concrete.* London: British Standards Institution, 2005, 22 pp.

41. British Standards Institution. *BS 812-123:1999: Testing Aggregates— Method of Determination of Alkali–Silica Reactivity: Concrete Prism Method.* London: British Standards Institution, 1999, 18 pp.

42. The Concrete Society/Corrosion Prevention Association/Institute of Corrosion. *Repair of Concrete Structures with Reference to BS EN 1504.* Camberley, United Kingdom: The Concrete Society, 2009, 18 pp.

43. British Standards Institution. *BS 1504-3:2005: Products and Systems for the Protection and Repair of Concrete Structures—Definitions, Requirements, Quality Control, and Evaluation of Conformity: Part 3. Structural and Nonstructural Repair.* London: British Standards Institution, 2005, 30 pp.

44. Mailvaganam, N. P. *Effective Use of Bonding Agents.* Construction Technology Update 11. Ottawa, Canada: National Research Council of Canada, 1997, 4 pp.

45. British Standards Institution. *BS EN 1504-4:2004: Products and Systems for the Protection and Repair of Concrete Structures—Definitions, Requirements, Quality Control, and Evaluation of Conformity: Part 4. Structural Bonding.* London: British Standards Institution, 2004, 32 pp.

46. British Standards Institution. *BS EN 1504-5:2013: Products and Systems for the Protection and Repair of Concrete Structures—Definitions, Requirements, Quality Control, and Evaluation of Conformity: Part 5. Concrete Injection.* London: British Standards Institution, 2013, 44 pp.

47. British Standards Institution. *BS EN 1504-6:2006: Products and Systems for the Protection and Repair of Concrete Structures—Definitions, Requirements, Quality Control, and Evaluation of Conformity: Part 6. Anchoring of Reinforcing Steel Bar.* London: British Standards Institution, 2006, 24 pp.

48. British Standards Institution. *BS EN 1504-7:2006: Products and Systems for the Protection and Repair of Concrete Structures—Definitions, Requirements, Quality Control and Evaluation of Conformity: Part 7. Reinforcement Corrosion Protection.* London: British Standards Institution, 2006, 20 pp.

49. British Standards Institution. *BS EN 1504-9:2008: Products and Systems for the Protection and Repair of Concrete Structures—Definitions, Requirements, Quality Control, and Evaluation of Conformity: Part 9. General Principles for Use of Products and Systems.* London: British Standards Institution, 2008, 32 pp.

50. Odgers, D. *Practical Building Conservation: Concrete.* London: English Heritage, 2012, 308 pp.

51. Concrete Repair Association. *Electrochemical Rehabilitation of Steel-Reinforced Concrete Structures.* Advice Note 4. Aldershot, United Kingdom: Concrete Repair Association, 2009, 7 pp.

52. Building Research Establishment. *Digest 444, Part 3: Corrosion of Steel in Concrete—Protection and Remediation.* Watford, United Kingdom: Building Research Establishment, 2000, 12 pp.

53. Orellan, J. C., G. Escadeillas, and G. Arliguie. Electrochemical chloride extraction: Efficiency and side effects. *Cement and Concrete Research*, v. 34, 2004, pp. 227–234.

54. Bennett, J., T. J. Schue, K. C. Clear, D. L. Lankard, W. H. Hartt, and W. J. Swiat. *Electrochemical Chloride Removal and Protection of Concrete Bridge Components: Laboratory Studies.* Washington, D.C.: National Academy of Sciences, 1993, 188 pp.

55. Fajardo, G., G. Escadeillas, and G. Arliguie. Electrochemical chloride extraction (ECE) from steel-reinforced concrete specimens contaminated by "artificial" seawater. *Corrosion Science*, v. 48, 2006, pp. 110–125.

56. Marcotte, T. D., C. M. Hansson, and B. B. Hope. The effect of the electrochemical chloride extraction treatment on steel-reinforced mortar: Part 1. Electrochemical measurements. *Cement and Concrete Research*, v. 29, 1999, pp. 1555–1560.

57. Tong, Y. Y., V. Bouteiller, E. Marie-Victoire, and S. Joiret. Efficiency investigations of electrochemical realkalisation treatment applied to carbonated reinforced concrete: Part 1. Sacrificial anode process. *Cement and Concrete Research*, v. 42, 2012, pp. 84–94.

58. Andrade, C., M. Castellote, J. Sarría, and C. Alonso. Evolution of pore solution chemistry, electroosmosis, and rebar corrosion rate induced by realkalisation. *Materials and Structures*, v. 32, 1999, pp. 427–436.

59. González, J. A., A. Cobo, M. N. González, and E. Otero. On the effectiveness of realkalisation as a rehabilitation method for corroded reinforced concrete structures. *Materials and Corrosion*, v. 51, 2000, pp. 97–103.

60. Ngala, V. T., C. L. Page, and M. M. Page. Corrosion inhibitor systems for remedial treatment of reinforced concrete: Part 1. Calcium nitrite. *Corrosion Science*, v. 44, 2002, pp. 2073–2087.

61. Holloway, L., K. Nairn, and M. Forsyth. Concentration monitoring and performance of a migratory corrosion inhibitor in steel-reinforced concrete. *Cement and Concrete Research*, v. 34, 2004, pp. 1435–1440.

62. Morris, W. and M. Vázquez. A migrating corrosion inhibitor evaluated in concrete containing various contents of admixed chloride. *Cement and Concrete Research*, v. 32, 2002, pp. 259–267.

Index

Milton Keynes UK
Ingram Content Group UK Ltd.
UKHW021844071024
449327UK00021B/1537

9 780367 865856